AGRICULTURE

RELEVANCE FOR TRANSGENIC CROPS

BCPC Symposium Proceedings No. 72

SYMPOSIUM PROCEEDINGS NO. 72

Gene Flow and Agriculture

Relevance for Transgenic Crops

Chaired by P J W Lutman

Proceedings of a Symposium held at the
University of Keele, Staffordshire
12 - 14 April 1999

BRITISH
CROP
PROTECTION
COUNCIL

49 Downing Street
Farnham, Surrey GU9 7PH

Tel: 01252 733072
Fax: 01252 727194
Email: md@bcpc.org
Web: www.bcpc.org

British Library Cataloguing in Publication Data
A catalogue record for this book is available from the British Library

British Crop Protection Council
Gene Flow and Agriculture: Relevance for Transgenic Crops
(Proceedings/Monograph Series, ISSN 0306-3941; No 72

ISBN 1 901396 72 X

Every effort has been made to ensure that the recommendations and statements made in these proceedings are correct, but the British Crop Protection Council cannot accept responsibility for any loss, damage, or any other accident arising from carrying out the methods advocated.

Nothing in these proceedings shall be taken as a warranty that any substances or mixture of substances mentioned herein is not the subject of patent rights and the Council does not hold itself responsible for any infringement of said rights.

Cover design by Major Design & Production Ltd, Nottingham
Printed in Great Britain by Major Design & Production Ltd, Nottingham

CONTENTS

Page

Preface VIII
Symposium Organising Committee IX
Abbreviations X

SESSION 1
GENETICALLY MODIFIED CROPS – THE CURRENT SITUATION

Transgenic plants: field testing and commercialisation including a consideration of novel herbicide resistant oilseed rape (*Brassica napus L.*)
S Barber 3

Deliberate release of genetically modified organisms: the UK regulatory framework
P Burrows 13

The role of DNA technologies in crop breeding
G C M Sage 23

A long term perspective on Ag-biotech
W de Greef 33

SESSION 2
GENE FLOW – MEASUREMENT AND IMPLICATIONS

Defining and measuring gene flow
A F Raybould and R T Clarke 41

Gene flow and risk assessment
C J Gliddon 49

Gene flow at the landscape level
G R Squire, J W Crawford, G Ramsay *et al.* 57

Genetic pollution: concepts, concerns and transgenic crops
R E Daniels and J Sheail 65

SESSION 3
CONSEQUENCES OF GENE FLOW WITHIN SPECIES

Gene flow in genetically modified herbicide tolerant oilseed rape (*Brassica napus*) in the UK
E C Simpson, C E Norris, J R Law *et al.* 75

Gene flow between sugar beet and weed beet
Y Vigouroux, H Darmency, T G de Garambe and M Richard-Molard 83

GeneSys: a model of the effects of cropping system on gene flow from transgenic rapeseed
N Colbach, J M Meynard, C. Clermont-Dauphin and A Messéan 89

Regional patterns of gene flow and its consequence for GM oilseed rape
C E Thompson, G Squire, G R Mackay *et al.* 95

Modelling the potential for gene escape in oilseed rape via the soil seedbank: its relevance for genetically modified cultivars
C Pekrun, P W Lane and P J W Lutman 101

SESSION 4
CONSEQUENCES OF GENE FLOW BETWEEN SPECIES – 1

Gene flow and rape – the Canadian experience
R K Downey 109

Gene flow from oilseed rape (Brassica napus) to related species
R B Jørgensen 117

Gene flow from oilseed rape to weeds
A M Chèvre, F Eber, M Renard and H Darmency 125

Gene flow from transgenic canola to wild radish – a model system to determine the risks
M A Rieger, C Preston, T Potter and S B Powles 131

Fitness costs associated with transgenic glufosinate tolerance introgressed from *Brassica napus* ssp *oleifera* (oilseed rape) into weedy *Brassica napa*
A A Snow and R B Jørgensen 137

Gene flow from oilseed rape to *Sinapis arvensis*: variation at the population level
C L Moyes, J Lilley, C Casais and P J Dale 143

SESSION 5
CONSEQUENCES OF GENE FLOW BETWEEN SPECIES – 2

Predicting the environmental impact of transgene outcrossing to wild and weedy rices in Asia
M B Cohen, M T Jackson, B R Lu *et al.* 151

The production of herbicide-resistant jointed goatgrass (*Aegilops cylindrica*) x wheat (*Triticum aestivum*) hybrids in the field by natural hybridization and management strategies to reduce their occurrence
S S Seefeldt, F L Young, R Zemetra and S S Jones 159

Potential for gene flow between wheat (*Triticum aestivum*) and jointed goatgrass (*Aegilops cylindrica*) in the field
C A Mallory-Smith, J Snyder, J L Hansen *et al.* 165

Assessing potential risks of transgene escape from fiber plantations
S P DiFazio, S Leonardi, S Cheng and S H Strauss 171

SESSION 6
CONSEQUENCES OF GENE FLOW BETWEEN HIGHER PLANTS AND OTHER ORGANISMS

Insecticidal transgenes into nature: gene flow, ecological effects, relevancy and monitoring
C N Stewart 179

Gene flow from virus-resisitant transgenic crops to wild relatives or to infecting viruses
P-Y Teycheney and M Tepfer 191

Interactions between insect tolerant genetically modified plants and natural enemies
T H Schuler, G M Poppy, R P J Potting *et al.* 197

Non-target effects of proteinase inhibitors expressed in potato as an anti-nematode defence
S E Cowgill, D Coates and H J Atkinson 203

Honeybees as vectors of GM oilseed rape pollen
G Ramsay, C E Thompson, S Neilson and G R Mackay 209

SESSIONS 7 AND 8
GENE FLOW – THE FUTURE

Concerns about gene flow and the implications for the development of monitoring protocols
J E Hill 217

Molecular aspects of multiple transgenes and gene flow to crops and wild relatives
I J Senior and P J Dale 225

Management of transgenic crops within the cropping system
J Champolivier, J Gasquez, A Méssean and M Richard-Molard 233

Assessing the impact and consequences of the release and commercialisation of genetically modified crops
J B Sweet, C E Norris, E Simpson and J E Thomas 241

Gene flow and the practical management of genetically modified crops in the UK
J H Orson and J F Oldfield 247

POSTER PAPERS

Monitoring weediness and persistence of genetically modified oilseed rape (*Brassica napus*) in the UK
C E Norris, E C Simpson, J B Sweet and J E Thomas 255

Millet as a model-crop to assess the impact of gene flow toward weed populations
H Darmency, L Assémat and T Wang 261

Origin of wild beet and gene flow between *Beta vulgaris* and *B. macrocarpa* in California
D Bartsch, J Clegg and N C Ellstrand 269

Evaluating the risk of transgene spread from *Brassica napus* to related species
R Pinder, N Al-Kaff, M Kreike and P Dale 275

Estimation of allele frequencies for *Bacillus thuringiensis* resistance in diamondback moth, *Plutella xylostella* and cotton bollworm, *Helicoverpa armigera*: an isofemale line (F_2) approach
M Ahmad and R Roush 281

PREFACE

Over the last two years, since this Symposium was initially proposed by the British Crop Protection Council (BCPC), the issue of genetically modified (GM) crops has received greater and greater attention in the media. Moral issues, food safety issues and environmental issues have all been raised. There is a clear need for scientifically based, open and independent discussion of the subject. This Symposium provides such a forum.

When this Symposium was conceived, it was felt that an event which discussed the issue of gene flow in relation to GM crops, was likely to be one of appreciable scientific and public interest. It was also one that was appropriate for the BCPC funding, in relation to its objectives of promoting environmentally sound crop protection, as the majority of GM crops approaching the market place were either herbicide or insecticide resistant. There was a need for good scientific research on the potential environmental impact of GM crops and for a forum for the discussion of the issues arising from the work. In 1996/7 only a limited amount of research had been completed but more was in progress and new work was just starting. Thus, 1999 seemed a good time to hold this Symposium. In 1996, the areas, world-wide, sown with GM crops were quite limited. The explosion in the uptake of these crops, particularly in North America, compounded by the upsurge in public concern in Europe about their cultivation, has put considerable pressure on the scientific community to provide sound scientific data on many aspects of these crops, not least their potential environmental impact.

The Smposium has been structured so that the initial sessions provide background to GM crops and to gene flow. The core of the event consists of invited and offered papers on gene flow, with the Symposium ending with a discussion forum on the future. This latter session is particularly relevant to Europe where the future planting of GM crops is still unclear.

I believe that this Symposium will provide a forum for the balanced and informed discussion of the likely impact of gene flow on the cultivation of GM crops. We have the opportunity to present objective and scientifically sound data that will provide a good basis for assessing aspects of their potential impact on the environment. Bearing in mind the exceedingly high political profile of.this subject in Europe in 1999, and the desire of pressure groups to exploit data for their own goals, I would encourage authors to weigh their comments with particular care. In conclusion, I hope that the Symposium will provide some 'light' on a subject that is currently characterised nationally by a predilection to generate heat but little light!

P J W Lutman
IACR-Rothamsted, Harpenden

SYMPOSIUM ORGANISING COMMITTEE

Programme Chairman

Dr P Lutman	IACR-Rothamsted, Harpenden, Herts AL5 2JQ

Session Organisers

Dr P Dale	John Innes Centre, Norwich , Colney Lane, Norwich NR4 7UJ
Professor A Gray	Institute of Terrestrial Ecology, Furzebrook Research Station, Wareham, Dorset BH20 5AS
Mr J Kilpatrick	ADAS Rosemaund, Preston Wynne, Hereford HR1 3PG
Mr J Orson	Morley Research Centre, Wymondham, Norfolk NR18 9DB
Dr H Smith	IACR-Broom's Barn, Higham, Bury St Edmunds, Suffolk IP28 6NP
Dr J Sweet	NIAB, Huntingdon Road, Cambridge CB3 0LE
Dr I Williams	IACR-Rothamsted, Harpenden, Herts AL5 2JQ

Administration Manager

Mr C Todd	BCPE, 49 Downing Street, Farnham, Surrey GU9 7PH

Editor-in-Chief & Press Manager

Miss F McKim	BCPE, 49 Downing Street, Farnham, Surrey GU9 7PH

ABBREVIATIONS

Where abbreviations are necessary the following are permitted without definition

acceptable daily intake	ADI
acid equivalent	a.e.
active ingredient	a.i.
approximately	*c.*
body weight	b.w.
boiling point	b.p.
British Standards Institution	BSI
by the author last mentioned	*idem.*
centimetre(s)	cm
Chemical Abstracts Services Registry Number	CAS RN
compare	cf.
concentration x time product	ct
concentration required to kill 50% of test organisms	LC_{50}
correlation coefficient	*r*
cultivar	cv.
cultivars	cvs.
day(s)	d
days after treatment	DAT
degrees Celsius (centigrade)	°C
dose required to kill 50% of test organisums	LD_{50}
dry matter	d.m.
Edition	Edn
editor	ed.
editors	eds
emulsifiable concentrate	EC
freezing point	f.p.
for example	e.g.
gas chromatography-mass spectrometry	gc-ms
gas-liquid chromatography	glc
gram(s)	g
growth stage	GS
hectare(s)	ha
high performance (or pressure) liquid chromatography	hplc
hour	h
infrared	i.r.
integrated crop management	ICM
integrated pest management	IPM
International Standardisation Organisation	ISO
in the journal last mentioned	*ibid.*
Joules	J
Kelvin	K
kilogram(s)	kg
least significant difference	LSD
litre(s)	litre(s)
litres per hectare	litres/ha
mass	*m*
mass per mass	*m/m*
mass per volume	*m/V*
mass spectrometry	ms
maximum	max.
melting point	m.p.
metre(s)	m
milligram(s)	mg
milligrams per litre	mg/litre
milligrams per kg	mg/kg
millilitre(s)	ml
millimetre(s)	mm
minimum	min.
Ministry of Agriculture Fisheries and Food (England & Wales)	MAFF
minute (time unit)	min
molar concentration	M
no observed adverse effect level	NOAEL
no observed effect concentration	NOEC
no observed effect level	NOEL
nuclear magnetic resonance	nmr
number average diameter	n.a.d.
number median diameter	n.m.d.
organic matter	o.m.
page	p.
pages	pp.
parts per million	ppm
pascal	Pa
percentage	%
post-emergence	post-em.
power take off	p.t.o.
pre-emergence	pre-em.
pre-plant incorporated	ppi
probability (statistical)	*P*
relative humidity	r.h.
revolutions per minute	rev/min
second (time unit)	s
standard error	SE
standard error of means	SEM
soluble powder	SP
species (singular)	sp.
species (plural)	spp.
square metre	m^2
subspecies	ssp.
surface mean diameter	s.m.d.
suspension concentrate	SC
systemic acquired resistance	SAR
technical grade	tech.
temperature	temp.
that is	*i.e.*
thin-layer chromatography	tlc
time for 50% loss; half life	DT_{50}
tonne(s)	t
ultraviolet	u.v.
United Kingdom	UK
United States Department of Agriculture	USDA
vapour pressure	v.p.
variety (wild plant use)	var.
volume	*V*
weight	*wt*
weight by volume	*wt/v*
(mass by volume is more correct)	(*m/V*)
weight by weight	*wt/wt*
(mass by mass is more correct)	*(m/m)*
wettable powder	WP

less than		<
more than		>
not less than		≮
not more than		≯
Multiplying symbols-		Prefixes
mega	($\times 10^{6}$)	M
kilo	($\times 10^{3}$)	k
milli	($\times 10^{-3}$)	m
micro	($\times 10^{-6}$)	μ
nano	($\times 10^{-9}$)	n
pico	($\times 10^{-12}$)	p

SESSION 1

GENETICALLY MODIFIED CROPS – THE CURRENT SITUATION

Chairman & Session Organiser: Dr P Lutman
IACR-Rothamsted, Harpenden

Transgenic plants: field testing and commercialisation including a consideration of novel herbicide resistant oilseed rape (*Brassica napus* L.)

S Barber
Organisation for Economic Co-operation and Development (OECD), Paris

ABSTRACT

From the first experimental field release of a transgenic plant (TP) in 1986, over 60 different species have been transformed using recombinant DNA techniques and released in confined experimental field trials. Following the first commercial TP release in 1992 the cultivated area of TP's has increased dramatically. In 1998 at least 69 million acres of TP's with novel herbicide tolerance and insect resistance were grown, predominantly in North and South America. In Europe, few TP's have received regulatory approval and cultivation of these crop varieties is very limited. Of the TP's already grown on a large acreage, *Zea mays* (maize/corn), *Brassica napus* (oilseed rape/canola) and *Gossypium hirsutum* (cotton) naturally outcross within the species, and of these *B. napus* is known to be able to outcross with certain related species that occur in some production areas. Because of this, the incorporation of novel herbicide resistance into *B. napus* using rDNA technology is viewed by many people as a major concern.

EXPERIMENTAL RELEASES OF TRANSGENIC PLANTS

Summary statistics from the OECD Field Trial Database (see Internet URL's) show that 98.3% of experimental releases of genetically modified organisms in Member countries are transgenic plants. These are made up of 38% *Zea mays*, 13% *Brassica* sp., 12% *Solanum tuberosum*, 10% *Lycopersicon esculentum*, 9% *Glycine max*, 7% *Gossypium hirsutum*, 5% *Nicotiana tabaccum*, 2% *Beta vulgaris*, with the remaining 4% made up from the other species listed in Table 1. The first experimental release[1] took place in 1986 and the number rose steadily to a maximum of just over 1400 in 1995. Releases declined to 1300 and 900 respectively in 1996 and 1997 as transgenic plant material in regulated "confined" trials in previous years were approved (deregulated) for commercial use. By far the greatest number of experimental releases amongst the OECD Member countries (70.5%) have been in the United States followed by Canada (11.9%), France (4.7%), Belgium (2%), UK (1.8%), Italy (1.7%), Holland (1.5%), Spain (1.2%) and Japan (1.2%) with 3.5% of releases in the remaining OECD Member countries. China has an active programme for testing and commercialising TP's (James, 1997), however, this data is not included in this paper.

Numerous novel traits have been introduced into TP's. Many early confined field releases were of plants that expressed marker proteins, in particular those expressed by the GUS gene, NPTII antibiotic resistance gene and herbicide resistance genes. These field tests were primarily to determine whether the growth, development and reproductive biology of the genetically modified host plant had been altered by the actual process or rDNA insertion. Apart from the delayed ripening tomato that provided benefits directly to the

[1] A release is an approval/permit to test a specific modification in a specific plant in one <u>or more</u> locations.

consumer, modifications were intended to be of benefit to the crop production and processing industries. TP's with novel herbicide resistance, insect resistance, virus resistance and a male sterility system for hybrid seed production, together with tomato lines with altered ripening qualities, made up the bulk of experimental field tests. More recent experimental releases have included TP's with tolerance to abiotic stress, resistance to fungal pathogens, modified quality (such as oil, starch and livestock feed components in the harvested part of the plant) and those producing specialty chemicals such as vaccines, pharmaceutical compounds and "feed" compounds for industrial processes such as plastics production.

Table 1. Plant species subject to rDNA modification and experimentally tested in confined field trials. (OECD Field Trial Database)

Actinidia deliciosa	Kiwi fruit	*Ipomea batatas*	Sweet potato
Agrostis stolonifera	Creeping bentgrass	*Juglans* sp.	Walnut
Allum cepa	Onion	*Lactuca sativa*	Lettuce
Arabidopsis thaliana	Thale Cress	*Linum usitatissimum*	Flax
Arachis hypogea	Peanut	*Liquidambar* sp.	Sweetgum
Asparagus officinalis	Asparagus	*Lupinus angustifolius*	Lupin
Atropa belladona	Belladonna	*Lycopersicon esculentum*	Tomato
Beta vulgaris	Sugar beet	*Malus domestica*	Apple
Betula pendula	Silver Birch	*Medicago sativa*	Alfalfa
Brassica carinata	Ethiopian mustard	*Nicotiana benthamiana*	Tobacco
Brassica juncea	Mustard	*Nicotiana tabaccum*	Tobacco
Brassica napus	Canola/Oilseed rape	*Oryza sativa*	Rice
Brassica oleraceae	Cabbage/Broccoli etc.	*Pelargonium* sp.	Pelargonium
Brassica rapa	Canola/Oilseed rape	*Picea abies*	Norway spruce
Capsicum annuum	Pepper	*Picea* sp.	Spruce
Carica papaya	Papaya	*Pinus sylvestris*	Scots Pine
Castanea sp.	Chestnut	*Pisum sativum*	Pea
Cichorium intybus	Chicory	*Populus* sp.	Poplar
Citrullus lanatus	Watermellon	*Prunus domestica*	Plum
Cucumis melo	Melon	*Rosa* sp.	Rose
Cucumis sativus	Cucumber	*Saccharum officinarum*	Sugar cane
Cucurbita pepo	Squash	*Sinapis alba*	White mustard
Daucus carota	Carrot	*Solanum melongena*	Eggplant
Dianthus caryophyllatus	Carnation	*Solanum tuberosum*	Potato
Eucalyptus sp.	Eucalyptus	*Tagetes* sp.	Marigold
Fragaria sp.	Strawberry	*Trifolium subterraneum*	Clover
Gladiolus sp.	Gladiolus	*Triticum aestivum*	Wheat
Glycine max	Soybean	*Vaccinium oxycoccus*	Cranberry
Gossypium hirsutum	Cotton	*Vitis vinifera*	Grape
Helianthus annuus	Sunflower	*Zea mays*	Corn/Maize
Hordeum vulgare	Barley		

COMMERCIAL RELEASES OF TRANSGENIC PLANTS

The regulatory procedures for approving TP's for commercial release and marketing fall into three safety assessment and approval areas: release into the environment; human food use; and, animal feed use. They follow a step wise approach that has evolved over 20 years of collaborative work among experts from governments, international government organisations, non-governmental organisations, academia and industry associations, and although different regulatory agencies in different countries use different legal instruments for the purpose of regulation, the procedures and information required for safety assessments are remarkably similar (OECD 1995a, 1995b).

The OECD Product Database, part of the OECD "BioTrack Online" internet site (see Internet URL's) provides information on commercial TP approvals for growing, food and feed use in Member countries (Table 2). These now include some TP's with stacked genes resulting in two or more novel traits in the same plant cultivar.

Table 2. Transgenic plant species with novel traits approved for commercial release by at least one regulatory agency (OECD Product Database)

Beta vulgaris	novel herbicide[1,2,3] resistance
Brassica napus	novel herbicide[1,2,3] resistance, male sterility and novel herbicide[2] tolerance stacked, modified fatty acid profile
Brassica rapa	novel herbicide[1] resistance
Carica papaya	virus resistance
Cichorium intybus	male sterility and novel herbicide[2] resistance stacked
Cucurbita pepo	virus resistance
Dianthus caryophyllus	novel flower colour, extended vase life
Glycine max	novel herbicide[1,2] resistance
Gossypium hirsutum	lepidopteran insect resistance, novel herbicide[1,3,4] resistance lepidopteran insect resistance and novel herbicide[3] resistance stacked
Linum usitatissimum	novel herbicide[4] soil residue resistance
Lycopersicon esculentum	modified ripening, lepidopteran insect resistance
Nicotiana tabaccum	novel herbicide[2,3] resistance
*Oryza sativa**	novel herbicide[2] resistance
Solanum tuberosum	coleopteran insect resistance, virus resistance, coleopteran insect resistance and virus resistance stacked
Zea mays	novel herbicide[1,2,3,4] resistance, lepdipopteran insect resistance, lepidopteran insect resistance and novel herbicide[1,2] resistance stacked, male sterility and novel herbicide[2] tolerance stacked

* Under review by USDA (see Internet URL's)
[1] Glyphosate herbicide resistance
[2] Phosphinothricin (glufosinate ammonium) herbicide resistance
[3] Oxynil herbicide resistance
4 Sulfonyl urea herbicide resistance

CULTIVATION OF TRANSGENIC CROPS WORLD-WIDE

James (1997, 1998) at the International Service for the Acquisition of Agri-Biotech Applications has collated data on the acreage of commercial TP production. In 1996 approximately 7 million acres of 7 transgenic crops were grown in the USA, China, Canada, Argentina, Australia and Mexico. In that year the principal crop was *Nicotiana tabaccum* (tobacco in China) followed by *Gossypium hirsutum* (cotton), *Glycine max* (soybean), *Zea mays* (corn), *Brassica napus* (oilseed rape/canola), *Lycopersicon esculentum* (tomato) and *solanum tubersoum* (potato). Of the transgenic crop area, 57% was in industrial countries and 43% in developing countries (principally China with 2.7 million acres). By trait, virus resistance (40%, again almost entirely in China) was followed by insect resistance (37%), herbicide [novel] resistance (23%) and quality traits (<1%). Data for 1996, 1997 and 1998, are presented in Tables 3 and 4, and country production for 1997 and 1988 is shown in Table 5. The adoption of TP culture, particularly in North and South America, has been rapid. (NOTE that data for China are not presented in the tables.)

Table 3. Acreages of TP's by species (James, 1997 and 1998), excluding China

By crop plant species	1996		1997		1998	
	10^6 Acres	%	10^6 Acres	%	10^6 Acres	%
G. max (soybean)	1.3	31	12.8	46	36.3	52
Z. mays (maize/corn)	0.7	17	8.0	30	20.8	30
G. hirsutum (cotton)	1.9	45	3.5	13	6.3	9
B. napus (oilseed rape/canola)	0.3	7	3.0	11	6.0	9
S. tuberosum (potato)	<0.1	<1	<0.3	<1	<0.3	<1
Total	4.2	100	27.5	100	69.50	100

Table 4. Acreages of TP's by novel trait (James, 1997 and 1998), excluding China

By novel trait	1996		1997		1998	
	10^6 Acres	%	10^6 Acres	%	10^6 Acres	%
Novel herbicide resistance	1.6	38	17.3	63	49.5	71
Insect resistance	2.6	62	10.0	36	19.3	28
Novel herbicide resistance and insect resistance stacked	-	-	<0.3	<1	0.8	1
Quality traits	<0.1	<1	<0.3	<1	<0.3	<1
Total	4.2	100	27.5	100	69.50	100

Table 5. Acreages of TP's by country (James, 1997 and 1998) excluding China, and, 1996 FAO data showing total arable land

Country	1997		1998		1996
	acres x 10^6	%	acres x 10^6	%	Total arable acreage x 10^6
USA	20.3	74	51.3	74	432.098
Argentina	3.5	13	10.8	15	61.728
Canada	3.3	12	7.0	10	112.000
Australia	0.3	1	0.3	1	123.484
Mexico	<0.3	<1	0.3	1	62.444
Spain	0	0	<0.3	<1	37.615
France	0	0	<0.3	<1	45.156
South Africa	0	0	<0.3	<1	37.000
United Kingdom	-	-	-	-	15.037
Total	27.5	100	69.5	100	926.562

TRANSGENIC HERBICIDE TOLERANT PLANTS

Data for 1998 (Table 4) show that of the novel traits grown world-wide, novel herbicide resistance had the largest acreage (71%), followed by *Bacillus thuringiensis* (Bt) ∂endotoxin derived insect resistance (28%), combined Bt insect resistance and herbicide resistance (1%) and quality traits (<1%).

For some years the popular press has given much attention to the issue of transgenic herbicide resistant crop plants, though generally not identifying this as "novel" herbicide resistance, and has emphasised the potential for development of superweeds (not clearly defined), generating much public concern and scientific and political debate. This is particularly true in the case of transgenic herbicide resistant oilseed rape/canola (*B. napus*). because of its natural ability to exchange genes, as either father or mother, with certain of its related species (OECD, 1997).

HERBICIDES - *BRASSICA NAPUS*: THE EXISTING PARADIGM

The development of new crop plant varieties results from a collaborative effort among several interested parties, and the growth of the Canadian oilseed rape/canola industry over the past 40 years is a good example of this. Food retailers in many countries provide information of their, and their consumers', preferences to the food processors and commodity shippers who in turn advise the plant breeders and the farming community of the quality requirements of the oilseed. Farmers also provide the plant breeders with the agronomic quality characteristics they require. This close collaboration among plant breeders, growers, processors and oilseed marketers has resulted in the regular setting of ever higher standards which have to be met by new canola cultivars (*B. napus* and *B. rapa*) if they are to become registered varieties. Processors have required higher seed oil and protein with low to zero levels of erucic acid in the oil and glucosinolates in the meal. Growers have required higher seed yield, improved disease and insect resistance, reduced shattering and lodging, and uniform plant growth and time to maturity (Downey and Röbbelen, 1989).

During this same time selective herbicides have been used by growers to control weeds with little or no damage to the crop plant, increasing yields for farmers and reducing weed seed "dockage" for processors. Selective herbicides have been a key part of the Canadian canola industry success story. Quality for the oilseed processors was further improved with the introduction of selective herbicides controlling species closely related to canola e.g., *Sinapis arvensis* (wild mustard) that when harvested with the crop could elevate levels of both erucic acid and glucosinlates in the final product. This was the existing canola variety breeding paradigm when rDNA technologies first became available to plant breeders.

HERBICIDE RESISTANT WEEDS: THE EXISTING PARADIGM

It is doubtful that all of the almost 1 billion acres of arable land shown in the countries listed in Table 5 have herbicide applied to crops grown on them, yet it is questionable that any of those acres are sown to crops that don't have tolerance[2] to one or more selective herbicides, i.e., their billion acres already have had herbicide tolerant crops growing on them. A major goal of herbicide producers over 50 years has been to develop "selective" herbicides that control as many weed species as possible without damaging crop plant species, and in this there has been good success. However, as predicted by Harper (1957) ill conceived management practices that relied on the same herbicidal mode of action in the same fields for several years resulted in the selection of those rare, naturally occurring, individuals within weed species populations that are resistant. The Weed Science Society of America (WSSA) provides data on weed species that have developed populations resistant (novel resistance) to specific herbicidal modes of action (Heap: see Internet URL's). WSSA reports 216 instances worldwide; 120 dicots and 216 monocots. Triazine herbicides have the most resistant weed species (60) the acetolactate synthase (ALS) disrupting herbicides (53), bipyridliums (26) and ACCase inhibitors (19). Some species, especially grasses, now have multiple herbicide resistances. These naturally occurring, but rare herbicide resistances can also be found in somaclonal variants, and in weed species closely related to crop plants. Some of these resistances have been incorporated into commercially available *B. napus* and *B. rapa* varieties. In Canada, plants with novel herbicide resistance (plants with novel traits) are subject to safety assessment no matter how they are developed, however, in other countries they are not subject to the same regulation and safety assessment as their rDNA derived counterparts, even should they have resistance to the same herbicide family.

---

[2] The Weed Science Society of America uses the term "resistance" to describe the inherent ability of a plant to survive and reproduce following exposure to a dose of herbicide normally lethal to the wild type. In a plant, resistance may be naturally occurring or induced by such techniques as genetic engineering or selection of variants produced by tissue culture or mutagenesis.
Herbicide "tolerance" is the inherent ability of a species to survive and reproduce after herbicide treatment. This implies that there was no selection or genetic manipulation to make the plant tolerant; it is naturally tolerant.

HERBICIDE RESISTANT *BRASSICA NAPUS*: THE NEWER PARADIGM

Maltais and Bouchard (1978) discovered triazine tolerant birds rape (*B. campestris* [=*B. rapa*]) in corn (*Z. mays*) fields in Québec where triazine herbicides had been applied over many years. This maternally inherited triazine resistance was incorporated into cultivated lines of *B. napus* and *B. rapa* oilseed rape (Beversdorf *et al.* 1980) and it was suggested the acquired triazine resistance could facilitate new and additional methods of weed control in oilseed rape. Triazine tolerant oilseed rape/canola varieties (*B. napus* and *B. rapa*) were developed, registered and have been grown in Canada especially where there are severe problems with related weed species, in particular *Thlaspi arvense*. Acreages have never been great since the reduced photosynthetic ability of the triazine resistant varieties resulted in a penalty, associated with a mutant photosystem II protein that prevented triazine herbicide binding, which lowered seed yielding ability.

WSSA data show that *Brassicaceae* weed species related to the oilseed rapes that have developed (been selected for) populations with novel herbicide resistance are *Brassica campestris* [=*rapa*] (oilseed rape/canola, bird rape, turnip), *B. tournefortii* (wild turnip), *Capsella bursa-pastoris* (shepherd's purse), *Raphanus raphanistrum* (wild radish), *Sinapis arvensis* (wild mustard), *Sisymbrium orientale* (Indian hedge mustard) and *S. thellungii* (African turnip weed). Of these, *B. rapa* and *R. raphanistrum* may naturally exchange genes with *B. napus*, although the potential for an exchange resulting in fertile offspring with *B. rapa* is very much greater that for *R. raphanistrum,* the latter being remote. As *B. napus* is an interspecific amphidiploid (n=19) sharing chromosomes from its two parents, *B. oleraceae* (n=9) and *B. campestris* [=*rapa*](n=10) this is expected (OECD, 1997). The probability of selecting mutant herbicide resistant *R. raphanistrum* compared with the probability of introgressing a transgene from *B. napus* conferring the same herbicide resistance is not known. These issues are discussed in detail in later papers.

Table 6 shows estimated acreages of Canadian registered varieties of *B. napus* resistant to glyphosate, phosphinothricin and imidazolinone herbicides. All have received regulatory approval (see Canadian Food Inspection Agency: Internet URL's).

Table 6. Estimated acreages of novel herbicide resistant oilseed rape/canola (*B. napus* and *B. rapa*) grown in Canada, 1996 – 1998. (Zeph: Pioneer HiBred International, pers. comm.)

Plant type	1996		1997		1998	
	10^6 Acres	%	10^6 Acres	%	10^6 Acres	%
B. napus varieties	6.21	69.0	10.41	86.0	12.24	92.0
B. rapa varieties	2.60	62.0	1.69	14.0	1.06	8.0
TOTAL	9.00	100.0	12.10	100.0	13.30	100.0
Glyphosate resistant	0.05	0.5	0.73	6.0	3.19	24.0
Phosphinothricin resistant	0.40	4.5	2.06	17.0	2.00	15.0
Imidazolinone resistant*	0.54	6.0	1.57	13.0	2.13	16.0
TOTAL novel herbicide resistant	0.99	11.0	4.36	36.0	7.32	55.0

*Imidazolinone resistance is NOT transgenic, but has been developed through somaclonal variation.

ENVIRONMENTAL SAFETY ASSESSMENT PARADIGMS FOR TRANSGENIC HERBICIDE RESISTANT *BRASSICA NAPUS*

In North America, the environmental safety assessment paradigm for TP's is based on a comparison of the novel varieties with their traditionally developed counterparts. This follows the environmental safety assessment model developed by the OECD Member countries (OECD 1993). The considerable "familiarity" gained from variety development and agronomic management practices have been used as a baseline in evaluating the potential risk of hazard of crops with novel herbicide resistance. Of course, the potential for different environmental interactions resulting from the presence of the novel proteins/enzymes conferring the novel trait are also considered. In the case of oilseed rape/canola it is accepted that gene flow between *B. napus* and certain related species, depending on the location of the release, might occur and that even though there is a remote possibility of introgression of herbicide resistance genes into these species, other existing cultural and chemical means can be used to control them. A more probable outcome is that gene exchange among oilseed rape/canola crops and its volunteers will lead to stacked transgenes in certain individual volunteer plants (see Canadian Food Inspection Agency Decision Documents: Internet URL's). Nonetheless it is considered that sound management practices will prevent serious problems from arising, whilst at the same time providing growers and oilseed rape/canola processors with improved "quality" oilseed rape/canola varieties.

In Europe there does not yet appear to be a consistent safety assessment and approval paradigm for oilseed rape/canola varieties with novel herbicide resistance. Recombinant DNA derived herbicide resistant varieties are subject to EC 90/220 regulation and safety assessment, whereas wide cross and somaclonal variant derived varieties with novel herbicide resistances are not. This suggests that the existing paradigm for breeding and growing oilseed rape varieties using registered selective herbicides may not be considered adequate as a baseline for evaluating the new transgenic herbicide resistant varieties.

REFERENCES

Beversdorf W D; Weiss-Lerman J; Erickson L R; Souza-Machado V (1980). Transfer of cytoplasmically inherited triazine resistance from bird's rape to cultivated oilseed rape (*Brassica campestris* and *B. napus*). *Canadian Journal of Genetic Cytology* **22(2)**, 167-172

Downey R K; Röbbelen G (1989). Brassica Species. In: *Oil Crops of the World,* eds G Röbbelen, R K Downey & A Ashri, pp 339 - 362. McGraw-Hill: New York.

Harper J L (1957). The evolution of weeds in relation to resistance to herbicides. *Proceedings 3rd British Weed Control Conference*. **3**, 179-188.

Internet URL's:

Canadian Food Inspection Agency (1999): Decision Documents on Determination of Environmental Safety: http://www.cfia-acia.agr.ca/english/plant/pbo/home_e.html

Heap I (1999). International Survey of Herbicide Resistant Weeds: http://www.weedscience.com

OECD (1999). BioTrack Online: http://www.oecd.org/ehs/service.htm

OECD (1999). Field Trials:Database: http://www.oecd.org/ehs/summary.htm

OECD (1999). Products Database: http://www.olis.oecd.org/bioprod.nsf

USDA (1999). Status of Petition 98-329-01p: http://www.aphis.usda.gov/bbep/bp/petday.html

James C (1997) *Global Status of Transgenic Crops in 1997. ISAAA Briefs No. 5.* ISAAA, Ithaca NY. USA

James C (1998) *Global Review of Commercialized Transgenic Crops: 1998. ISAAA Briefs No. 8.* ISAAA, Ithaca NY. USA

Maltais B; Bouchard C J (1978). Une moutarde des oiseaus (*Brassica campestris* L.) resistante a l'atrazine. *Phytoprotection* **59**, 117 - 119

OECD (1995a). *Commercialisation of Agricultural Products Derived through Modern Biotechnology: Survey Results.* OECD Environment Monograph No. 99, Organisation for Economic Co-operation and Development, Paris, France

OECD (1995b). *Analysis of Information Elements Used in the Assessment of Certain Products of Modern Biotechnology.* OECD Environment Monograph No. 100, Organisation for Economic Co-operation and Development, Paris, France

OECD (1992). *Safety Considerations for Biotechnology: Scale-up of Crop Plants.* Organisation for Economic Co-operation and Development, Paris, France.

OECD (1997). *Consensus Document on the Biology of* Brassica napus *L. (Oilseed rape) Series on the Harmonization of Regulatory Oversight in Biotechnology No 7.* Organisation for Economic Co-operation and Development, Paris, France.

Deliberate release of genetically modified organisms: the UK regulatory framework

P Burrows
*Department of the Environment, Transport and the Regions, Ashdown House,
123 Victoria Street, London SW1E 6DE, UK.*

ABSTRACT

This paper reviews the development and operation of the regulations covering the release of genetically modified organisms (GMOs) in the UK. The regulatory framework aims to prevent or minimise damage to the environment by establishing a statutory system of risk assessment and prior consent, before any GMO can be released or marketed. Directive 90/220/EEC *on the deliberate release into the environment of genetically modified organisms* does much to harmonise the release of GMOs across the European Community. The Directive recognises two classes of release depending of the purpose; Part B releases for research and Part C releases for marketing. Individual Member States differ in how the Directive has been implemented and in the UK. No GMO may be released without the express consent of the Secretary of State. All UK applications for Part B (research) releases are submitted to the Department of the Environment Transport and the Regions, and the Advisory Committee on the Releases to the Environment (ACRE) advise whether the release should be allowed. In issuing consent the Secretary of State will also take account of views expressed by other Government Departments, the Statutory Nature Conservation Bodies and the general public. Applications for Part C (marketing) consent follow a different procedure and are submitted initially to any one of the 15 Member States, which then review the application and form an opinion. The other 14 States then evaluate the application. If there are no objections the lead Member State issues the marketing consent which applies throughout the European Community. If one or more State objects to the application then it falls to the Commission and Council to resolve. The current system, based on Directive 90/220/EEC, has been criticised mainly because of lengthy delays in obtaining marketing consent. Discussions are in progress to revise it in a number of key areas. An overview of the main proposed changes is given in the paper.

INTRODUCTION

No discussion of gene flow in agriculture can currently stray far from the challenges - real or imaginary - presented by the environmental release of genetically modified organisms (GMOs). The anti-GM pressure groups are quick to raise the spectre of gene flow from GM crops to wild relatives leading to 'superweeds' or, at the very least, the 'genetic pollution' of our botanical national heritage. In response it is often pointed out that herbicide tolerant weeds and hybridisation between conventional crops and their wild relatives or sexually compatible neighbouring crops are already common in the agricultural environment and not a feature unique to GM. Although they might be described as a nuisance in some agricultural situations they are not a significant threat to the environment.

Faced with the media hysteria that has surrounded GM crops and GM food in recent weeks, it comes as a surprise to many genuinely concerned observers that genetic modification is actually tightly regulated in the UK. In fact there are several layers of risk assessment and safety testing that take the GMO right from initial research and development in the laboratory through to testing in the environment and finally placing a product on the market.

This paper outlines briefly the regulatory framework in the UK[1] in as far as it controls the deliberate release of GMOs. The emphasis is placed on the *process* whereby the regulations are put into action and consent may be obtained to release GMOs either for experimental purposes or for commercialisation.

DEVELOPMENT OF THE DELIBERATE RELEASE REGULATIONS

The regulation of genetic modification has its roots in the early 1970s when scientists working at the forefront of the emerging technology recognised its potential power and called for a 'moratorium' until a number of safety concerns had been considered. In 1976 the Genetic Manipulation Advisory Group (GMAG)[2] was set up to consider proposals for work involving genetic manipulation. Soon afterwards, the *Health and Safety (Genetic Manipulation) Regulations 1978* came into force and required that any activity involving genetic manipulation [modification] should be notified to the Health and Safety Executive (HSE). These regulations only covered GM work in containment[3], and at this time the release of GMOs to the environment was controlled by a voluntary code of practice overseen by HSE.

The first specific controls over the environmental release of GMOs in the UK were provided by Part IV of the *Environmental Protection Act 1990*. At the same time the Environmental Protection Act was being drafted the European Commission was also preparing community legislation to control deliberate release, this later emerged as the now familiar *Council Directive 90/220/EEC on the deliberate release into the environment of genetically modified organisms*. Directive 90/220 was implemented in the UK by the *Genetically Modified Organisms (Deliberate Release) Regulations 1992*[4] (amended 1995 and 1997) and came into force on 1 February 1993.

AIMS AND PROVISIONS OF THE REGULATIONS.

The regulatory framework provided by these various statutory instruments aims to prevent or minimise damage to the environment by establishing a statutory system of 'prior informed consent' before any GMO may be released or marketed. In the UK this means that no GMOs may be released without the express consent of the Secretary of State. The regulations set out prescribed information (see below and Box 1) which must be supplied in an application for consent to do a deliberate release. Central to the approval process is an *environmental risk*

---

[1] The UK regulatory framework is intimately linked to the wider European Community controls over GMOs
[2] GMAG latter became the Advisory Committee on Genetic Modification (ACGM) which still advises Government today on the contained use of GMOs
[3] Contained Use of GMO refers to activities in, for example, research laboratories and industrial facilities
[4] These regulations also gave effect to the substantive provisions in Part IV of the *Environmental Protection Act 1990*

Figure 1. Flow diagram for obtaining deliberate release consent (Part B) in the UK

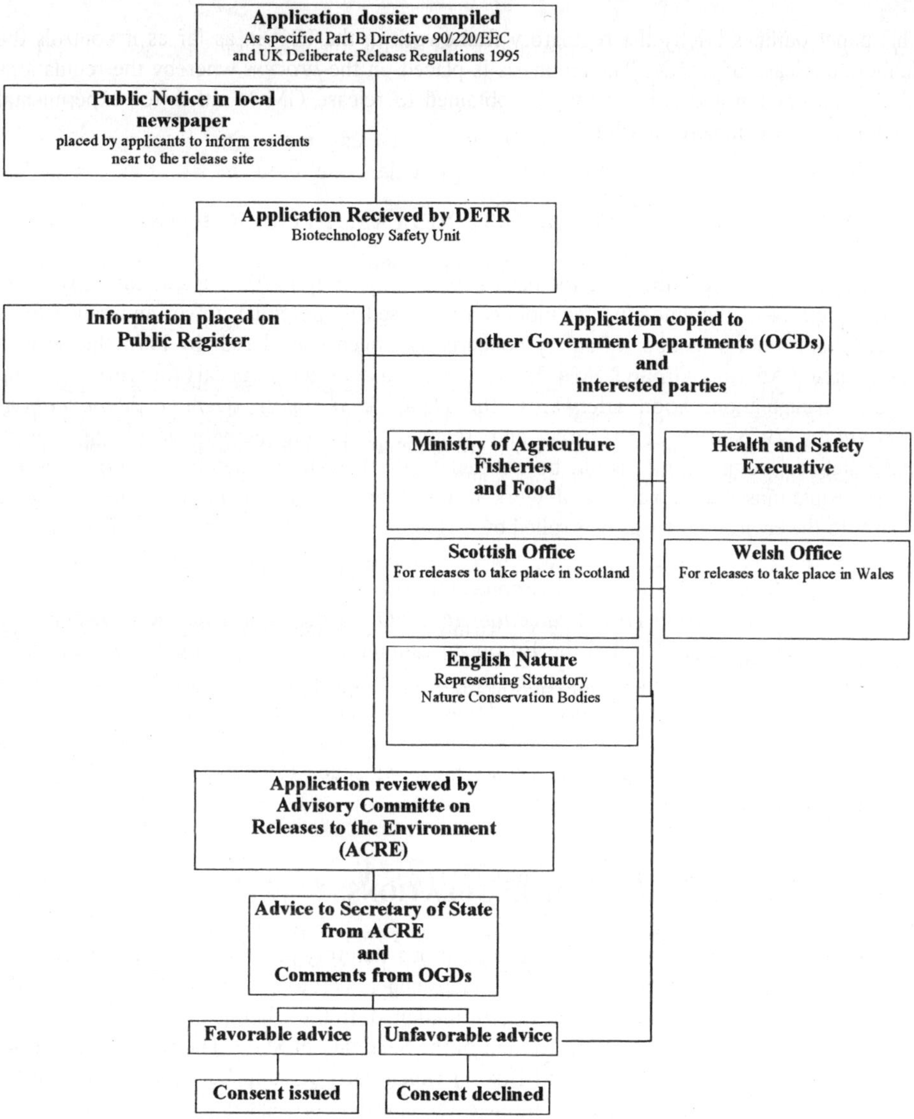

assessment which considers the potential risk to human health and the environment posed by releasing a particular GMO and, where necessary, identifies risk management procedures to avoid or minimise any damage. The regulations are framed in such a way that if it can be shown, to the best of current scientific understanding, that a release poses a low or negligible risk to the environment then it should be given the necessary consent to proceed.

Directive 90/220 recognises two broad categories for the deliberate release of GMOs[5] depending on the proposed purpose:

I. releases for *research and development*, which are made under *Part B* of the Directive, cover a number of activities but, in particular, are used mostly for conducting experimental field trials with GM crops.

II. releases for *placing on the market* are made under *Part C* of the Directive. Part C consent is necessary before a GMO may be used commercially or marketed within the European Community.

The principal procedural difference between these two categories of release is that Part B consents are granted at the national level on an individual Member State by Member State basis, whereas Part C marketing consents are granted at the European Community level. Once issued they apply across all Member States.

THE REGULATORY PROCESS: PUTTING THE REGULATIONS INTO ACTION

Although Directive 90/220 does much to harmonise the deliberate release of GMOs across the European Community there are nevertheless differences between the individual Member States in the ways in which the Directive has been implemented and the national procedures followed to obtain release consent. By way of illustration, the following section describes the process in the UK whereby Part B and Part C consent may be obtained. The example of seeking approval to release a GM plant will be used, but the process is essentially the same for all GMOs covered by the Directive.

Part B consent - releases for research and development

Applicants wishing to conduct an experimental field trial in the UK with GM crops must apply to the Secretary of State for consent to conduct the deliberate release (figure 1). The application is submitted to the Biotechnology Safety Unit of the Department of the Environment, Transport and the Regions (DETR). Applications consist of a dossier of information substantially composed of the responses to 41 prescribed questions (box 1) which are set out in schedule 1 to the 1995 regulations[6]. Taken together the questions and answers build up a detailed description of the GM plant, how it has been modified and the conditions of the proposed release. Importantly, each application dossier also contains a detailed environmental risk assessment which considers the potential harm that may be caused to

---

[5] Which are also reflected in our national deliberate release regulations

[6] *Genetically Modified Organisms (Deliberate Release) Regulations 1995*

Box 1. Information required under Schedule 1 of the 1995 Regulations for the Deliberate Release of Genetically Modified Higher Plants

General information

1. The name and address of the applicant.
2. The title of the project.

Information relating to the parental or recipient plant

3. The full name of the plant: family, genus, species, subspecies, cultivar,
4. Information on reproduction of plant: mode, generation time and the sexual compatibility with other cultivated or wild plant species.
5. Information on the survivability of plant: survival structures, dormancy etc.
6. Information concerning dissemination of plant: means, extent and factors affecting dissemination
7. The geographical distribution of the plant.
8. If the plant species is not normally grown in Member States, describe the natural habitat
9. Information on any significant interactions of the plant with organisms other than plants in the ecosystem where it is usually grown, including toxicity to humans, animals and other organisms.

Information relating to the genetic modification

10. A description of the methods used for the genetic modification.
11. The nature and source of the vector used.
12. The size, function and donor organism(s) of each DNA sequence intended for insertion.

Information relating to the genetically modified plant

13. A description of the trait(s) and characteristics of the GM plant which have been modified.
14. Information on sequences inserted or deleted: size/structure, copy number of insert, information on any vector sequences or foreign DNA remaining in the GM plant. The size/function of any deleted regions. Cellular location of insertion (e.g. chromosomal, mitochondria, chloroplast etc.).
15. Information on the expression of the insert: expression and parts of the plant where expressed.
16. How does GM plant differ from the recipient plant in: mode/rate of reproduction, dissemination, survivability
17. The genetic stability of the insert.
18. The potential for transfer of genetic material from the GM plants to other organisms.
19. Information on any toxic/harmful effects on human health and the environment arising from the genetic modification.
20. The mechanism of interaction between the genetically modified plants and target organisms.
21. Any potentially significant interactions with non-target organisms.
22. A description of detection and identification techniques for the genetically modified plants.
23. Information about previous releases of the genetically modified plants.

Information relating to the site of release

24. The location and size of the release site or sites.
25. A description of the release site ecosystem, including climate, flora and fauna.
26. Details of any sexually compatible wild relatives or cultivated plants present at the release sites.
27. The proximity of the release sites to officially recognized biotopes or protected areas.

Information relating to the release

28. The purpose of the release.
29. The foreseen dates and duration of the release.
30. The method by which the genetically modified plants will be released.
31. The method for preparing and managing the release site, prior to, during and after the release.
32. The approximate number of genetically modified plants (or plants per m^2) to be released.

Information on control, monitoring, post-release plans and waste treatment plans

33. A description of any precautions to minimize or prevent pollen or seed dispersal from GM plant.
34. A description of the methods for post-release treatment of the site or sites.
35. A description of post-release treatment methods for the GM plant material including wastes.
36. A description of monitoring plans and techniques.
37. A description of any emergency plans.

Information on potential environmental impact of the release of the genetically modified plants

38. The likelihood of the GM plant becoming more persistent or invasive than recipient plants.
39. Any selective advantage or disadvantage conferred to other sexually compatible plants species, which may result from genetic transfer from the genetically modified plant.
40. Potential environmental impact of the interaction between the GM plant and target organisms.
41. Any possible environmental impact resulting from potential interactions with non-target organisms.

human health and the environment and identifies ways in which the risks may be avoided or minimised.

On receipt of the application dossier, it is initially reviewed by specialist scientists (case officers) in the DETR Biotechnology Safety Unit who copy it for comment to other Government Departments which also have an interest. These Departments are the Ministry of Agriculture Fisheries and Food, the Health and Safety Executive, the Scottish Office[7] and the Welsh Office[8]. Details of each application and the environment risk assessment are also copied to English Nature, which in GMO matters represents the statutory Nature Conservation Bodies.

The deliberate release regulations also specify certain information[9] about each application which is required to be placed on a statutory public register[10] within 12 days of receipt of the application by DETR. This information includes, among other things, a summary of the application written in non-technical language, the location at which the release will take place and the environmental risk assessment. In practice, DETR go much further than is required by law and make all of each application available to the public, excluding the names of private individuals and any information agreed to be 'commercial in confidence'.

To aid transparency in the regulatory process the applicants are required to place an announcement of the proposed release in a newspaper circulating in the area where the release is planned. The announcement must be placed within 10 days of submitting the application to DETR and it gives concerned or curious members of the public the opportunity to write the Secretary of State to object or seek further information about the release.

Most applications when they are first submitted to DETR contain inadequate or ambiguous information and the case officer has the statutory power to request more information from applicants or clarify specific points. Once the case officer is content that the application contains sufficient detail and is compliant with the regulations then the dossier is put before the statutory Advisory Committee on Releases to the Environment (ACRE). ACRE is a scientific and technical committee made up of leading experts in subjects such as ecology, plant breeding, microbiology, plant and animal molecular biology and toxicology. The Committee advises the Government on the potential risks to human health and the environment from the release of genetically modified organisms. All members of ACRE are appointed by the Secretary of State for the Environment, together with the Secretaries of State for Scotland, Wales and Northern Ireland and the Minister for Agriculture Fisheries and Food. Members are appointed on the basis of their technical and scientific expertise. They do not represent any particular stakeholder interests such as the biotechnology industry or environmental pressure groups.

ACRE reviews all of the information in applications put before it with particular emphasis on the risk assessment and any proposed risk management procedures. Gene flow is always an

[7] If a release is to take place in Scotland
[8] If a release is to take place in Wales
[9] Information prescribed under Part V section 17 of the Genetically Modified Organisms (Deliberate Release) Regulations 1992
[10] Held at DETR in London. Copies of the register are also held at regional offices of the Environment Agency and at the Scottish Office in Edinburgh

important part of the risk assessment where both the potential for gene flow and its consequences are considered. Often the Committee will advise that the risk management is not sufficient and indicate what must be done to make it acceptable. Similarly the Committee may instruct the Secretariat (provided by DETR Biotechnology Safety Unit) to seek more information from applicants. When ACRE is content that the proposed release poses a low risk to human health and the environment, then the Committee will advise the Secretary of State that the consent may be granted. In reaching a final decision the Secretary of State will also take into account views expressed by other Government Departments, English Nature and any letters received from the general public in response to the initial Public Notice placed in a local newspaper. If ACRE have expressed a favourable opinion and there are no scientific objections from the other parties then consent will be granted.

Part C consent - for placing on the market

Marketing applications (also called 'notifications') are submitted initially to any one of the 15 Member States. The selected country then takes the regulatory lead on that particular application. The lead Member State has 90 days in which to form an opinion on the application. During this time additional information may be requested from the applicants and the clock is stopped while this is supplied. Marketing notifications that come to the UK, as the lead competent authority, go through the same strict evaluation process outlined for the Part B releases above, in that information is placed on the statutory public register and the dossiers are copied to other Government Departments, including Northern Ireland in the case of marketing, and English Nature. ACRE is asked to advise the Secretary of State whether the proposed commercialisation of a particular GMO poses a risk to human health and the environment.

After reviewing the dossier, if the lead competent authority is content for the GMO to be placed on the market in the European Community then the application dossier is forwarded to the Commission with a favourable opinion. The Commission must then circulate the dossier to the other 14 Member States who have 60 days in which to comment. During the 60 days Member States can request further information but there is no opportunity to stop the clock. If none of the Member States object to the application then the Commission instructs the lead country to issue the marketing consent which applies across all Members States.

If however one or more Member States objects then it falls to the Commission to make the decision. In practice, to do this the Commission seeks advice from the Scientific Committee on Plants (SCP) which serves a purpose much the same as ACRE. If the SCP reaches a favourable opinion on the marketing application the Commission will propose that it is given consent, and asks Member States to participate in a qualified majority vote (QMV) on whether to adopt the Commission's proposal to place the GMO on the market. If the QMV is in favour then the original lead country issues the consent. If the QMV is against placing the product on the market (i.e. the Member States reject the Commission's proposal) then the final decision is referred to the Council of Ministers. The Council have 90 days to reach a decision but at this stage according to the comitology (the voting and administrative procedures) it can only block the GMO from getting marketing consent by a unanimous decision. Otherwise the Commission's original proposal is accepted and the product is given consent.

THE DELIBERATE RELEASE REGULATIONS: FUTURE DEVELOPMENTS

With the current scepticism and mistrust of GM in Europe it is unlikely that there will be any major effort, at least in the next several years, to deregulate GMOs - even if increasing experience world wide in releasing and using GMOs begins to suggest that our worst fears are unfounded. Nevertheless something may be done to make the existing regulations work better.

The current Directive 90/220 has been in operation for several years and has attracted some criticism mainly because the procedures for getting marketing consent are slow and lack transparency. Furthermore, like all legalisation of this sort it is only when it has been operating for some time that areas in which there is room for improvement are revealed. Discussions are currently in progress to revise Directive 90/220 based on a text put forward by the Commission early in 1998. The main elements[11] of the proposed revision are set out in the paragraphs below. However, it should be emphasised that there is still great deal of negotiating to be done within the EU system and it is not at all certain what form the revisions will take in the final text. There have already been several rounds of talks at Member State level during the Austrian presidency and the European Parliament has recently commented on the proposal. It is already clear that some of the proposed changes are significantly more 'problematic' than others.

The main proposed changes to Directive 90/220 are:

I. procedural time limits - clear time limits are set out in which particular actions must be taken during consideration of marketing applications. Even if all of the proposed procedural steps are taken to the allowed time limit it should still be possible to get marketing consent within one year from submitting the dossier to the lead Member State.

II. risk assessment - clarification of the risk assessment and harmonisation cross Member States. The scope of the risk assessment is to include 'direct, indirect, immediate and delayed' effects.

III. monitoring and time limited marketing consents - specific provision is included for mandatory monitoring of commercial releases post marketing consent. Marketing consents will have a time limit after which they are again reviewed. The time limit suggested by the Commission is seven years.

IV. transparency and labelling - the Commission will make the content of the marketing applications available for public comment and there will be greater transparency at Community level. There is commitment for GMOs to be labelled 'in accordance with Community policy'. Current requirements are that the GMO product is labelled where the gene or gene product can be detected.

V. ethics - the Commission will be able to consult a committee that it appoints on the ethical issues raised by biotechnology.

---

[11] for a more detailed consideration and comment on the proposed changes to Directive 90/220 see *EC Regulation of Genetic Modification in Agriculture*, House of Lords Select Committee on the European Communities 2nd report, 15 December 1998. London; the Stationary Office.

VI. comitology - this is essentially a change in the regulatory committee procedures followed by the Commission[12] in reaching a decision. The main effect is to strengthen the role of the Council of Ministers in agreeing or blocking proposals for marketing put forward to it by the Commission. For example, under the proposed new comitology the Council may agree a Commission proposal to give marketing consent by a QMV, or reject it by a simple majority.

OTHER CHANGES

When Directive 90/220/EEC was first produced it covered the environmental release of all GMOs, but over the past few years the primary regulatory responsibility for some GMOs has been removed. GM medicines and vaccines[13] and novel foods[14] consisting of or containing GMOs are now covered by separate regulations. There are future plans to remove responsibility for other GMOs such as GM seeds and animal feed containing GMOs. Eventually it is expected that Directive 90/220 will cease to exist, as all of its elements are covered by other regulations.

CONCLUDING REMARKS

The deliberate release of GMOs is tightly regulated in the UK and intimately linked to the wider European Community. It is accepted that the regulations may not be completely perfect and there is always room for improvement, but despite what is claimed in sensational stories in the press the regulations do provide a very effective safeguard for human health and the environment. The UK is generally regarded as being a 'Gold Standard' in Europe and other Member States look to us for a lead on key policy issues.

It is difficult to imagine a greater contrast in acceptance of GMOs between that which currently exists in the USA and that in Europe. It is important that those who have responsibility for regulating Biotechnology do not themselves get swept along in the highly charged debate. We have a science based regulatory system in the UK and science/safety decisions should not be made on emotional grounds. That said, science alone cannot address all of the concerns of society over Biotechnology and there is a good argument for a wider debate on ethics and, more generally, what society actually wants from agriculture.

It is often said that we do not know the long term consequences of releasing GMOs. This is true, but no field of human endeavour is absolutely free from risk, nor can long-term consequences ever be foreseen at the outset. This is as much true for new advances in engineering or medicine as it is for agriculture. Given that there will always be some uncertainty, then the long-term unpredictability of a new technology should not be used as a reason for not taking the first cautious steps. Genetic modification may bring great benefits for the environment but we must recognise that there is also potential for harm. The challenge for us all is to get the good bits without the bad.

---

[12] technically this is a change from IIIa to IIIb regulatory committee procedures

[13] GM medicines regs

[14] Novel foods and food ingredients

The role of DNA technologies in crop breeding

G C M Sage
Plant Breeding International, Cambridge, CB2 2LQ, UK

ABSTRACT

DNA technologies are bringing together, in life science companies, genetic understanding of biological systems which will support the production of new agricultural and medical products. In crop breeding these technologies potentially extend the range of both the biodiversity that can be accessed and the traits that can be delivered. Some varieties bred using DNA technologies will require more screening and regulation, depending on the trait involved, than conventionally bred varieties. The first such varieties are being taken up by farmers very rapidly in many parts of the world.

BACKGROUND

Conventional crop breeding, as carried out historically by farmers and, from the early part of this century, also by increasingly science based plant breeders, is predominantly a process of visual selection among a range of phenotypes produced by artificial cross-pollination. Its purpose is to offer to society new biological options, in the form of varieties, which are then taken up or not depending on society's assessment of their usefulness relative to the varieties already available. Usefulness is never easily quantified because it encompasses profitability, reliability, fitness for purpose, safety and adaptation to site and current husbandry. Conventional commercial plant breeders finance themselves, investing income from seed sales and, with inbred crops also royalties, in long-term, incremental improvements in both crop performance and selection techniques.

PLANT BREEDING

The DNA technologies, popularly referred to as Genetic Engineering or Genetic Modification, allow scientists to influence heredity at the molecular level. They form the basis of both Genomics and Plant Transformation, two new possibilities for crop breeders. Genomics provides ways in which all genes, not only those of large phenotypic effect, can be mapped and selected and Plant Transformation allows genes to be built into constructs which can be moved rapidly across species barriers. Structural Genomics is the process of finding out where all the genes are and has led to the remarkable discovery of the co-linearity or synteny, that is the conservation of gene order on chromasomes, that exists across different species within crop plant families. Functional Genomics is the process of finding out what individual genes control at the biochemical level and is leading to a greater understanding of many new and potentially economically important traits. An understanding of, and capacity in, Genomics gives the breeder the ability to track the genes controlling comparable traits in different species. If the crop breeder also has capacity in plant transformation he has the potential to combine the genes for a trait from different species and thus to exploit a much wider range of crop biodiversity than before. Conventional crop breeding is limited by its dependence on sexual compatibility and genes of large effect. Crop breeding using DNA technologies avoids these limitations and, since defined genetic constructs are used, also offers increased precision and speed (Figure 1).

Figure 1 Difference between conventional breeding and genetic engineering

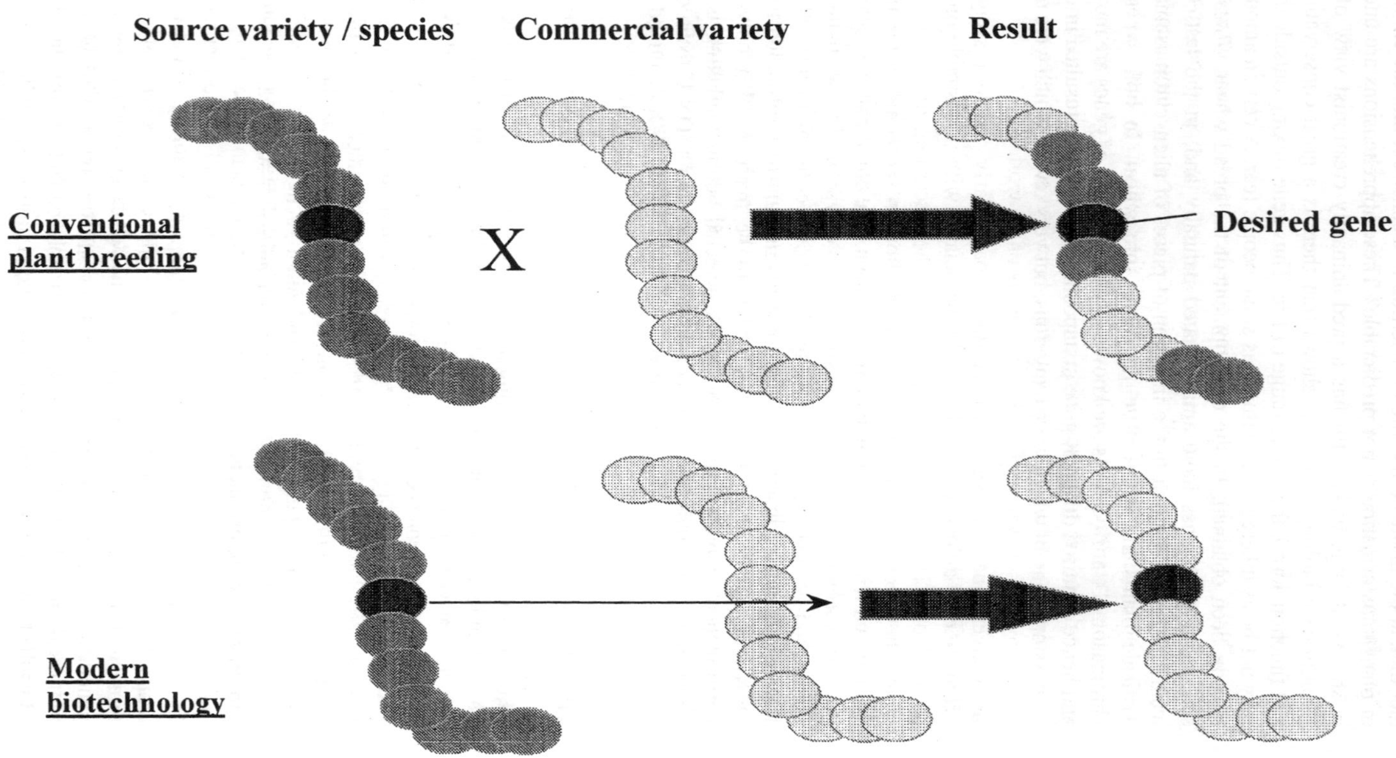

It is useful to consider two questions when comparing conventional breeding with genetic modification. 1) Are they different because the juxtaposition of genes from distinct species might lead to dangerous novel gene interactions? and 2) Is this possibility exacerbated by the fact that, at the moment, the site of insertion of a construct into the chromosomes of the host in plant transformation is not predictable? These apprehensions are understandable enough in society at large which still has a predominantly creationist view of the origins of life. Evolutionary biology, however, shows that there is a great conservation of gene structure and function across the whole range of life. Thus gene mixing outside the species gene-pool may not be as inherently 'unnatural' as some people fear. A certain amount of experience has already been obtained. In the development of the bread wheat, *T.aestivum*, three distinct species of goat grass have amalgamated naturally and, in the last 30 years cytological techniques have resulted in the insertion of pieces of alien chromasome, and in the case of Triticale plants the whole Rye geneome, into wheat. In both wheat and barley many chromasomal translocations are known in which blocks of genes are moved about the geome and hence acquired different gene juxtapositions. All of this resulted in combinations of trait expression, none of which were dangerous, but very few of which were useful and stable.

In conventional crossing programmes, particularly back-crossing programmes, the acquisition of a desired gene is usually associated with undesirable ones which have to be selected out. This takes time and involves the discarding of very large numbers of lines which the breeder would not risk his reputation on by taking to the marketing stage. The same is and will be true of genetic modification. The combination of genes from very widely different species, animals and plants for example, will need to be studied with greater effort, until a body of experience is built up, than would combinations of well understood genes. Because the site of insertion of any construct in plant transformation is unpredictable many individual transformant plants will have to be made so that only those with optimal performance, stability and safety can be selected before marketing is considered. Variety production, whether conducted conventionally or with DNA techniques, will remain a numbers game in which only a very few lines, which have survived extensive screening tests, ever reach the market.

DNA TECHNOLOGY AND THE PLANT BREEDING INDUSTRY

The costs of conventional breeding are already so high that it has to be conducted, if it is to be profitable, on a climatic region basis rather than in individual countries. Thus during the past fifteen years governments in the developed world have withdrawn from funding crop breeding to be replaced by multinational seed companies. These in turn have tended to become aligned with agrochemical companies in order to exploit synergies in marketing to growers. We have now reached the stage where a number of global life science companies (Dupont, Monsanto, Novartis and most recently Aventis) are beginning to be formed, from the combination of pharmaceutical, seed and agrochemical companies, where the high cost of investment in genomics and plant and animal transformation can be spread across a greater number of products.

In the last three years these life science companies have launched the first wave of GM crops. These have genetically modified agronomic traits and are designed to help growers do more with less. But these will very soon be followed by a second wave which will have genetically modified quality traits and, eventually, by a third wave where the genetic modification will turn the plants into sustainable biofactories producing specific molecules. Thus the target is a whole range of value-added traits or specialty crops. Not all of these will be involved in the food chain and, indeed, not all of them will be offered widely to growers if only limited areas of specialist, identity preserved, production are required to meet demand.

REGULATION

The crop breeders inside life science seed companies want to utilise DNA technologies because they give, in a commercially highly competitive market, more efficient means to the end of offering new biological options. But, as we all know, society, particularly in Europe, is not yet convinced that it wants to take up these particular options. Public apprehensions centre around either the fear that by blurring the dividing lines between species we may be doing something that is unnatural or the fear that the new technologies are being introduced too fast with inadequate regulation by over-powerful multinational companies. Society at large, understandably, has difficulty in coming to terms with any new technology but particularly with a technology that affect its food. In Europe it is an irony of history that DNA technology is being introduced with more legal regulation in place than any previous technology just at a time when the BSE disaster has wrecked public confidence in regulation.

Regulation of GMOs in Europe is based on new, purpose designed legislation concerned predominantly with human and environmental safety. In contrast, in the rest of the world that has GMO regulation this has come about by the adaptation of pre-existing legislation. The European system is admirably thorough but it takes longer in Europe than in the rest of the world, not only to get applications approved, but also to get changes made to the application process. Despite this, European regulation of GMOs is evolving. On the one hand, as experience is accumulated, fast track product, as opposed to process, based systems of approval are being introduced. At the same time, however, an element of ethical consideration is being recommended for example by the recent UK House of Lords report.

Obtaining society's approval to use GMOs by passing the regulatory hurdles is, however, not the end of the need for rules concerning modified crops. Experience in other parts of the world has shown that the advantageous effects of a GM trait may not only affect the performance of the variety into which it is placed but may also influence what happens in subsequent crops in the rotation. A possible UK example of this may come to pass when Roundup Ready sugar beet are grown in rotation with non-herbicide resistant potatoes. It is probable that in this situation volunteer potatoes will be better controlled by glyphosate than by the more expensive, less environmentally benign, herbicide regimes currently used. The agricultural divisions of Life Science companies see themselves, from now on, not as providers of either varieties or agrochemicals, but as providers of sustainable solutions to farming problems.

INTELLECTUAL PROPERTY

For several hundred years western societies have encouraged new technology by granting intellectual property rights. These are a deal done between national governments and inventors whereby, in exchange for the inventor declaring details of an invention according to strict rules, society grants the inventor exclusive rights to the commercial development of the invention for a fixed time period after which the invention becomes freely available to all. World Trade Organisation rules currently require participating countries to move towards the intellectual property protection of plants and plant varieties.

Plant varieties can by protected in 37 countries using Plant Variety Rights based on the international conventions of UPOV, the Union for the Protection Of new Varieties. Plant Variety Rights have been extensively used, particularly for inbreed crops in Europe, since the mid 1960's. In fewer countries some plant varieties and novel gene constructs, which control particular traits involved in plant transformation, can be protected by patents.

The level of protection given by Plant Variety Rights and by patents differs in that with Rights the holder cannot prevent other breeders from using his protected variety as a parent in crosses aimed at the selection of further varieties but with patents he can. Thus on the face of it, particularly if patents with broad claims are granted, the holders of patents are in a powerful position and this possibility causes at least part of the disquiet that some members of the public feel about multinational companies. However it must be remembered that patents, unlike Plant Variety Rights, are expensive to both obtain and to defend from challenge. Also, given the great genetic complexity especially of many second and third wave traits, competitor inventors will always find ways of developing their own similar but not identical patents. Thus with sensible adjustment of the laws the consumer will continue to be offered choices. As long as this happens then, if innovation is to continue, society should see to it that a potential investor in life science activities can choose a level of intellectual property protection that is appropriate to the level of risk taken in making the investment.

CULTURAL GUIDANCE AND UPTAKE OF GM CROPS

It is important to ensure that understanding of the best husbandry practice relevant to a particular sustainable solution is rapidly and widely disseminated to farmers growing GM crops. This can be encouraged in two ways as is already happening in the USA. First when a farmer there signs up to buy GM seed he not only agrees to pay a technology charge but is also instructed in and agrees to carry out particular husbandry procedures. Secondly the supply chain also agrees on methods of labelling seed and produce and controlling their exchange so that the crop can be passed along the food chain with all concerned knowing what they are handling. Last summer in the UK, an organisation called SCIMAC (The Supply Chain Initiative on Modified Agricultural Crops) (Figure 2), which consists of the Trade Associations associated with crop production, was formed and is already proposing 1) Terms of Inter-Professional Agreements, 2) Administration of Independent Audit and 3) The Ensurance of Compliance.

Figure 2 Who is SCIMAC?

SCIMAC membership includes:

- Farmers Union (NFU)
- British Society of Plant Breeders (BSPB)
- British Agrochemicals Association (BAA)
- United Kingdom Agricultural Supply Trade Association (UKASTA)
- British Sugar Beet Seed Producers Association (BSBSPA)

Launched publicly in June 1998

The importance of a meeting concerned with gene flow within and between species lies not only in that it will provide information relevant to the evolution of regulatory requirements but, probably more importantly in the long run, it will help to specify best practice on farms. The use of the phrase 'in the long run' implies that we have time to work out the systems before we grow the crops. Unfortunately, in practice, because of the speed with which GM crops are being taken up around the world, we will have to do both simultaneously.

THE PACE OF CHANGE

To emphasise this, figures indicating for GM crops the diversity of traits, the speed of development, the areas grown and their potential value are presented below:

It took Monsanto sixteen years from initiating its own research to getting its first GM product onto commercial farms in 1996. The first field releases to the environment of GM crop material began in 1987. By 1998 there were ten Monsanto GM products available. (Figure 3)

Figure 3 Monsanto's Biotechnology Timeline

Started R&D efforts in corporate R&D	1980
First engineered plants	1983
Programme transferred to commercial unit	1985
First field tests	1987
Demonstrated commercial performance	1993
First Regulatory submissions	1993
First Commercial approval	1995
First Commercial products introduced	1996
10 products available in the marketplace	1998

Figure 4 Extent of field tests of GM crops

- More than 25,000 field tests globally
- In 45 Countries
- With 60 different crops
- Most frequent categories 1987-1998

Herbicide Tolerant	29.0%
Insect Resistant	24.1%
Product Quality	21.3%
Viral Resistant	10.1%
Fungal Resistant	4.1%
Agronomic Properties	4.1%
Other	7.3%

(USDA Field Releases Permits Issued and Notifications Acknowledged)

Between 1987 and 1998 there have been more than 25.000 field tests in 45 Countries with GM material in sixty different crops. The most frequent trait categories have been herbicide resistance and insect resistance. (Figure 4)

In 1998 Monsanto's share of the area planted was over 50 million acres. This involved five crops, soyabean, maize, cotton, oilseed rape and potatoes and seven countries. (Figure 5).

Figure 5 Total 1998 Plantings of Commercialised Monsanto Biotechnology Crops (acres in thousands)
* = First year of commercialisation

Region / Product	U S A	CANADA	CHINA	MEXICO	ARGENTINA 1998-99	AUSTRALIA 1998-99	SOUTH AFRICA 1998-99
Roundup Ready Soybeans	25,000+	100			10,000		
Bollgard/Ingard Cotton (B.t)			130*	100	20*	200	30*
Roundup Ready Cotton	5,000+ (45% of cotton)						
Bollgard/R R Cotton Stacked Gene				2			
Roundup Ready Corn	950*						
YieldGard Corn (B.t)	10,000+	300+			42*		
Roundup Ready/YieldGard Corn Stacked	30*						
Roundup Ready Canola		Well over 1,000					
NewLeaf Potatoes (B.t)	50	10					
Laurate Canola (oil modification)	50+						
BXN Cotton (herbicide-resistant)	1,000						
Approximate TOTALS	42,000+	1,400+	130	100+	10,000+	200	30

This very rapid uptake of varieties derived using genetic modification technology has come about because farmers want these varieties. Some understanding of why can be shown by the cost benefit experience of on-farm use of Monsanto GM crops in 1997 given in Figure 6. In each of five different crops a yield benefit over previous practice was clearly demonstrated. This translated into appreciable additional value potential which, in turn, produced high levels of farmer satisfaction and repurchase intention.

Figure 6 Monsanto cost benefit experience of on-farm use of GM Crops in 1997

Crop	Yield Benefit %	Value Potential $/acre	Farmer Satisfaction %	Repurchase Intention %
Roundup Ready Soyabeans	5	12-20	90	90
Yield Gard Bt Maize	8	20-35	90	80+
Bollgard Bt Cotton	10	35-70	80	70+
Roundup Ready Cotton	5-6	8-20	90	95
New Leaf Potatoes	5	40-100	85	90

The setting of the technology charge at approximately half the expected additional value potential ensures that the farmer and technology provider roughly share the financial benefit. This wider picture of the uptake of GM crops is in stark contrast to the often parochial and introspective debate that is going on in Europe. In our search for understanding of the environmental biology of GM crops we must not lose sight of the fact that, after a very large number of tests covering an eleven year period, most non-Europeans agree with our House of Lords' conclusion that the benefits are considerable and the technology has the potential to create and sustain revenue and jobs. This is reinforced by estimates of the value of the plant biotechnology industry by 2005 (Figure 7). This shows, as I suggested earlier, that the predominance of Herbicide Tolerance will fade as novel quality characteristics come on stream.

The most likely GM crop developments in the UK, and their estimated time of introduction, are listed in Figure 8. Beet and oilseed rape varieties are going to reach the market before wheat and potatoes. This means that our first commercialised GM crops, unlike the maize and soyabean of the USA, are precisely those where a knowledge of gene flow is going to be important and where we will not have the potential to learn from extensive experience from outside Europe.

Figure 7. Future development of the Biotechnology Industry by 2005, the Plant Biotechnology Industry will generate revenues of over $6 Billion

$ Billion	Year	
Annual industry revenues from	2000	2005
Herbicide Tolerance	0.8	1.1
Other Agronomic Characters	0.8	2.1
Quality Characters	0.6	3.4

Figure 8 Biotech development projects and possible timelines for the UK

Roundup Tolerant crops

Fodder beet	spring	2000
W OSR	Aut	2000
Sugar beet	spring	2001
Fodder maize	spring	2002

Hi-Laurate Spr OSR — spring 2002?

<u>Beyond 2002</u>

Fusarium-resistant wheat
Virus resistant potatoes
Blight resistant potatoes
Septoria resistant wheat
Other speciality oils in OSR

A long term perspective on Ag-biotech

W de Greef
Novartis Seeds, Basle, CH4002, Switzerland

ABSTRACT

The prediction of the future direction of the Ag-biotech industry is fraught with difficulties, yet it is a valuable exercise. Such assessments of the impact of biotech on a global scale needs to concentrate on direct effects including, an evaluation of societal needs as well the requirements for food, feed, raw material and energy production. The world has to move towards a sustainable food production system that is capable of feeding an increasing population on less land. There are two targets for biotechnology, input and output traits. The former mainly concern those that affect plant nutrition and adaptation to biotic and abiotic stresses. The target is to increase yields with reduced external inputs. Herbicide and insecticide resistant crops are already being grown and there has been some progress with fungus resistance. Future advances will concentrate on abiotic stress factors such as drought, heat and salt tolerance. The 'holy grail' of developing nitrogen-fixing cereals is being explored but will take time as it is a complex system. Anticipated advances in the acquisition of 'output' traits are, in the short term, the modifications of the starch, protein, oil and sugar contents of plants. In the longer term the development of 'functional' foods (*eg*. enriched with vitamins), 'cured' foods (*eg*. removal of allergenic factors) is predicted, along with modifications that affect food digestibility. Future advances in genomics will deliver an increasing range of genes with potential to develop totally new biochemical pathways in plants.

INTRODUCTION

Futurology is a very inexact science. That makes it quite safe to make long term prognoses on technology development. It also threatens to make the exercise seem futile.

Does it make sense to try to look 30-40 years ahead in a sector where most of the products coming to the market were not even conceived 10 years ago? Worse, most of the **technologies** needed to make those products did not exist ten years ago! Worse still, even among the technologies we work with today, there are lurking those that will transform the sector, and we are likely to miss them in any attempt to pick among the key new developments of today, those that will transform our lives in a decade or two.

We would be in good company in such a case. Alvin Toffler, arguably the most famous futurologist in the world, totally failed to see the coming of the Internet in a book he wrote on communications and power in 1989!

And yet, it is important to do this type of exercise. It is the only way we have to take stock of what we have achieved in the sector, and it forces us to reflect not on what will be (which is impossible indeed), but on what may be in different scenarios. This is precisely what this paper sets out to do.

Talking about the impact of biotech on a global scale only makes sense if the following premises are taken into account:

Be modest : concentrate on direct impacts.

- Identify areas of activity where biotech can have a direct impact.
- Evaluate societal needs in the broadest sense, as they are likely to be affected by demography, the environment, the economy, and emerging policies on sustainability and on health protection.
- Concentrate on the needs in food and feed production, industrial raw material production and energy production. Assess these needs in quantitative and qualitative terms.
- Assume that in the long run, key components of society such as food security are demand driven.

Within this framework, it does become possible to identify areas of activity in which biotechnology is likely to deliver a significant contribution. It is indeed likely that when a difficult problem related to agricultural production emerges that does not have a satisfactory solution through other technologies, then biotechnology will be asked to address it, and is likely to come up with a solution. In other words, what we will look at is the evolution of demand for the technology, not the evolution of its supply.

So what do we expect agriculture to achieve in the next 3-5 decades ? The nineties have been a decade of intensive reflection on the ways in which humankind has provided for its basic needs. Out of it comes a broad consensus that every major step forward in the future will have to be subjected to the « sustainability test ». In other words : does it have the potential to contribute to a more sustainable world ?

The modalities, and indeed what constitutes sustainable technology are still far from clear. There are large groups in society that have a fundamental problem with the association between high technology, cutting edge science and sustainability. In the long run though, knowledge driven society will come to terms with using technology to promote sustainability, not opposing it as a matter of principle (or as a matter of fear).

The challenge before agriculture between now and 2050 is to feed at least 3 billion more people on less land, while reducing a number of non-sustainable inputs. In addition, agriculture is also expected to allow space for the introduction of new crops (and the transformation of existing ones) and for the production of a much more diversified range of raw materials for a wider variety of industries. These are only the « material » objectives.

On top of that comes a much broader array of socio-economic and cultural considerations. These tend to be dismissed or marginalised in technical discussions, even though it is by now clear that technological innovation without an appraisal of its societal impact is likely to be less effective at least, and can even lead to the possible death of the innovation.

This neglect of the non-technical aspects of agricultural innovation has led to a gigantic backlash, to the point where technical elements are becoming absent from decision making. A good example is provided by the EU policy for agriculture (Agenda 2000). This so-called strategic document on EU Common Agriculture Policy does not even address innovation and technology as relevant!

TARGETS FOR AG-BIOTECH AT PRESENT AND IN THE FUTURE

Today, the biotechnologies considered most relevant for the next decades are those associated with IPM, those associated with seed technologies and those associated with the genetic make-up of the crop (and its associated micro-organisms). The industry most likes to divide the application fields in input traits and output traits.

Input traits

Input traits include those that affect plant nutrition and the adaptation of crops to biotic and abiotic stresses. These traits generate value by providing yield increases, or crop security, or replacement of other inputs. The value chain is short. Seed companies develop the product, and the farmer gets the immediate benefit.

Whatever the present controversies over consumer benefit from the first generation of GMO crops (almost all of which contain modifications of input traits), the most important impact of GM crops over the next decades may still be at this level. The reason is simple: over a 30-50 year perspective, increasing yield with reduced external inputs remains the biggest challenge for agriculture world-wide. The agricultural production glut of this decade is achieved at the cost of unsustainable levels of fertiliser, pesticide and energy inputs, and it uses categories of land for production that should not be farmed. It also uses unsustainable amounts of water.

All these factors have to be gradually corrected to make them sustainable, and this has to be done while increasing world basic food production by a minimum of 30 million tonnes per year. It looks like a challenge that will require every bit of human ingenuity and innovation available. It cannot be stressed enough that this remains the real challenge for world agriculture. It requires long term commitment from governments, academia and industry to keep developing the necessary tools: insights, knowledge, technology, products and policies.

Instead of this, we find that governments are gradually reducing their commitments for training and research, and ignoring the need for production increases in their policies. In fact, the flavour of the day in policy making is production decrease, and extensification. That is not a major long-term problem, as long as it remains reversible. The agricultural sector is sufficiently technically competent to introduce new production technologies for higher yield in a very short time. However, if R&D is slowed down, and if overly restrictive legislation covering new technologies is put in place, that effectively makes it much more difficult to reverse the trend when (not if) the present food glut changes to become a shortage. A food shortage due to short-term policies such as pricing and set-aside programmes is conjunctural, and easily reversed. A shortage due to an empty R&D pipeline is structural, and likely to lead to much longer periods of tight food supply.

Can genetic engineering deliver on the input traits? For biotic stresses, the answer is by now a clear yes. For abiotic stresses and for nutrition, it is too early to tell, but the answer should be there within the next few years.

Insect and herbicide resistant crops are by now fully integrated in agriculture, at least technically. This does not mean that there is nothing more to be done. We have not even started looking at how to integrate these traits in IPM schemes, although they could transform our capabilities there. We also are still all working with Bt crystal proteins. Given what is in the R&D pipeline now, it is most likely that within 20 years, we will look back on Bt as the Stone Age of insect control. 20 years may look like a long time, but let us keep in mind that even PCs needed more than 20 years to get where they are now.

Although technological feasibility has been demonstrated by now for fungus resistance, it is probably going to take another decade before we see widespread adoption of the technology. For this set of traits, breeders are going to be confronted with a new challenge : how to integrate in the most synergistic way « conventional » and « engineered » disease resistance genes in a finished variety. The same goes for nematode resistance, one of the key long-term objectives of crop genetic engineers.

Major advances can still be made in resistance to unfavourable weather conditions : drought, water logging, cold, heat. It is not clear at this moment whether these improvements will come mainly from genetic engineering or from genomics though. The improvements required here are often of a quantitative nature, and there tends to be a lot of genetic diversity available within the gene pool of each species for such traits.

There has been recent and quite spectacular progress in salt tolerance and in resistance to Aluminium toxicity. Both traits are of particular significance for tropical agriculture, and could well transform farming in parts of the developing world.

One of the holy grails for biotechnology has been, since day 1, the idea of developing nitrogen-fixing cereals. In fact, much of the early investment in the technology was « sold » on that promise. By now we know it is not going to be simple of course. However, our understanding of gene function and plant physiology is moving very fast indeed. Possibly the most interesting route will ultimately prove to be the one by which we create cereal plants that attract nitrogen-fixing micro-organisms, and feed them. In any event, this is not likely to happen in the next couple of decades (at least at the commercial level).

In the meantime though, our rapidly expanding knowledge of what goes on at the interface between plant roots and the community of micro-organisms that live on (and in) them, is slowly beginning to yield clues on how to direct these interactions more to the advantage of the farmer.

Output traits

It is generally believed that most of the potential added value of transgenic crops lies in the modification of traits that change the composition of the end product. The changes already achieved or in advanced research stage include dramatic composition changes of starches, proteins and oils in the final crop. Less widely publicised are changes in sugar composition (e.g. fructo-furanose producing sugar beet), changes in baking quality (wheat with altered gluten profiles), and changes in malting quality of barley. Into the same category belong traits such as long shelf life of fruits (and less known: flowers).

These products only present a very primitive first generation. At best they give a glimpse of how our understanding of the quality determinants of food and other agricultural products allows us to modify crops further to our advantage.

The next generation of products will certainly include two types of food products: functional foods, in which specific health promoting substances have been enriched (e.g. vitamins and specific types of fibres), and « cured » foods, in which toxic or allergenic components have been eliminated. Depending on the manufacturing costs, plants may also become more frequent bioreactors for the production of proteins with pharmaceutical applications. In this field they face stiff competition from milk produced proteins though.

In the long run, we should expect to see further improvements especially in factors that influence digestibility of food, especially in crops that are grown predominantly for animal feed. One of the most effective ways to increase available food and feed with reduced environmental impact is to ensure that the food is entirely available for the animal to metabolise. Early efforts include the same compositional changes of the macro-components mentioned above, but research now has moved on to the elimination of the numerous anti-nutritional factors that most plants contain, and to the addition of factors that improve the availability of nutrients. An elegant example of the latter is the expression of phytase in plants. Phytase increases the availability of phosphorus in diets by solubilising polyphosphates. The enzyme is used with considerable success as a feed additive, and now companies are moving on to expressing it in the storage organs of crops.

The elimination of anti-nutritional factors often brings its own problems though. Many of them are in fact defence mechanisms of the plant, especially against insects, but also often against fungi. So stripping the crop of these factors can leave them wide open to pest and disease attack. The future will probably see many projects where a native anti-nutritional factor is eliminated, and another gene inserted for the replacement of its defence function but without the side effects.

Even though the use of plants as sources of drugs is widely overestimated as a consequence of the politically correct reference to « natural products », plants do remain interesting targets for the pharmaceutical industry as bioreactors for the synthesis of raw materials or intermediates of drugs. Several crops have for decades been the preferred sources of oestrogen hormone intermediates, but now it becomes possible to take a species that produces an intermediate, and sometimes add new genes for further reaction steps. A vivid example of this process is on the market, albeit in a totally different field: it is the blue carnation. In this plant, enzymes that can metabolise the pigments that are responsible for the red colours of carnation have been added. The source of the genes was delphinium, which can naturally do the next biosynthesis steps.

The rapid advances in genomics are delivering a rapidly increasing store of genes for all kinds of metabolic steps, and it should not take more than another decade before we see many uses of this resource in the development of entire new biosynthetic pathways in plants.

SESSION 2

GENE FLOW – MEASUREMENT AND IMPLICATIONS

Chairman & Session Organiser: Professor A Gray
Institute of Terrestrial Ecology, Furzebrook

Defining and measuring gene flow

A F Raybould, R T Clarke
ITE Furzebrook Research Station, Wareham, Dorset, BH20 5AS, UK

ABSTRACT

We describe developments of regression-based methods for estimating the relationship between gene flow and distance using allelic variation at marker loci. These new methods allow the correct testing of the difference between regression slopes, and the assignment of confidence intervals to estimates from the regression equation, when the regression is between matrices of pairwise data, rather than between independent values. We show that there are significant differences in gene flow estimates from RFLPs and isozymes in sea beet. Estimates of the distance at which populations exchange one migrant per generation are lower for wild cabbage than for sea beet.

INTRODUCTION

A full assessment of the environmental impact of GM crops will consider the movement of transgenes into and among natural populations of crop relatives. Regardless of variation in sexual compatibility and selection, variation in gene flow among crop relative populations might cause some transgenes to spread and become common very quickly, whereas others might remain rare and be confined to areas very close to the source crop. Because transgenes are inherited in a mendelian way, gene flow among non-transgenic natural populations is an 'appropriate biological model' (Gliddon, 1994) for predicting the movement of transgenes within and among populations of crop relatives.

Gene flow is the movement of genetic information among individuals, populations or taxa. In plants, "potential" gene flow is the movement of seed and pollen as a function of distance (i.e. dispersal). "Actual" gene flow is the amount of fertilisation (in the case of pollen) and establishment of reproductive individuals (in the case of seeds) as a function of distance from a source (Levin & Kerster, 1974). Clearly, not all pollen will effect fertilisation, and not all seeds will establish reproductive plants, therefore actual gene flow can be much lower than potential gene flow. Also, the shapes of the pollen and seed dispersal curves may not predict the rate of change of gene flow with distance because the probability of fertilisation/establishment by a pollen grain/seed may vary with distance (e.g. Levin, 1981).

Gene flow is measured in two ways: by 'direct and' 'indirect' methods. The most common direct method for plants is the observation of seed and pollen movement, which gives an estimate of potential gene flow (dispersal). Other direct methods use genetic markers to estimate actual gene flow. A simple method is to introduce or identify a plant in a population with a unique genetic marker (e.g. an isozyme allele) and to follow the appearance of the marker in the next generation (e.g. Latta *et al.*, 1998). A more sophisticated approach uses markers to identify the fathers of half-sib families. If the markers are highly variable (e.g. microsatellites) and the number of potential fathers is relatively small, the father of each seed can be identified unambiguously (e.g. Dow & Ashley, 1998).

Indirect methods use the distribution of genetic variation to infer actual amounts of gene flow. The most powerful type of data is allele frequencies at one or more discrete loci. These data can be treated in a number of ways, but essentially high variation in allele frequency between populations or patches of plants indicates that gene flow between the populations is low, whereas similar allele frequencies across populations imply high gene flow.

Direct methods, whether estimating potential or actual gene flow, only measure gene flow at the time of the observations. Indirect genetic methods, on the other hand, measure average amounts of actual gene flow, by reflecting the cumulative effects of temporal variation in the spatial distribution of dispersal and establishment over preceding years, including rare, unpredictable events (e.g. Slatkin, 1985). If rare long-distance dispersal events (i.e. founder effects) have shaped the genetic structure of a species, direct methods may give lower estimates of gene flow than indirect methods (e.g. Campbell & Dooley, 1992). On the other hand, direct estimates can be higher than indirect estimates, for example where genetic drift has removed immigrant alleles from populations (e.g. Rasmussen & Brødsgaard, 1992). If possible, it is desirable to use both types of method.

In this paper, we describe indirect methods for estimating gene flow in two crop relatives: *Beta vulgaris* ssp. *maritima* (sea beet, the wild relative of sugar beet) and *Brassica oleracea* ssp. *oleracea* (wild cabbage, a wild relative of cultivated cabbage and oilseed rape). Our aims are to derive robust estimates of gene flow for these species, and to draw attention to the assumptions that underlie these estimates.

ESTIMATION OF GENE FLOW FROM ALLELE FREQUENCY DATA

The starting point for indirect estimates is data on the distribution of a genetic polymorphism. Neutral, co-dominant allelic variation at discrete loci gives the most powerful type of information. Dominant markers can be used, but very large sample sizes are needed to achieve the same statistical power (Lynch & Milligan, 1994). Since the mid-1960s, isozymes have provided plant population geneticists with a ready source of co-dominant markers. Microsatellites are now becoming the markers of choice because the number of polymorphic loci and the number of alleles per locus tend to be higher than isozymes. However, the mutation mechanisms at microsatellite loci are poorly understood, which presents problems in deciding the most appropriate estimator of genetic structure (and hence gene flow) at these loci.

The most common method of inferring gene flow from allele frequency data uses the 'Infinite Island' model of Wright (1931), in which an infinite number of finite populations ('islands') produce and receive migrants. The importance of an infinite number of islands is that allele frequencies in the system as a whole do not change. In each generation, islands have an effective population size of N, of which a proportion, m, are migrants. Wright showed that the product Nm is related to a parameter F_{ST} (see below), such that

$$F_{ST} \approx 1/(4\text{Nm} + 1)$$

Nm is the amount of gene flow - the number of migrants per population per generation averaged over all islands.

The parameter F_{ST} has been variously formulated. Perhaps the most widely adopted is the analysis of variance approach of Weir & Cockerham (1984). Consider a single locus with two alleles A and a with overall frequencies p and (1-p) respectively. Arbitrarily assign values of X=1 for A and X=0 for a. If X_{ijk} denotes the value (0 or 1) for the *i*th allele in the *j*th individual in the *k*th population, then the total variance of X can be partitioned as follows:

$$X_{ijk} = p + a_k + b_{jk} + w_{ijk}$$

where a_k represents the difference (from overall p) in the frequency of A in population k (with variance of the set of a_k equal to σ_a^2), the b_{jk} denote differences between individuals within populations (variance σ_b^2), and the w_{ijk} represent differences between alleles within individuals (variance σ_w^2). The total variance of X = p(1-p) $= \sigma_T^2 = \sigma_a^2 + \sigma_b^2 + \sigma_w^2$. F_{ST} is the proportion of the variance that is due to differences in the mean allele frequency among populations. In other words

$$F_{ST} = \sigma_a^2 / \sigma_T^2$$

There are different ways of treating loci with more than two alleles and for combining data from several loci. Weir & Cockerham's (1984) suggested approach, which is widely used, is to estimate the overall F_{ST} as

$$F_{ST} = \sum_r \sum_s \sigma_{a(rs)}^2 / \sum_r \sum_s \sigma_{T(rs)}^2$$

which is effectively an average of the individual F_{ST} values ($F_{ST(rs)} = \sigma_{a(rs)}^2 / \sigma_{T(rs)}^2$) for each allele s of each locus r, weighted by respective total variances ($\sigma_{T(rs)}^2$).

It is relatively straightforward to estimate F_{ST} for a group of populations, and so derive the average amount of gene flow among them. However, one may be more interested in whether gene flow declines with increasing geographic distance between populations, a phenomenon termed "isolation by distance" (Wright, 1943). Using computer simulations, Slatkin (1993) showed that under a variety of demographic models, isolation by distance was represented by a linear relationship between log Nm and log distance (D) estimated between all pairs of populations. Normal tests of significance of a correlation or regression of log Nm and log D are not valid because the number of points (pairs) is higher than the number of independent pieces of information (populations). If there is no isolation by distance present in the study (null hypothesis), then each of the n populations of plants could have come from any of the n geographic positions. Therefore the significance is obtained from a Mantel randomisation test (Mantel, 1967), which effectively repeatedly randomises the positions of the n populations. The test significance is the proportion of correlations between Log Nm and Log D from say 10000 randomised data sets that are equal to or less than the observed correlation (Nm and distance are negatively correlated when there is isolation by distance).

A regression relationship between log Nm and log D allows us to derive an indirect estimate of the actual gene flow between a pair of populations a given distance apart. It also enables us to estimate the distance (which we term 'isolation distance') at which gene flow is reduced to some critical level. However, in both cases, without confidence intervals such estimates are of little value. The pairwise nature of the data means that valid confidence intervals cannot be derived from the formula for standard errors of parameter estimates and predictions available for normal linear regression of independent observations. We are developing a more appropriate regression model that incorporates terms for the variability between populations in their levels of gene flow for a given distance. Maximum likelihood estimates of model parameters and their standard errors are used to derive confidence limits for the regression line and for the 'isolation distance'. We call this approach the 'Maximum likelihood population-effects' (MLPE) method.

GENE FLOW AMONG SEA BEET POPULATIONS

We estimated gene flow among ten sea beet populations in Dorset (see Raybould *et al.*, 1996b; 1997). Fifty plants per population were analysed for variation at 7 isozyme and 6 restriction fragment length polymorphism (RFLP) marker loci. The data were used to estimate all pairwise F_{ST} values separately for the isozyme and RFLP data. F_{ST} estimates were converted to log Nm values and regressed on log D. The results are displayed in Figure 1 and Table 1.

Table 1. Details of regressions (log Nm = a + b log distance) among 10 populations of sea beet in Dorset. P = Mantel one-sided test probability ($b < 0$).

Marker Type	b	P	R^2
RFLP	-0.479	0.0007	33.2%
Isozyme	-0.068	0.4910	1.0%

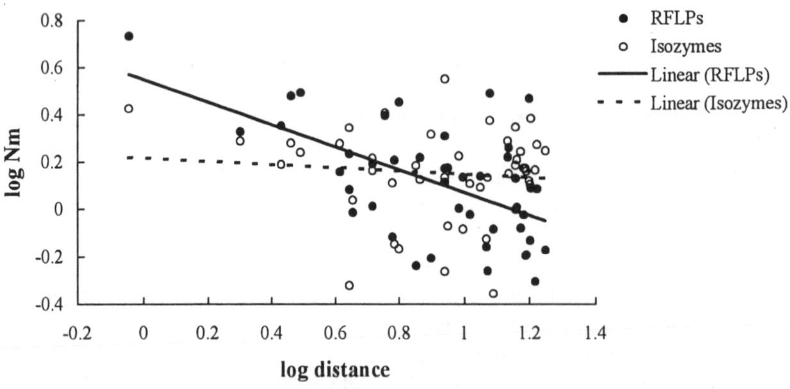

Figure 1. Regressions between log Nm and log distance among 10 populations of sea beet in Dorset. Nm estimated separately from isozymes and RFLPs.

The RFLP data suggest a strong isolation by distance effect (significant negative slope and high R^2), whereas the isozyme data show no relationship between gene flow and distance (non-significant negative slope and low R^2). Because of the non-independence of the points in the regression analyses, a formal test of the significance of the difference between the RFLP and isozyme slopes cannot be made using a conventional t-test. However, if there is no significant difference between the slopes, there will be no correlation (and regression relationship) between the difference in log Nm values for the two types of marker (i.e. log $Nm_{(RFLPs)}$ – log $Nm_{(isozymes)}$) and log distance. The regression slope is significantly different from zero (Mantel two-sided test $P = 0.0064$), showing that there is a statistically significant difference between the isozyme and RFLP regression slopes (Figure 2).

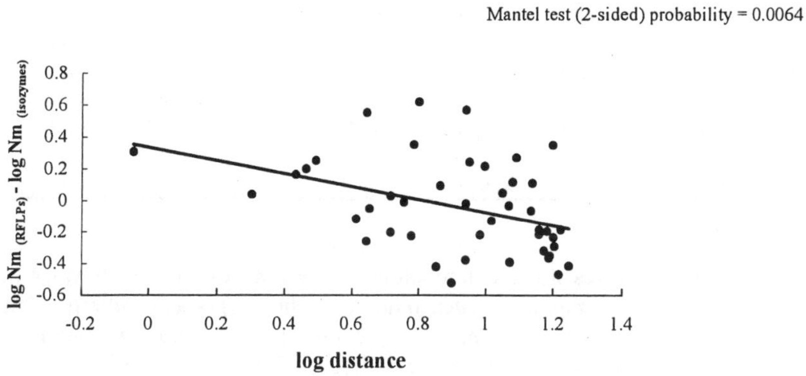

Figure 2. Regression between (log $Nm_{(RFLPs)}$ – log $Nm_{(isozymes)}$) and log distance among 10 populations of sea beet in Dorset.

The reason for the difference is uncertain at present. However, it is possible that the difference is due to the effects of a small subset of loci. If each RFLP locus is excluded in turn from the analysis, the correlation between log Nm and log distance for the 5 remaining loci becomes slightly less negative in each case, apart from *R1* which makes the correlation slightly more negative (Table 2). In all cases the correlation is significant. The results suggest a consistent pattern of isolation by distance at each RFLP locus.

In general, when an isozyme locus is excluded, the remaining 6 loci give a slightly more positive correlation compared with the full data set (Table 2). However, when either locus *got-3* or *pgi* is excluded, the correlation becomes negative, though non-significant. When both loci are excluded, the correlation becomes negative and significant ($P = 0.0077$) and the regression slope for log Nm based on the remaining five isozyme markers is not significantly different from that using the six RFLPs (Mantel two-sided test $P = 0.1085$).

The analysis is preliminary and needs to be confirmed by other tests. For example we need to obtain correlations for single loci. Single locus correlations are a problem at present because often a number of pairwise F_{ST} values are undefined because the two populations are fixed for the same allele and our current procedures for Mantel tests do not handle missing data. Nevertheless, we can speculate that the differences between the RFLP and isozyme data are mainly due to the effects of two loci, *got-3* and *pgi*. The case of *pgi* is particularly interesting because a large number of studies have suggested that variation in PGI enzymes is adaptive (Riddoch, 1993).

Table 2. Changes in the correlation between log Nm and log distance when single loci are excluded from the analysis.

Isozymes			RFLPs		
Locus excluded	r	P (1-sided)	Locus excluded	r	P (1-sided)
None	0.003	0.4913	None	-0.631	0.0007
Acph	0.085	0.7410	*L3*	-0.614	0.0006
Est	0.030	0.5674	*L9*	-0.564	0.0013
Got-3	-0.080	0.2398	*R1*	-0.658	0.0004
Got-4	0.030	0.5758	*R4*	-0.611	0.0004
Per	0.009	0.5049	*R7*	-0.598	0.0009
6-pgdh	0.013	0.5324	*R13*	-0.567	0.0007
Pgi	-0.168	0.0927			
Got-3 & pgi	-0.318	0.0077			

For risk assessment purposes, it is useful to know the precision of our gene flow estimates. In conservation genetics, one migrant per generation (i.e. Nm = 1) is a useful rule of thumb for the minimum amount of gene flow necessary to prevent populations becoming fixed for different alleles through the effects of genetic drift (e.g. Mills & Allendorf, 1996). Therefore the distance at which Nm = 1 might be a useful way to compare the pattern of gene flow among species, and serve as a very rough measure of 'isolation distance' among natural populations. Again, because of the pairwise nature of the data, confidence intervals for estimates of the 'isolation distance' cannot be obtained by ordinary least-squares (OLS) regression techniques. We use our 'maximum likelihood population-effects' (MLPE) method to overcome this problem. For the beet RFLP data, an ordinary regression assuming independent points, gives Nm = 1 at 13.9 km with under-estimated 95% confidence intervals of 10.3-24.4 km. The MLPE method gives Nm = 1 at 14.5 km with wider, but in this case more accurate 95% confidence intervals of 6.7-39.3 km.

We have also carried out a study of seven *Brassica oleracea* (wild cabbage) populations using microsatellites (Raybould *et al.*, 1999), for which the estimated distance at which Nm = 1 was 6.1 km with best-estimate 95% confidence limits of 1.5-10.4 km. The lower 'isolation distance' was perhaps expected as wild cabbage pollen is insect dispersed, whereas beet pollen is dispersed by wind. In addition, beet seed may be dispersed by tides from some populations.

ASSUMPTIONS INVOLVED IN THE F_{ST}-BASED APPROACH

The F_{ST} based approach to estimating gene flow outlined above is very simple, but has many potential problems with both the interpretation of indirect estimates and with the population genetic models used to make these estimates (Bossart & Prowell, 1998). An essential assumption of the approach is that the observed distribution of genetic variation is the result of equilibrium between gene flow and genetic drift. This assumption can be violated for many reasons, such as recent colonisation, selection acting on markers (e.g. through environmental effects) and high mutation rates. To some extent these problems can be overcome by using a wide variety of markers. A perhaps more serious problem is that the infinite island model is probably a gross oversimplification in most natural populations. Some models consider a finite number of islands, whereas others assume migration between neighbouring populations only (stepping stone models). A 'general' model would consider a matrix of migration rates between all populations (considered together, rather than as separate 2-population systems), however analysis based on such a model would be too complex to provide useful estimates of gene flow (e.g. see Nagel, 1997).

It is now possible to analyse data on population gene frequencies using migration models that are more realistic than the island model. For example, Tufto *et al.* (1998) present an analysis of data on 21 sub-populations of sea beet spread over two transects along the shore of Furzey Island in Poole Harbour, on the South Coast of England (Raybould *et al.*, 1996a). In the models of Tufto *et al.* (1996, 1998), explicit distance distribution functions are specified for pollen dispersal (assumed to be two-dimensional) and seed dispersal (assumed to be one-dimensional along the shore) based on perceived underlying physical movement processes. These functions are used with the assumed, known or estimated effective population sizes to derive a matrix of equations for the migration transition probabilities between each pair of populations. This matrix can be used to repeatedly calculate the probability distribution (i.e. multivariate normal variance-covariance matrix) for the observed population gene frequencies using a range of parameter values for the dispersal functions until the likelihood is maximised. This maximum likelihood approach can also be used to distinguish between alternative models and dispersal functions. Tufto *et al.* (1998) found that the Furzey Island sea beet data was best fitted by a model assuming the sub-populations formed an isolated meta-population with the standard deviation for both pollen and seed dispersal distances equal to 75m. This type of method therefore provides estimates of potential gene flow.

CONCLUSIONS

New markers and statistical approaches to the indirect estimation of gene flow are being developed continuously. The traditional approach based on F_{ST} estimates and the infinite island model may eventually be replaced by methods that are more realistic and as easy to use. However, we agree with Bohonak *et al.* (1998) who 'believe that the limitations of traditional approaches are generally understood and that they still provide a valuable first approximation in many cases'. Comparisons of gene flow among species have previously been based on differences in mean F_{ST} or Nm values regardless of the spatial scale of the study (e.g. Hamrick & Godt, 1996). Comparisons based on the above regression and modelling approaches are more informative, especially for GMO risk assessment, in that they incorporate the effects of scale and inter-population distances.

REFERENCES

Bohonak AJ; Davies N; Roderick GK; Villablanca FX (1998). Is population genetics mired in the past? *Trends in Ecology and Evolution* **13,** 360.

Bossart JL; Prowell DP (1998). Genetic estimates of population structure and gene flow: limitations, lessons and new directions. *Trends in Ecology and Evolution* **13,** 202-206.

Campbell DR; Dooley JL (1992). The spatial scale of genetic differentiation in a hummingbird-pollinated plant – comparison of models of isolation by distance. *American Naturalist* **139,** 735-748.

Dow BD;Ashley MV (1998). High levels of gene flow in bur oak revealed by paternity analysis using microsatellites. *Journal of Heredity* **89,** 62-70.

Gliddon C (1994). The impact of hybrids between genetically modified crop plants and their related species: biological models and theoretical perspectives. *Molecular Ecology* **3,** 41-44.

Hamrick JL; Godt MJW (1996). Effects of life history traits on genetic diversity in plant species. *Philosophical Transactions of the Royal Society B* **351,** 1291-1298.

Latta RG; Linhart YB; Fleck D; Elliot M (1998). Direct and indirect estimates of seed versus pollen movement within a population of ponderosa pine. *Evolution* **52,** 61-67.

Levin DA (1981). Dispersal versus gene flow in plants. *Annals of the Missouri Botanical Garden* **68,** 233-253.

Levin DA; Kerster HW (1974). Gene flow in seed plants. *Evolutionary Biology* **7,** 139-220.

Lynch M; Milligan BG (1994). Analysis of population genetic structure with RAPD markers. *Molecular Ecology* **3,** 91-99.

Mills LS; Allendorf FW (1996). The one-migrant-per-generation rule in conservation management. *Conservation Biology* **10,** 1509-1518.

Nagel JE (1997). A comparison of alternative strategies for estimating gene flow from genetic markers. *Annual Review of Ecology and Systematics* **28,** 105-128.

Rasmussen IR; Brødsgaard B (1992). Gene flow inferred from seed dispersal and pollinator behavior compared to DNA analysis of restriction site variation in a patchy population of *Lotus corniculatus* L. *Oecologia* **89,** 277-283.

Raybould AF; Goudet J; Mogg RJ; Gliddon CJ; Gray AJ (1996a). Genetic structure of a linear population of sea beet revealed by isozyme and RFLP analysis. *Heredity* **76,** 111-117.

Raybould AF; Mogg RJ; Clarke RT (1996b). The genetic structure of *Beta vulgaris* ssp. *maritima* (sea beet) populations: RFLPs and isozymes show different patterns of gene flow. *Heredity* **77,** 245-250.

Raybould AF; Mogg RJ; Gliddon CJ (1997). The genetic structure of *Beta vulgaris* ssp. *maritima* (sea beet) populations. II. Differences in gene flow estimated from RFLP and isozyme loci are habitat-specific. *Heredity* **78,** 532-538.

Raybould AF; Mogg RJ; Clarke RT; Gliddon CJ; Gray AJ (1999). Variation and population structure at microsatellite and isozyme loci in wild cabbage (*Brassica oleracea* L.) in Dorset (UK). *Genetic Resources and Crop Evolution* (in press).

Riddoch BJ (1993). The adaptive significance of electrophoretic mobility in phosphoglucose isomerase. *Biological Journal of the Linnean Society* **50,** 1-17.

Slatkin M (1985). Gene flow in natural populations. *Annual Review of Ecology and Systematics* **16,** 393-430.

Slatkin M (1993). Isolation by distance in equilibrium and non-equilibrium populations. *Evolution* **47,** 264-279.

Tufto, J.; Engen S; Hindar K (1996). Inferring patterns of migration from gene frequencies under equilibrium conditions. *Genetics* **144,** 1911-1921.

Tufto J; Raybould AF; Hindar K; Engen S (1998). Analysis of genetic structure and dispersal patterns in a population of sea beet. *Genetics* **149,** 1975-1985.

Weir BS; Cockerham CC (1984). Estimating F-statistics for the analysis of population structure. *Evolution* **38,** 1358-1370.

Wright S (1931). Evolution in Mendelian populations. *Genetics* **16,** 97-159.

Gene flow and risk assessment

C J Gliddon
School of Biological Sciences, University of Wales, Bangor, Gwynedd, LL57 2UW, UK

ABSTRACT

The primary aim in assessing the amount of gene flow from higher plants, in the context of risk assessment of the release of GMO's, is to obtain an estimate of the distribution of the dispersal stages (e.g. pollen and seeds). A knowledge of this distribution will then permit estimation of the probability of the dispersal stage travelling further than a specified distance from its source. Such cumulative dispersal probabilities, can then be used to design both any appropriate confinement measures for a release and monitoring procedures to ensure the efficacy of the confinement.

A standard statistical method for analysing gene-flow data is described and applied to data obtained from crop plants or their relatives, previously published in the scientific literature. The concept of relative risk is developed and applied to this data. Attention is drawn to the inherent variability of both gene flow and its estimates in these experiments and to the consequent problems in attempting to quantify risk to the environment.

INTRODUCTION

In Europe, we are approaching a possible transition between the confined use in the environment of genetically modified organisms (GMO's), primarily in small-scale field trials for the purpose of research and development, and a scaling-up of releases (farm-scale trials), prior to moving on to their unconfined commercial release. While there is still no harmonisation of regulations relating to the release of GMO's on a global scale, there is a large degree of conformity in the aims of the regulations, namely prevention or minimisation of environmental damage. To this end, the assessment of risk associated with the release of GMO's into the environment is required during the research and development stage with the intention that salient information obtained from monitoring those small and medium scale releases will be used to improve the assessment of risk associated with commercial (unconfined) releases.

The spread of genes due to spatial dispersal of pollen has been of great interest to agronomists, population geneticists and environmentalists. Contamination of seed crops by an undesirable source of pollen - such as a different cultivar, wild or weedy relative - led to many field studies in order to evaluate isolation distances required to maintain a satisfactory level of genetic purity of seeds. Results of such trials have often been used in order to establish, modify or contest isolation requirements for Certified Seed production (e.g. Stringam, 1978). Agronomists have also been looking for the best strategies to minimise isolation distances: barren zones, removal of border lines or pollinators.

More recent concerns relating to pollen flow have arisen from the development of GM crops. The nature of these concerns often varies, depending on the stage of development of the GM variety. At early stages of development there is often a perceived need to prevent, or at least minimise, the escape of 'novel' genes from a field trial into wild or feral relatives (Ellstrand & Hoffman, 1990; Dale, 1992; Raybould & Gray, 1993). This need represents, in essence, the converse of the problems faced in traditional seed multiplication, in that contamination of other plants is to be prevented rather than contamination of the trial by them. At later stages of development, it must be realised that escape of genes from the GM crop in medium to large scale trial is inevitable when sexually compatible relatives exist in the same environment. Nevertheless, it would be useful to be able to quantify likely rates of spread into the environment to enable the design of appropriate monitoring protocols and/or mitigation procedures should the risk assessment necessitate them. A further need for quantification of levels of gene flow arises in late stages of development where a single crop species has been modified for a variety of end-uses. For example, if oil seed rape varieties are being developed for use as human food, in industrial processes and for production of pharmaceuticals, it is clearly desirable to have a knowledge of the magnitude of gene flow which may occur, to permit the appropriate segregation in agriculture of the different forms of the crop.

RISK ASSESSMENT

Mackenzie & Henry (1990) described the present paradigm in evaluating risk as follows:

$$risk = exposure \text{ x } hazard$$

In the application of this conceptual relationship to GMO's released into the environment, *exposure* is a measure of the organism's (insert's) ability to escape from its place of deliberate introduction. The subsequent fate of any escaped organisms (inserts) then needs to be quantified in terms of the likelihood of persistence, increase and spread in the environment. The third term in the equation, *hazard*, refers to the impact of the escaped organism on the existing ecosystem.

In order for information to be useful in addressing the problem of quantification of *exposure* for use in risk assessment, it is necessary that it produce data allowing the estimation of the probability of escape as a function of distance from the intentional introduction, together with some estimate of the effect of size of the introduction and some measure of the likelihood of persistence.

The quantification of *hazard* or impact of the escaped organism on the existing ecosystem has rarely been addressed by monitoring of field trials, not least because field trials have been designed primarily to preclude the possibility of escape and, hence, of any *hazard* occurring. Furthermore, the quantification of *hazard* should involve both biological and socio-economic criteria since, for example, the local extinction of a species certainly has an impact on the local ecosystem but the question of whether this extinction constitutes damage rather than benign change requires a value judgement for its answer.

EXISTING ANALYSES

Virtually all of the sampling methods and monitoring protocols described in the literature fail to describe the minimum levels of detection which could be achieved using their particular protocol. This problem is exacerbated by the designs of the experiments - in the vast majority of cases using higher plants, the marked organisms are in a small minority of total organisms in the design. This results in the experimental design making it difficult to detect the spread of the marker in relation to the probability of recovering the non-marked gene. For example, if a marker is represented by 1% of the total organisms, even if its spread is uniform across the entire experimental area, it will only be recovered in 1% of samples. This fault of experimental design could well account for the very small distances that have been reported for the spread of GMO's (e.g. Scheffler *et al.*, 1993).

The vast majority of reported monitoring results fail to try to fit a distribution to the data. The results are usually presented as a simple histogram of raw data with either no attempt at further analysis or a simplistic analysis consisting of calculation of a mean dispersal distance (occasionally with errors which have been erroneously calculated assuming an underlying normal distribution). Furthermore, results are usually presented for marker genes at a given distance as a percentage of total genes sampled. This is inappropriate as it is scale dependent, the correct form being of marker genes at a given distance as a percentage of the total number of marker genes recovered. This method removes the dependence on size of the source of marker genes and correctly emphasises the rate of decrease of marker genes recovered with distance (see Kareiva *et al.*, 1994).

While undoubtedly there has been useful information collected from monitoring of the release of GMO's, the absence of appropriate analysis of the data makes it virtually useless, in its present form, for the purpose of risk assessment.

ALTERNATIVE APPROACHES

Gene flow may be studied by both indirect and direct methods. Indirect methods involve the use of techniques developed in population genetics theory to estimate rates of gene-flow in natural populations (Goudet *et al.*, 1994; Raybould *et al.*, 1996; Raybould *et al.*, 1997). Such indirect methods are problematic in that they need natural populations. However, in cases where the perceived risk involves spread of escaped genes through populations of crop relatives, these methods are ideal since they combine the effects on rate of spread of both gene-flow and population structure (i.e. connectivity). Direct methods involve the estimation of the parameters of dispersal distributions from actual field experiments. Traditionally, dispersal was assumed to follow a bivariate normal distribution (Wright, 1943; Haldane, 1948). However, dispersal distributions from plants have, in the main, been found to be strongly leptokurtic (e.g. Levin & Kerster, 1974) and this led to the suggested use of an exponential power function of the form e^{-ax^b} (Bateman, 1947; Kareiva *et al.*, 1994). Rather than use this essentially descriptive distribution, Lavigne *et al.* (1996) and Tufto *et al.* (1997) have proposed using methods based on a consideration of Brownian Motion in three dimensions to describe pollen deposition. It is clear that under some conditions, for example wind strength varying in direction during an experiment, this mechanistic method gives a superior fit compared to the descriptive exponential power function (Tufto *et al.*, 1997).

Nevertheless, this paper will apply the exponential power function to a range of published data on dispersal in crop plants.

The methodology used follows and extends that suggested by Kareiva *et al.* (1994), using a dispersal reliability function, $R(x)$, associated with an exponential power function for the density of exposure:

$$R(x) = e^{-ax^b} \qquad (1),$$

where a and b are constants which are estimated from the data, defining the rate of decay (shape) of the distribution. $R(x)$ then gives the relative density of genes as a function of distance, x, from the source. Kareiva *et al.* (1994) suggest using a maximum likelihood approach in which the data can be realised as a binomial sampling process in which the binomial probabilities of obtaining a seed carrying a marker gene are proportional to $R(x)$. In order to estimate a and b using this method, a prior estimate of the contamination, c, at distance 0 from the source is required. In situations where the published data was not sufficient to allow the use of this maximum likelihood approach, the function $c.R(x)$ was fitted using the method of least squares. The parameter c defines the level of contamination within the source of pollen (zero distance), b affects the convexity of the curve relative to simple exponential decay (the lower b, the greater the convexity) and a affects the rate of exponential decay. It should be noted that a, b and c are highly correlated, exacerbating problems of estimation.

For use in assessing risk, an expression describing the probability of a gene travelling further than a specified distance, x, is required. This is a cumulative density function (c.d.f.) with shape (rate of decay) described by the parameters a and b estimated above.

The appropriate c.d.f., derived from equation (1), is as follows:

$$\Gamma(\alpha,\beta) \,/\, \Gamma(\beta)$$

where α is $a.x^b$,
β is $2/b$ for dispersal in two dimensions (all directions),
$\Gamma(\bullet,\bullet)$ is the incomplete Gamma function and $\Gamma(\bullet)$ is the complete Gamma function

The c.d.f. above describes the decrease in exposure as a function of distance from the source, relative to the exposure with no isolation. Since, in any given release, the hazard is independent of isolation distance, the c.d.f. is also a measure of relative risk. That is, the probabilities obtained estimate the magnitude of risk, relative to the risk measured at a distance of zero from the release site.

RE-ANALYSIS OF EXISTING DATA

Data was obtained from the scientific literature by computer-searching bibliographies. From several hundred publications, those papers which provided suitable raw data on effective gene flow as a function of distance were selected and the data analysed in a consistent manner.

Source references for these papers and brief details of the experiments are given in Gliddon *et al.* (1999).

The estimates of parameters *a* and *b*, together with their 95% confidence intervals for a selected set of crops are given in Table 1. For completeness, the 'contamination rate' describing the percentage of marked genes recovered at distance zero is also given although, as described above, it plays no further part in the subsequent analysis. It can be seen from Table 1 that less than half of the experiments have estimates of *a* and *b* that do not include zero in their confidence interval (bold italics in the Table). This indicates that the experimental data for these experiments alone shows a SIGNIFICANT decrease in recovery of marked pollen as a function of distance. In the main, this failure to describe a decay in contamination with distance for many crops is regarded as an indication of the lack of appropriate data collected in the experiments rather than as a biological attribute of the crop.

Table 1: Estimates of *a* and *b*

Crop	a	a min	a max	b	b min	b max	Contamination	Range (metres)	
ALFALFA	0.099	-0.330	0.528	0.501	-0.272	1.274	0.620	50	1610
ALFALFA	0.007	-0.036	0.049	1.191	-0.367	2.748	0.201	5	105
MAIZE	***0.522***	***0.337***	***0.707***	***0.958***	***0.624***	***1.291***	***0.838***	***0.5***	***24***
MAIZE	***1.413***	***1.058***	***1.768***	***0.974***	***0.160***	***1.788***	***0.046***	***1***	***34***
POTATO	0.523	-2.514	3.560	0.942	-1.619	3.504	0.031	0.75	21.5
POTATO	0.577	-15.549	16.703	0.647	-22.105	23.398	0.001	0.75	9.5
RADISH	***1.643***	***0.703***	***2.583***	***0.761***	***0.215***	***1.306***	***0.855***	***0.2***	***14***
RADISH	0.274	-0.064	0.612	0.613	0.294	0.932	1.000	6	180
RADISH	0.214	-0.074	0.501	0.555	0.221	0.889	0.522	2.5	150
RAPE	0.835	-2.400	4.071	0.773	-1.856	3.401	0.028	1	70
RAPE	***0.750***	***0.554***	***0.946***	***0.206***	***0.102***	***0.156***	***0.019***	***n/a***	***n/a***
RAPE	***0.832***	***0.693***	***0.971***	***0.304***	***0.236***	***0.374***	***0.035***	***n/a***	***n/a***
RAPE	***1.332***	***1.053***	***1.611***	***0.535***	***0.451***	***0.619***	***0.051***	***3***	***55***
RAPE	***0.752***	***0.526***	***0.978***	***0.652***	***0.541***	***0.763***	***0.026***	***3***	***55***
RYEGRASS	0.018	-0.021	0.057	0.816	0.262	1.371	0.060	3	76
RYEGRASS	***0.133***	***0.073***	***0.194***	***0.774***	***0.593***	***0.955***	***0.348***	***0.5***	***915***
RYEGRASS	0.003	-0.001	0.007	1.555	1.191	1.919	0.454	0	100
RYEGRASS	0.031	-0.104	0.165	1.169	-0.023	2.361	0.454	10	70
SUGAR BEET	***1.310***	***0.330***	***2.290***	***0.524***	***0.097***	***0.952***	***0.141***	***0.2***	***23***
WHEAT	***0.483***	***0.385***	***0.581***	***0.872***	***0.684***	***1.060***	***0.254***	***0.025***	***30***
WHEAT	0.038	-0.020	0.096	0.870	0.411	1.329	0.107	3	48

a min, b min: lower 95% confidence limit of *a* and *b* respectively
a max, b max: upper 95% confidence limit of *a* and *b* respectively
Range: The distances from the souce over which the experiment was carried out

DISCUSSION

Variability within crops

A cursory glance at Table 1 shows that there can be immense variability in rates of decay estimated from different experiments for particular crops. The relative risk for the five experiments carried out with oil seed rape is shown graphically in Figure 1. In part, this variability may be ascribed to the lack of good data and, therefore, the lack of statistical power in the estimates. However, an alternative explanation is that this variability is an inherent feature of the pollination biology of certain crop species.

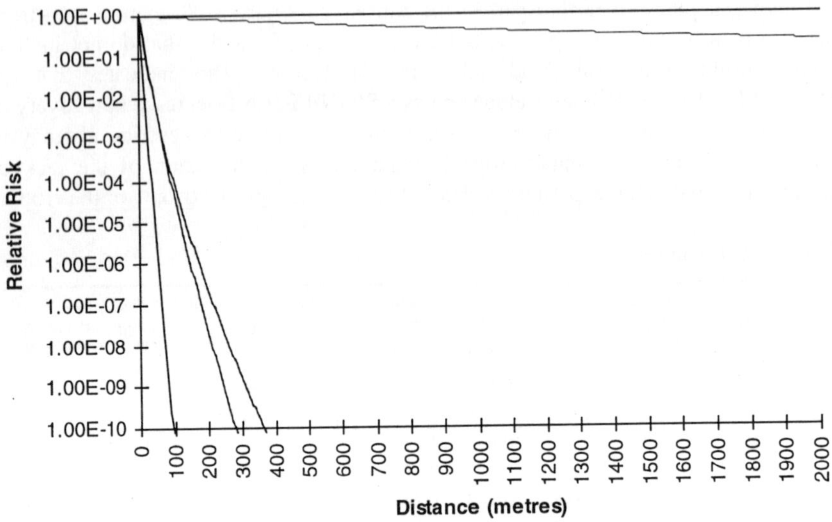

Figure 1. Relative risk in five different experiments using oil seed rape

Unfortunately, the existing data is not sufficient to discriminate between these two alternatives. The large body of literature on pollination biology certainly suggests that, at least for insect pollinated species, the crop species itself is unlikely to be a sufficient descriptor of expected pollen movement. Factors including the type and density of surrounding vegetation, flowering stage of other vegetation and meteorological conditions are likely to influence the distance that pollen is carried to a far greater extent than the species of pollen donor. It is obvious that such variability needs to be taken into account if carrying out a risk assessment as a precursor to the possible release of a GMO.

It is clear that for insect pollinated species in particular, there can be a very strong effect of experimental conditions determining the amount of gene flow measured. One striking component of the experimental design that has often been shown to have a major effect is that of barren zones or borders separating potential pollen recipients from donors. Goplen *et al.* (1972) reported on the effects of isolation distance on contamination in sweetclover (*Melilotus alba*). They used a recessive gene marker, *cu*, for low coumarin content. In one of their experiments, a source plot of 0.4 ha was located 46 m and 389 m from two sink plots of a similar size. Both these plots showed a similar average level of contamination (0.14 and 0.13 respectively). However, within each of these plots, the contamination declined rapidly with distance from the source. An explanation for the above observations is that pollinators arrive in the recipient plots loaded with pollen from the donors. They then unload their pollen as they move within the recipient plots. However, the likelihood of arriving at one or other recipient plots is not determined, to any major extent, by the distance from the source.

In conclusion, there is a need to be extremely careful in using published data on isolation with distance as an element of risk assessment. This due to two factors: first, the data in the literature have frequently not been analysed appropriately for the purposes of risk assessment; and, secondly, there is often great variability among experiments describing pollen movement in a single crop. In general, the variability within crops can be assumed to be greater in insect-pollinated species than in wind pollinated species. However, there is often insufficient data in the literature to allow reasonably confident predictions to be made about the levels of contamination to be expected in almost all crops.

The statistical method described above provides a basic standard for the analysis of experiments designed to measure the decay of contamination by pollen as a function of distance for use in risk assessment. It should be noted that although only three parameters need be estimated in this method, the majority of existing data-sets are insufficient to describe significant decay of contamination with distance. This is a clear indictment of the utility of many existing descriptions of pollen movement in crop plants in the context of risk assessment. The development of more complex methods of statistical description, taking into account, for example, the effects of wind strength and direction on pollen movement (e.g. Lavigne *et al.*, 1996; Tufto *et al.*, 1997), will not render existing data more useful as they will require estimation of more parameters from very sparse data-sets

REFERENCES

Bateman A J (1947). Contamination in seed crops. III. Relation with isolation distance. *Heredity*, **1**, 303-336.

Dale P J (1992). Spread of engineered genes to wild relatives. *Plant Physiology*, **100**, 13-15.

Ellstrand N C; Hoffman C A (1990). Hybridization as an avenue of escape for engineered genes. *BioScience*, **40**, 438-442.

Gliddon C; Boudry P; Walker D S (1999). Geneflow – a review of experimental evidence. In: *Environmental impact of genetically modified crops*, eds A J Gray, F Amijee & C J Gliddon. Genetically Modified Organisms Research Report 10, pp. 67-81. DETR: London

Goplen B P; Cooke D A; Pankiw P (1972). Effects of isolation distance on contamination in sweetclover. *Canadian Journal of Plant Science*, **52**, 517-524.

Goudet J; De Meeüs T; Day A J; Gliddon C (1994). The different levels of population structuring of the dogwhelk, *Nucella lapillus*, along the south Devon coast. In: *Genetics and evolution of aquatic organisms*, ed. A R Beaumont, pp. 81-96. Chapman & Hall: London.

Haldane J B S (1948). The theory of a cline. *Journal of Genetics*, **48**, 277-284.

Kareiva P; Morris W; Jacobi C M (1994). Studying and managing the risk of cross-fertilization between transgenic crops and wild relatives. *Molecular Ecology*, **3**, 15-21.

Lavigne C; Godelle B; Reboud X; Gouyon P H (1996). A method to determine the mean pollen dispersal of individual plants growing within a large pollen source. *Theoretical and Applied Genetics*, **93**, 1319-1326.

Levin D A; Kerster H (1974). Gene flow in seed plants. *Evolutionary Biology*, **7**, 139-220.

Mackenzie D R; Henry S C (1990). Towards a consensus. In: *The biosafety results of field tests of genetically modified plants and microorganisms*. Proceedings of the Kiawah Island Conference, eds D R MacKenzie & S C Henry, pp. 273-283. Agricultural Research Institute: Bethesda, Maryland, USA.

Raybould A F; Gray A J (1993). Genetically modified crops and hybridization with wild relatives: a U.K. perspective. *Journal of Applied Ecology*, **30**, 199-219.

Raybould A F; Goudet J; Mogg R J; Gliddon C J; Gray A J (1996). The genetic structure of a linear population of sea beet (*Beta vulgaris* ssp. *maritima*) revealed by isozyme and RFLP analysis. *Heredity*, **76**, 111-117.

Raybould A F; Mogg R J; Gliddon C J (1997). The genetic structure of *Beta vulgaris* ssp. *maritima* (sea beet) populations .2. Differences in gene flow estimated from RFLP and isozyme loci are habitat-specific. *Heredity*, **78**, 532-538.

Scheffler J A; Parkinson R; Dale P J (1993). Frequency and distance of pollen dispersal from transgenic oilseed rape (*Brassica napus*). *Transgenic Research*, **2**, 356-364

Stringam G R; Downey R K (1978). Effectiveness of isolation distance in turnip rape. *Canadian Journal of Plant Sciences*, **58**, 427-438.

Tufto J; Engen S; Hindar K (1997). Stochastic dispersal processes in plant populations. *Theoretical Population Biology*, **52**, 16-26.

Wright S (1943). Isolation by distance. *Genetics*, **28**, 114-138.

Gene flow at the landscape level

G R Squire, J W Crawford, G Ramsay, C Thompson
Scottish Crop Research Institute, Invergowrie, Dundee DD2 5DA, UK

J Bown
School of Computing, University of Abertay, Dundee DD1 1HG, UK

ABSTRACT

The distribution of fields and fragmentation of habitat in agricultural landscapes has a dominant effect on the rate of gene flow among crops and feral or wild relatives. An approach to quantifying the salient factors is illustrated with reference to experimental and theoretical studies of oilseed rape crops and feral brassicas. Existing models of long distance pollen movement underestimated gene flow when compared with measurements of pollen concentration and pollination in an agricultural context. The evidence indicates that large pollen sources, such as crop fields, interacted on a regional scale to increase gene flow. The value of modelling landscape-scale gene flow for synthesis and forming hypotheses is illustrated. Existing approaches, based on pollen-trap measurements from what were presumed to be single source fields need to be modified to account for multi-source interference.

INTRODUCTION

In the study of gene flow and its consequences for agriculture, the main feature of the landscape is the fragmented or distributed pattern of its crop fields and the associated populations of feral or wild relatives. This fragmentation is due in part to variation in the soil and topography, and in part to the variety of agricultural practice. Interactions both between and within fragments are important (Husband & Barrett, 1996). Movement of pollen and propagules determines the potential rate of spread of a gene, while local interactions for space or resources determine the rate at which individuals containing the gene increase in number or become extinct. While mechanisms of dispersal and pollination set the upper spatial limit to possible exchanges in any reproductive cycle, the local population dynamics determine the consequences of those exchanges.

However, the rate of gene transfer from one plant to another and the residence time of the gene in the population are unlikely to be independent factors. Suppose, for example, that a new genotype was grown in an agricultural region for a few life cycles and left a number of feral populations containing its genetic material. Any property or factor conferring stability on the individuals in a fragment would increase the residence time of the genetic material but tend to reduce its spread to new individuals within the fragment. Also, dynamics that maximise the rate at which a gene increases in frequency at the landscape scale are unlikely to coincide with *both* the maximum residence time in a fragment and the maximum rate of gene movement between fragments. Rather each would be in some optimum state. If gene flow is undesirable, knowledge of such optimum states should allow policy and management to avoid them. The challenge is to define them from imperfect knowledge of processes operating over a wide range of spatial and temporal scales.

There is a need for both conceptual and mathematical frameworks to enable scaling between local and regional processes and to incorporate specific observational and experimental detail on factors such as the physiological properties of individuals, their competitive interactions, the mating system, competition between self and foreign pollen and the properties of the pollen vector. Research towards developing such a framework has been in progress for several years and is now summarised. Many of the issues considered are generic and will be widely applicable in open pollinated systems.

GENE FLOW IN FERAL BRASSICAS AS A MODEL SYSTEM

Crop fields of oilseed rape and feral descendants of these crops exist in close proximity throughout the arable areas of the UK. Here *feral* includes both volunteer weeds within fields and other populations on field margins, soil dumps and roadsides. The crops and ferals are mainly *Brassica napus*, accompanied in certain areas by crops and ferals of *Brassica rapa*, and by wild *Brassica rapa*. Ferals are a common and widespread weed in cereal rotations, anecdotal evidence in many areas indicating they reside for several years in the seedbank (Lutman, 1993). The commercial lifetime of most cultivars is only a few years, so many seedbanks are likely to contain the descendants of several cultivars. Within arable fields, the volunteers and crops mingle intimately, but the plant habit of the volunteers is still governed by the cycles of cultivation and spraying, so remains similar to what it was in the original crop. The feral populations outside fields experience a wide range of selection pressures, leading to diverse forms including individuals that flower when very small (<5 cm height) to those that flower in at least two successive years. Many of the out-of-field feral populations are not routinely controlled, while phenotypic plasticity allows others that are controlled to avoid or recover from cutting or herbicide treatment. Seed is spilled while in transport and moved around in farm machinery. Pollination can occur by both wind and insect but the extent of each is uncertain in the arable context. Potentially, therefore, the feral oilseed rape in a region can receive genetic material from crops, keep it for several generations, and pass it on to other crops and ferals even when the cultivar source is no longer extant.

Much of the evidence of population dynamics and gene flow in oilseed rape has been obtained in a context that could not directly inform regional scale studies of feral populations. So, for instance (Bilsborrow *et al.*, 1998; Scheffler *et al.*, 1993), estimates of rather low levels of gene flow from moderate-sized plots to a somewhat larger surrounding planted area were more representative of gene flow from a feral population to a nearby or surrounding crop than the reverse. On population establishment, the work of Crawley *et al.* (1993) on the dynamics of oilseed rape populations artificially established in a range of habitats was mostly relevant to the risk of oilseed rape invading semi-natural vegetation. The increasing evidence of the extensive persistence and long-distance gene flow among crops and feral populations has developed from three other lines of investigation, all carried out in an agricultural context. First, demographic surveys in areas of Tayside and north east Fife defined the density of fields and feral populations in the landscape (DETR, 1999). About 25% of the populations recurred at the same site over at least two years, while others disappeared for a year or more then reappeared. Since the recurring populations might have been caused by new seed spilled each year, the second line of research developed DNA markers to distinguished previous and extant crop cultivars (Charters *et al.*, 1996). Use of these markers gave unequivocal evidence that a cultivar could persist for several years at a site, even after it had ceased to be grown in the area; that populations could consist of more than one cultivar; and that outcrossing occurred within feral populations and from crops to ferals, sometimes leading to a genetically heterogeneous feral population. In one population studied, the descendants of a crop cultivar have persisted for at least ten years. The third line of investigation on transects of pollen density and direct measurements of gene flow indicated that early work with small sources might have underestimated the distance of pollen movement (Timmons *et al.*, 1995). Although the evidence for actual hybridisation was still very sparse and uncertain at distances over 0.25 km, many feral populations in the region lived within the hybridisation range of at least one field.

The application of modelling

None of this work in isolation could inform the crucial question of what would happen if many transgenic fields were distributed in the landscape. Clearly, if pollen concentrations decay rapidly with distance, such that fields are (in effect) isolated from each other, then regional considerations are not important; but if the pollen fronts from fields coalesce, then landscape scale behaviour – which might be unpredictable from work with single, small sources - is possible. A preliminary landscape-scale model was therefore developed as a means of synthesis and hypothesis-forming (Crawford *et al.*, in press; Timmons *et al.*, 1996). Equations based on field data represented pollen movement by air and deposition with distance from a field to give a two-dimensional pollen density profile over a simulated landscape. Simple assumptions were that there was enough pollen to fertilise every flower, and that self and foreign pollen were equal in their efficacy to pollinate. The percentage hybridisation between, say, transgenics and ferals was simply in proportion to the relative concentrations of the two types of pollen. When transgenic and non-transgenic fields were distributed together with feral populations at realistic frequencies found in the demographic survey, landscape-scale behaviour was implicated, specifically in that the gene flow from transgenics and ferals was influenced by the regional aggregation pattern of transgenic and non-transgenic fields. Although the model predicted substantial hybridisation in some areas, there were still major uncertainties over the existence and effect of long distance airborne transport (>1 km), and the balance between wind and insects as pollen vectors.

The principal purpose of new theoretical and experimental work in 1998 was to test the hypothesis that the distribution of fields and ferals on a landscape scale plays a significant role in gene flow. An experiment was constructed in two parts, as described later in this volume (Thompson *et al.*, 1999). The first used male-sterile bait populations as biological pollen traps, glass slides covered in double-sided sticky tape to measure surface deposition of pollen and a series of pollen suction traps, measuring airborne concentrations, all distributed along a line from one relatively isolated field, starting in the field itself then at 5 m, 170 m and 400 m. The results from this transect would provide back-reference to previous pollen transects on which the model was based. In the second part, bait plants and sticky slides were distributed in the region during the flowering period of oilseed rape. The aim was to determine pollen transport parameters from the line transect, and use these to predict the distribution of aerial and surface deposited pollen. A comparison with the new regional-scale data would confirm or reject the hypothesis.

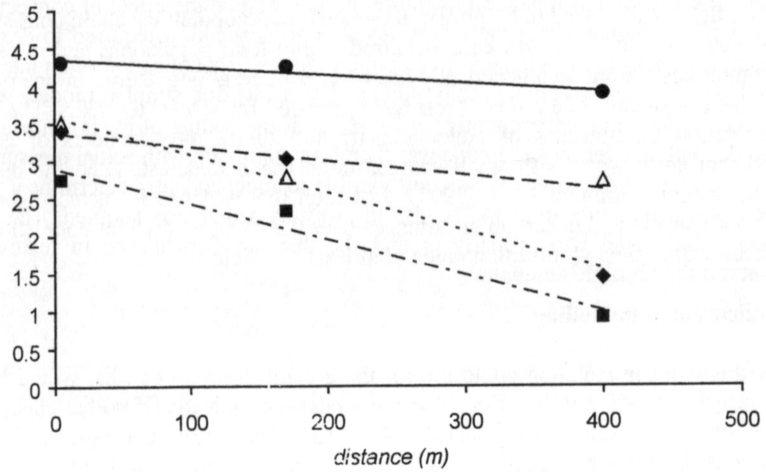

Figure 1. Plot of the logarithms of (filled squares) seed number produced per flower, (triangles) pollen concentration ($cm^{-2}d^{-1}$) on the glass slides, (diamonds) pollen concentration in arbitrary units recorded in suction traps, (circles) pods produced per flower i.e. % hybridisation, with distance from the field margin.

Clear trends were found in aerial and deposited pollen concentration, seeds/flower and % hybridisation with distance along the line transect (Fig. 1). A feature of the trends is the nature of the distance dependence of the airborne and deposited pollen concentrations. Both are consistent with an exponential decay with distance (r^2=0.94 and r^2=0.72 respectively). However the decay rate is different: the airborne pollen falls off more quickly with a characteristic dispersal distance, α_a, (equal to the distance at which pollen drops to 1/e=0.37 of the value at the margin) of 200 m compared with the deposited pollen which has a characteristic dispersal distance, α_d, of approximately 500 m. The difference is most likely due to the fact that pollen is lost from the plume sampled by the traps both through deposition and upward convection. This confirms the need to understand the consequences of the sampling technique for the interpretation of measurements. The value of α_a is consistent with the values recorded in previous years in a different part of the region (Timmons *et al.*, 1996) but significantly larger than reported from plot experiments (e.g. McCartney & Lacey, 1991). Slide data collected from a similar transect adjacent to another field of oilseed rape in the current experiment showed the same spatial dependence and value for α_d. By contrast, slide data collected from a transect adjacent to a field of turnip rape (*B. rapa*) showed a lower value of α_d around 200m. Such large distances imply that many fields in the landscape will be in wind-borne pollen contact with feral populations.

As a first means of scaling between field and landscape, the existing model for airborne pollen transport (Timmons *et al.* 1996, Crawford, Squire & Burn, in press) was calibrated from the transect containing the pollen traps. The model includes a term for deposition and therefore we could calculate both airborne and deposited pollen levels. However, the model was essentially two-dimensional and therefore did not include the effect of convected losses of pollen into the higher atmosphere. The level of sophistication required to deal with 3-dimensional transport is substantially greater and requires data collection of an intensity that is beyond the scope of regional-scale studies. By using our simpler model, we ignore upward convection and consequently $\alpha_a = \alpha_d$ and the value of α_d is underestimated. Predicted deposition over the landscape was therefore compared with actual measurement of pollen on the glass slides. Since it was not possible to independently determine a value for the pollen production rate from the fields, an arbitrary value was adopted. The resulting output from the model will properly predict the distance dependence in relative pollen levels, but not the absolute amounts.

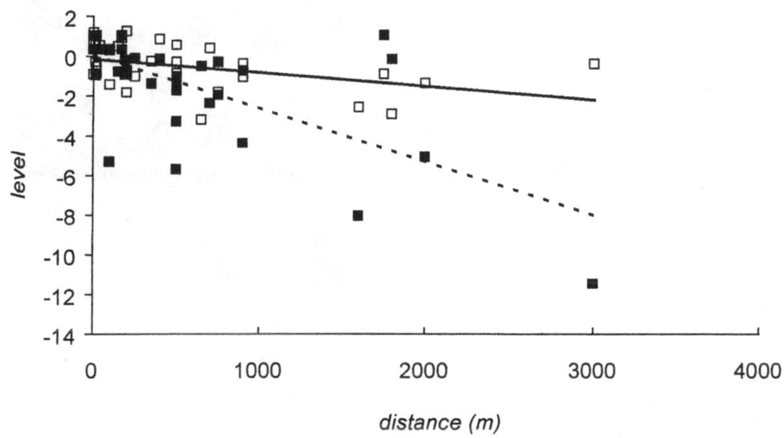

Figure 2. Plot of predicted (filled squares) and measured (open squares) levels (arbitrary units) of deposited pollen as a function of distance to nearest oilseed rape field (m). The absolute values have been renormalised for graphical clarity, a process that does not affect the slope of the distance relation.

Examples of the approach and indicative output (Fig. 2, Fig. 3) illustrate generic problems in scaling. Measured deposition was more variable at the landscape scale (Fig. 2). The value for α_d estimated from the slide data is 1500 m, three times larger than that estimated from comparable data on the transect. The measurements over a 400 m line extending from a single field were not therefore representative of pollen deposition over the landscape. The modelled values of deposition show even more variation between sites (Fig. 2); the value of α_d is 370 m, a similar order to that measured in the transects but about a quarter of that measured at the landscape scale. These results indicate the original hypothesis is correct: regional context has a significant role in determining the amount of pollen transferred by wind from fields to the environment. In calibrating our model, we chose a relatively isolated field. However, until the results of the simulation and analysis were complete, we could only have provided a subjective and qualitative definition of 'isolated'. The transect probably suffered significant interference from surrounding fields (Fig. 3), even although these were the order of 1 km or more distant. Therefore our calibration was context dependent. The discrepancies between measured and modelled values appear to have a complex origin. As discussed above, the estimates at distance are very sensitive to the slope measured in the transect (Fig. 1), while the two-dimensional nature of our pollen transport model means that it will underestimate α_d. Furthermore, distance to the nearest field is not an appropriate variable at the landscape scale: only locations in the more isolated parts of the study area would have received wind-born pollen from mainly one source. The analysis shows the uncertainties that arise from scaling just one variable, airborne pollen. Scaling other salient factors, such as insect-borne pollen, would add further complexity.

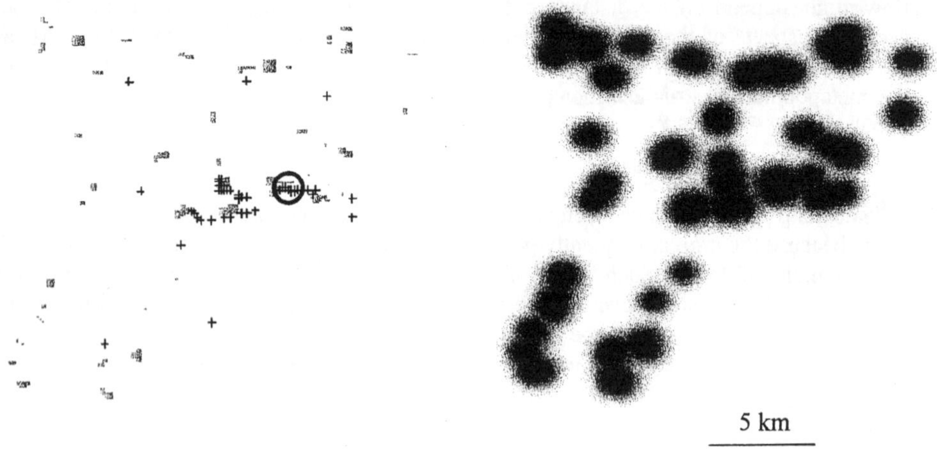

Figure 3. Left: map of fields (grey areas) in the 20 km x 20 km study region. Crosses indicate the locations of bait plants and glass slides for collecting deposited pollen. Circle identifies the main transect. Right: contour map of deposited pollen levels predicted by the model indicating the degree of mixing across the region after 2 simulated days of pollen release.

DISCUSSION

Crop and feral oilseed rape in a region such as Tayside should be considered as a metapopulation governed by a combination of local population dynamics at numerous sites, seed movement and gene exchange by pollen. Many other distributed, sexually compatible populations are likely to behave in a similar manner, differing only in degree from that of the feral brassicas. Therefore aspects of local stability and regional connectivity described here will be relevant to other systems. Further definition is required, both with respect to oilseed rape and generally, of both long-distance and local processes.

The effects of the spatial arrangement and size of fragments on pollen transfer are still uncertain and will be influenced by the type and features of the pollen vector as well as local landscape. For example, variation in pollination in the study with bait plants was traced, among other things, to local features such as woodlands and the occurrence of insect pollinators. Molecular techniques allow paternity to be identified in feral oilseed rape, but only where the pollen sources are sufficiently distinct. In some feral populations we studied, up to 50% of the progeny of some individuals contained genetic material that came from other plants as a result of outcrossing in that season. Given the landscape-scale gene movement from crops to ferals, however, several identical source cultivars will usually be within pollination distance of any feral population, making precise detection of the source impossible in many circumstances. Elsewhere, more definition has been obtained. In fragmented populations of the insect-pollinated tropical tree, *Swietenia humilis*, source-individuals were identified by paternally transmitted markers. Gene flow within moderate or large fragments was mainly between individuals in the fragment, while gene flow between fragments (up to 5 km) was greatest from small to large fragments (White & Powell, 1997). More generally, the relative extents and direction of between- and within-fragment gene flow might depend on the distance between fragments, the flight behaviour of the insect vector or aspects of the airborne pollen profile, as well as sexual compatibility. All are likely to be context-specific, but the possibility exists that a small degree of isolation within the metapopulation structure increases the likelihood of gene movement out of isolated populations.

The causes of establishment and spatial heterogeneity in the distribution of feral populations are uncertain. Areas of high occupancy by feral brassica populations are known to exist, partly because the crop is frequently grown there, but for other reasons that must favour the establishment of ferals. Such areas allow build up of populations from many different cropping and harvesting events, thereby allowing gene exchange over many years. Moreover, the stability over time of local populations has rarely been examined in detail, though great differences exist in their longevity. While the observational evidence for brassicas agrees with experiments that feral oilseeds are not becoming established in perennial scrub, grassland or woodland, evidence is accumulating that ferals can be stable and long-lived within and around the margins of agricultural land. Molecular techniques coupled with local knowledge have demonstrated persistence for over ten years in our study region. The contributions of source-cultivar physiology, and the effect of environment on local sorting and selection need to be much better defined.

The attempt to model regional gene flow has shown the need for a deeper understanding of the underlying principles, as opposed to a fragmentary and occasional look at topical issues. The purpose of modelling as used here is not so much prediction of absolute events but to produce testable hypothesis based on existing information, much of which is still at a small scale. Modelling and experimentation are therefore complementary and cyclical. Reaching an understanding of the broad picture will guard against unsound generalisations arising from specific experiments out of context.

ACKNOWLEDGEMENTS

We thank the Ministry of Agriculture, Fisheries and Food and the Scottish Office Agriculture, Environment and Fisheries Department for their support. JB is funded by a studentship from the University of Abertay, Dundee.

REFERENCES

DETR (1999). *Investigations of feral oilseed rape populations. Research Report No. 12.* Department of the Environment, Transport and the Regions: London.

Bilsborrow P E; Evans E J; Bowman J; Bland B F. (1998). Contamination of edible double-low oilseed rape crops via pollen transfer from high erucic cultivars. *Journal of the Science of Food and Agriculture* **76**, 17-22.

Crawford J W; Squire G R; Burn D (in press) Modelling spread of herbicide resistance in oilseed rape. In: *Environmental impact of genetically modified crops. Research Report No. 10,* eds A J Gray, F Amijee & C J Gliddon, pp. 101-112. Department of the Environment, Transport and the Regions: London.

Crawley M J; Hails R S; Rees M; Kohn D; Buxton J. (1993). Ecology of transgenic oilseed rape in natural habitats. *Nature* **363**, 620-623.

ChartersY M; Robertson A; Wilkinson M J; Ramsay G (1996). PCR analysis of oilseed rape cultivars (*Brassica napus* L. ssp. *oleifera*) using 5' anchored simple sequence repeat (SSR) primers. *Theoretical and Applied Genetics* **92**, 442-447.

Husband B C; Barrett S C H. (1996) A metapopulation perspective in plant population biology. *Journal of Ecology* **84**, 461-469.

Lutman P J W. (1993). The occurrence and persistence of volunteer oilseed rape *(Brassica napus*). *Aspects of Applied Biology* **35**, *Volunteer crops as weeds*, pp 29-36.

McCartney H A; Lacey M E. (1991). Wind dispersal of pollen from crops of oilseed rape (*Brassica napus* L.). *Journal of Agriculture Science, Cambridge* **107**, 299-305.

Scheffler J A; Parkinson R.; Dale P J. (1993). Frequency and distance of pollen dispersal from transgenic oilseed rape (*Brassica napus*). *Transgenic Research* **2**, 356-364.

Squire G R; Marshall B; Dunlop G; Wright G (1997). Genetic basis of rate-temperature characteristics for germination in oilseed rape. *Journal of Experimental Botany* **48**, 869-875

Thompson C E; Squire G R; Mackay G R; Bradshaw J; Crawford J W; Ramsay G. (1999). Regional patterns of gene flow and its consequence for GM oilseed rape. In*: Gene flow and agriculture: relevance for transgenic crops.* British Crop Protection Council.

Timmons A M; Charters Y; Crawford J W; Burn D; Scott S; Dubbels S J; Wilson N U J; Robertson A; O'Brien E T; Squire G R; Wilkinson M J (1996). Risks from transgenic crops. Nature **380**, 487

Timmons A M; O'Brien E T; Charters, Y M; Dubbels S J; Wilkinson M J (1995). Assessing the risks of wind pollination from fields of genetically modified *Brassica napus* ssp. *oleifera. Euphytica* **85**, 417-423.

White G; Powell W. (1997). Isolation and characterisation of microsatellite loci in *Swietenia humilis* (Meliaceae): an endangered tropical hardwood species. *Molecular Ecology* **6**, 851-860.

Genetic pollution: concepts, concerns and transgenic crops

R E Daniels
ITE, Furzebrook Research Station, Wareham, Dorset, BH20 5AS, UK

J Sheail
ITE, Monks Wood, Abbots Ripton, Huntingdon, Cambridgeshire, PE17 2LS, UK

ABSTRACT

Genetic pollution implies the loss of identity of wild plant species as a result of transfer of genes from crop plants, with engineered fitness genes causing special concern. However, significant reproductive barriers exist between most wild species and crop plants. Gene flow is expected to have its greatest effect on weed species which are closely related to crop plants, taxonomically, ecologically and in their reproductive biology. Such plants may already have a long history of gene exchange with their crop relatives.

INTRODUCTION

Recently, conservation interest has become increasing focussed on the need to maintain biodiversity, as witnessed by a number of national and international political initiatives (the 1992 Rio Convention on Biological Diversity being the best known). This carries with it the implication that it is necessary to retain not only a range of habitats and species, but also the separate gene pools giving rise to the singularity of species and distinctive variants at the infra-specific level. Under this philosophy, individual, isolated, populations become worthy of conservation because their genetic structure endows them with a suite of characteristics which may be unique. It may also provide both short term fitness and long term evolutionary potential which, together, act as a buffer against extinction. The desire to maintain the maximum possible amount of genetic diversity has, in turn, given rise to fears that the genetic individuality of local populations or rare species with small populations could be compromised following gene flow from external sources. Consequently, the management of risk to genetic identity has become a component in the design of conservation strategies for rare species.

Genetic drift and, to a much lesser extent, mutation will tend to produce variation between populations which are small and isolated reproductively, so that those populations come to differ in morphology, physiology, biochemistry or in life history and breeding system characteristics. Gene flow, even at very low rates, will counter such differentiation. The greater the amount of gene flow between populations, the less the opportunity for the stochastic development of genetically distinct populations by drift. Where selection strongly outweighs the counterbalancing effects of gene flow, and dominates any tendency for genetic drift, it will dictate the evolution and genetic architecture of plant populations. It is the balance between these three processes which determines the genetic character of populations and any changes that may take place in them. It is against this background that genetic pollution risks must be assessed

WHAT IS GENETIC POLLUTION?

The term "genetic pollution" (Butler, 1994) is one of a number of disparaging terms which have been used to define the mixing of gene pools by hybridisation and introgression, especially where a common, non-local species or variety has been brought into contact with a small local population or a rare species. Less emotive are expressions such as, "genetic assimilation" (Ellstrand, 1992) and "mixing" (Rhymer & Simberloff, 1996). Whatever the terminology, the reference is to a single process which raises three fundamental problems having philosophical, ecological and practical implications:

- that locally-adapted populations, or whole species will become extinct
- that natural patterns of diversity will become obscured within a gene pool common to all populations of that species, and that this will ultimately compromise the capacity to adapt to environmental change
- that the introduction of new fitness traits will have long-term implications for population genetics and community dynamics.

The consequences of gene flow producing such effects has been voiced, particularly in relation to the use of non-local material in wildflower seed mixes (Akeroyd, 1994; Handel, *et al.*, 1994) but may also be invoked in considering the effects of forage or arable crop species on close relatives in natural grasslands.

Many of the terms used imply that there is an inherent value in what are regarded as "pure" gene pools and that this value is lost following introgression of genes from outside this pool. This view is incorporated in various national and international conventions whose underlying ethos is to protect variety, including genetic variation between and within species. The need for such protection of variation is rooted in a belief that it forms part of a national or international heritage and, as such needs to be regarded in much the same way as York Minster or the pyramids of Egypt. These buildings contain within their structures evidence of their history, their purpose and the culture of the societies which constructed them or modified the original design. In the same way, plant populations also contain evidence of their history. Because of this, and because man feels a duty to pass on items of heritage value to his successors (for their intrinsic value or because they may have potential for future use) there are sound philosophical grounds for retaining as much genetic diversity as possible within species and populations.

Plant populations and the genetic structures of those populations are not static, fixed entities: they are dynamic. The processes of gene flow, isolation (and the consequent possibilities of inbreeding depression), genetic drift, selection and mutation are continuously operating and will operate to modify the amount of biodiversity found, whatever management options are exercised. In the long term it does therefore seem unreasonable to expect to be able to maintain the full range of combinations of genetic variation currently exhibited in diverse and isolated populations. Nevertheless the concern shown to maximise the possibilities of doing so is understandable.

Among animals there are several well known cases where the distinct identity of a particular species has been thrown into doubt, or has been lost following the breakdown of reproductive barriers, usually following the invasion or introduction of more common species into areas formerly occupied by their rarer relatives. Examples include that of the extensive hybridisation found to occur between mallard (*Anas platyrhnynchos*) and a number of rarer endemic duck species in various parts of the world (Rhymer & Simberloff, 1996). Similar hybridisation and introgression occurs in plants and may threaten rare, or endemic, species (Ouborg & Van Treuren, 1994). Introgression from domesticated forms of crop plants have also been implicated (directly or indirectly) in the extinction of wild populations of maize and rice (Gregorius & Steiner, 1993). The emphasis in all these cases has been on the loss of identity at the species or sub-species level because the rarities have been subsumed into their more common relatives. This is why there is a fear of genetic pollution.

Extended to an agricultural context the fear is that hybridisation may occur following the introduction of an agricultural crop into the vicinity of a population of a wild relative, so that some of the fitness genes selected as part of breeding programmes (particularly those which enable plants to establish sooner, grow more quickly or attain greater stature) may be transferred into wild populations, giving the hybrids such a competitive advantage that it may lead to their eventual extinction of the native species. Alternatively, the transferred genes may become dispersed throughout the whole population to produce a continuum between crop and wild species and, hence, loss of identity of the wild species. The development of transgenes, especially those conferring pest or disease resistance, is seen as cranking up this process, resulting in widespread modification of the genetic structure of wild populations and elimination of the variation they contain.

Gene movement between different species, or even genera, is not an unusual phenomenon as shown by the high proportion of hybrids present in the British and other floras (Stace, 1975; Ellstrand *et al*., 1996) but this general observation hides wide discrepancies in the frequency of hybridisation within different groups of plants. Among native British species, hybridisation is especially common in some families or genera, but rare in others (Table 1).

Table 1. Frequency of hybridisation in 7 genera of British plants (data from Stace, 1975)

Genus	Number of species	Number of interspecific hybrids	Number of intergeneric hybrids
Salix	23	123	0
Rumex (subgenus Rumex)	23	42	0
Geranium	17	7	0
Plantago	5	2	0
Dactylorhiza	7	27	22
Trifolium	21	2	0
Lolium (native species)	3	3	10

A contrasting situation is found in two genera which have a long history of agricultural use during which considerable infra-specific variation has been exploited or developed. Whilst there are strong breeding barriers between *Trifolium* species, *Lolium* species not only produce hybrids with each of their two congenors but are also capable of hybridising with *Festuca*, *Glyceria* and *Bromus* species. As the crop base in Britain (in terms of the original progenitor species) is relatively small, it is not surprising to find that even in crops with a long history of selective breeding and adaptation, the possibility of gene flow between them and wild relatives has not been eliminated. An increasing number of studies have shown specific cases where genes have passed to wild relatives from different crops and the whole topic has been reviewed by a number of authors, including De Vries *et al.*, (1992), Raybould & Gray (1993) and Gregorius & Steiner (1993). Assessments of relatedness may help determine the possibility of gene flow but it will not necessarily assist in predicting its actual incidence because gene flow rates between members of a crop-weed-wild plant complex may vary widely (Van Raamsdonk & Van der Maesen, 1996). Individual assessment on a species by species basis is necessary.

FACTORS INFLUENCING THE EFFECTIVENESS OF GENETIC POLLUTION

The process of gene flow has been discussed by many authors and extensively reviewed by Levin & Kerster (1967). All emphasise that a number of reproductive barriers have to be overcome before genes can be transferred successfully into new populations: barriers to dispersal, to compatibility and to post fertilisation processes. Gene flow from a crop to a relative, with which it can hybridise, is most likely to be accomplished directly by pollen dispersal because successful establishment following seed movement will depend on overcoming additional barriers to germination, establishment survival, flowering and seed set in an environment to which it may not be adapted.. Whichever route gene movement takes, provided the reproductive barriers are not insurmountable, a gene (or a transgene) may eventually be expected to become established in a close relative, even if dispersal rates are low.

As pointed out by Gregorius & Steiner (1993), once the transfer has taken place, subsequent consequences will be independent of the degree of relatedness between the crop and its wild relative. What will be more important are any effects that a transferred gene or gene complex may have in modifying fitness characteristics, and whether it does so independently of other fitness genes. Phenotypic plasticity may be able to compensate for some differences between growing conditions which exert selection pressure but, in crop plants artificial selection is designed to increase conformity to desired end points within a fixed environment, which reduces both genetic variation and plasticity, and so reduces the capacity for adaptation to non-agricultural conditions. In this respect crops differ from wildflower seed mixes, where the risk of "genetic pollution" is seen as being particularly high. There an intimate mixture of different species, each containing a range of genotypes is sown. Selection can act upon this variation, favouring those genotypes with more appropriate fitness traits at the expense of others. When plants containing adaptive genes developed to enhance agricultural production are transferred to situations where different selective pressures operate, the possession of those genes will not necessarily confer any selective advantage. Enhanced fitness of a weed/crop hybrid of radish (*Raphanus sativus*) has been demonstrated (Klinger & Ellstrand, 1994) but this was within an agricultural context and the results do not necessarily demonstrate a wider fitness capability in other environments. Whitton *et al.*, 1997 stress that the greatest opportunities for the movement of transgenes are within the crop-weed complex and factors which enhance fitness in weedy populations, such as increased invasive capacity, are likely to me the most significant.

Where on the genome a gene occurs will influence the chances of its transfer from a crop because different parts of the genome may introgress at different rates. Such differential transfer has been shown to occur in crosses between *Brassica napus* and *B. campestris/rapa*, and in subsequent backcrosses (Mikkelsen *et al*., 1996). It will also be influenced by whether the transgene in *B. napus* is on the A genome (derived from *B.rapa*) or the C genome (from *B. oleracea*). Once it is incorporated into a wild relative a gene closely associated with some other strong selectively-important gene or group of genes may also have a greater capacity for being retained and, ultimately, it could form part of what may become a highly conserved gene complex. Alternatively, if it behaves as a neutral gene, it will persist as a random component of the population gene pool, provided that mating within that population is also random. Raybould & Gray (1994) concluded that, indeed, many transgenes coding for proteins of no selective advantage in the wild (e.g. those conferring herbicide or antibiotic resistance, or used to produce pharmaceutical or industrial chemicals) will act as neutral genes. As there is no selective advantage, there will be no change in fitness.

THE HISTORICAL CONTEXT

Man has had a profound effect on evolutionary processes (selection, migration, drift, recombination, hybridisation and mutation) in many plant species since pre-historic times (Baker, 1972; Ledig, 1992). Indeed Brown (1992) goes so far as to claim that it would be difficult to find a species that has not been subject to some form of human-influenced selection. The value of certain plant species to man led him to impose his own selection criteria, initially collecting his crop and his seed from those individuals which were most suitable for his needs. Where this process occurred within natural populations, e.g. forests, there is evidence of considerable shifts in genetic structure depending on the harvesting technique used (Ledig, 1992). In primitive agricultural systems where the distinction between natural communities and planted areas was not clearly defined, and where wild and cultivated individuals remained essentially the same, opportunities for gene flow remained high. As a consequence Renno *et al.* (1997) assert that wild and cultivated forms of pearl millet (*Pennisetum glaucum*) have exchanged genes for the last 3000 years and have co-evolved over large parts of the Sahel.

In the course of agricultural development, some crop plants became increasingly differentiated from their wild relatives through domestication, defined by Van Raamsdonk (1995) as, "a process resulting in characteristics profitable for man [but] which generally reduces the fitness of plants in natural habitats." Reduction in fitness may be equated with loss of genetic diversity such that crop plants become increasingly confined to open environments where continuous disturbance is maintained by man. Through selection, plants are optimised for uniformity under these controlled conditions, whilst competition, diseases and pests are checked by a combination of selection, the use of effective agronomic procedures and the application of chemical agents. Where man has moved agricultural plants to new locations he has often narrowed the genetic base further by reducing the number of species used and their intra-specific diversity (Barrett, 1981). He has also taken his weeds with him. At the same time, if the introduced species have encountered novel ecological conditions, these may have been instrumental in changing life history and demographic characteristics (Eckert *et al*., 1996).

Such processes might be expected to reduce the effective rate of gene flow between crop plants and their wild relatives. However, by breaking down geographical or ecological boundaries, new opportunities for hybridisation may be presented.

As Thirsk (1997) emphasised, farmers have often been extremely innovative in responding to difficult market conditions and opportunities for gene flow from new crops may well have arisen from the adaptation of cropping patterns which resulted from this innovation. Given their geographical proximity, and similar soils and climate, new crops with wild relatives already present in Britain, were most likely to have spread, in the Middle Ages, from the Low Countries and northern France to East Anglia and southern England. Thus, there is documentary evidence of rape (*Brassica napus*) being grown on the monastic estates of east Norfolk in 1255. Although vetches (*Vicia* spp.) were introduced before farm records were kept, evidence from the Bishop of Winchester's estate and from manorial records of the late thirteenth and early fourteenth centuries record the spread, with a distance decay pattern, of cultivated vetches from Kent along the coast to Hampshire and thence, via the chalk, to Berkshire, the Chilterns and, eventually, Norfolk. Clover (*Trifolium* spp.), lucerne (*Medicago sativa*) and sainfoin (*Onobrychis viciifolia*) were first recorded as a substitute for grass in the 1620s, although indigenous forms of clover may have been consciously used as fodder long before. The most widely-practised response to depressed grain prices in the early nineteenth century was to convert arable fields to pasture and, in doing this, farmers were urged not to use promiscuous grass seeds collected from a hay crop, even where it had come from a high quality field. The imported substitutes, sown under guidance from seed merchants' publications, which became an increasingly important source of advice, would have contained not only genetically different varieties, but also pernicious weeds (Sheail, 1986). Thus, over many centuries crop and weed species in the agricultural landscape must frequently have been exposed to imported genotypes (Ambrosoli, 1997).

A significant effect of farming in the past has been reduction of variability in weed species (Warwick, 1991) either through gene movement from crop relatives (though, of course, not all weeds are closely related to crop species and would not be subject to such gene flow) or the imposition of a strong selection agent. For example, reduction in polymorphism has been linked to increase in triazine resistance (Warwick & Black, 1993), though the extent of such reduction may vary between species from small, e.g. *Brassica rapa* to very large e.g. *Chenopodium album*. If agricultural systems have already reduced genetic variability in crops and some of the associated weed species, as well as reducing the capacity for phenotypic plasticity, plants within any weed-crop complex may be expected to resemble crop plants in having a low capacity for permanent spread beyond the confines of the agricultural environment. Although the constant and widespread introduction of novel varieties of crop plants may increase variation it will not increase fitness beyond the limits of cultivation, nor lead to the formation of locally-adapted gene complexes of the type found in land races of crop species, and which make them especially susceptible to gene mixing (Renno *et al.*, 1997). Similar gene complexes may also be present in highly adapted local populations and rare species but the threat of breakdown of these complexes, leading to disruption of the adaptive mechanism and extinction of the populations containing them is likely to be minimal because of the strength of geographical, ecological, taxonomic and reproductive barriers between them and most crop species. A few close relatives of crops do not exhibit the classical features of weed species and may be components of more clearly defined natural or semi-natural communities. These include, for example *Daucus carota*, *Beta maritima* and some forage varieties of wild grasses and legumes.

CONCLUSIONS

Because the closest relatives of many crop plants are species with weedy characteristics, invading open or disturbed ground (and hence usually in close physical proximity to crops), barriers to introgression formed by reproductive and ecological isolation are small. Much larger barriers will exist where the relationship is distant, the populations are remote, the communities containing them are stable and closed, ecological conditions are very different and selection operates strongly in favour of locally-adapted genotypes. Under these circumstances, there is much less chance of incorporation of crop genes or transgenes into the genome of such plants. These factors lead to a conclusion that, because the mechanisms controlling gene flow from transgenes into wild relatives are much the same as those for other genes, there will be no significantly enhanced effect of transgenes on the genetic constitution of most components of the native flora, including many of the rarest species. The risks will be largely confined to close relatives which already share, through taxonomic links and past association with crops, a common history which is likely to have included extensive gene flow. Whether to improve understanding of the potential for gene flow in the past, or to help explain its present day significance, more extensive, yet detailed, study of the historic record might well repay the effort. Within this context, any incorporation of transgenes into weed species may be seen as a continuation of the developmental history of the crop/weed relative association, and not as an entirely new phenomenon.

REFERENCES

Akeroyd J (1994). *Seeds of destruction? Non-native wildflower seed and British floral diversity*. Plantlife, London.

Ambrosoli M (1997) The wild and sown. Botany and agriculture 1350-1850. Cambridge University Press: Cambridge.

Baker H G (1972). Human influences on plant evolution. *Economic Botany* **26**, 32-43.

Barrett J A (1981). The evolutionary consequences of monoculture. In: *Genetic consequences of man-made change*, eds J A Bishop & L M Cook, pp 209-248. Academic Press: London,.

Brown A H D (1992). Human impact on plant gene pools and sampling for their conservation. *Oikos* **63**, 109-118.

Butler D (1994). Bid to protect wolves from genetic pollution. *Nature* **370**, 497.

De Vries F T; Van der Meijden R; Brandenberg W A (1992). Botanical files: a study of the real chances for spontaneous gene flow from cultivated plants to the wild flora of the Netherlands. *Gorteria* Supplement **1**, Rijksherbarium/Hortus Botanicus, Leiden.

Eckert C G; Manicacci D; Barrett S C H (1996). Genetic drift and founder effect in native versus introduced populations of an invading plant, *Lythrum salicaria* (Lythraceae). *Evolution* **50**, 1512-1519.

Ellstrand N C (1992). Gene flow by pollen: implications for plant conservation genetics. *Oikos* **63**, 77-86.

Ellstrand N C; Whitkus R; Rieseberg L H (1996). Distribution of spontaneous plant hybrids. *Proceedings of the National Academy of Sciences of the USA* **93**, 5090-5093.

Gregorius H-R; Steiner W (1993). Gene transfer in plants as a potential agent of introgression. In: *Transgenic organisms: risk assessment of deliberate release*, eds K Wöhrmann & J Tomuik, pp. 83-107. Birkhäuser Verlag: Basel.

Handel S N; Robinson GR; Beattie A J (1994). Biodiversity resources for restoration ecology. *Restoration Ecology* **2**, 230-241.

Klinger T; Ellstrand N C (1994). Engineered genes in wild populations: fitness of weed-crop hybrids of *Raphanus sativus*. *Ecological Applications* **4**, 117-120.

Ledig F T (1992). Human impacts on genetic diversity in forest ecosystems. *Oikos* **63**, 87-108.

Levin D A; Kerster H W (1967). Gene flow in seed plants. In: *Evolutionary Biology vol. 7*, eds T Dobzhansky; M K Hecht; W C Steere, pp. 139-219. Plenum Press: New York.

Mikkelsen T R; Jensen J; Jørgensen R B (1996). Inheritance of oilseed rape (*Brassica napus*) RAPD markers in a backcross progeny with *Brassica campestris*. *Theoretical and Applied Genetics* **92**, 492-497.

Ouborg N J; van Treuren R (1994). The significance of genetic erosion in the process of extinction. IV. Inbreeding load and heterosis in relation to population size in the mint *Salvia pratensis*. *Evolution* **48**, 996-1008.

Raybould A F; Gray A J (1993). Genetically modified crops and hybridisation with wild relatives: a UK perspective. *Journal of Applied Ecology* **30**, 199-219.

Raybould A F; Gray A J (1994). Will hybrids of genetically modified crops invade natural communities? *Trends in Ecology and Evolution* **9**, 85-89.

Renno J-F; Winkel T; Bonnefous F; Bezançon G (1997). Experimental study of gene flow between wild and cultivated *Pennisetum glaucum*. *Canadian Journal of Botany* **75**, 925-931.

Rhymer J M; Simberloff D (1996). Extinction by hybridization and introgression. *Annual Review of Ecology and Systematics* **27**, 83-109.

Sheail J (1986). Grassland management and the early development of British ecology. *British Journal for the History of Science* **19**, 283-299.

Stace C A (1975). *Hybridisation and the flora of the British Isles*. Academic Press, London.

Thirsk J (1997). *Alternative agriculture: a history from the Black Death to the present day*. Oxford University Press: Oxford.

Van Raamsdonk L W D (1995). The effect of domestication on plant evolution. *Acta Botanica Neerlandica* **44**, 421-438.

Van Raamsdonk L W D; van der Maesen L J G (1996). Crop-weed complexes: the complex relationship between crop plants and their wild relatives. *Acta Botanica Neerlandica* **45**, 135-155.

Warwick S I (1991). Herbicide resistance in weedy plants: physiology and population biology. *Annual Review of Ecology and Systematics* **22**, 95-114.

Warwick S I; Black L D (1993). Electrophoretic variation in triazine-resistant and susceptible populations of the allogamous weed *Brassica rapa*. *Weed Research* **33**, 105-114.

Whitton, J; Wolf, D E;. Aria D M; Snow A A; Rieseberg L H (1997). The persistence of cultivar alleles in wild populations of sunflowers five generations after hybridisation. *Theoretical and Applied Genetics* **95**, 33-40.

SESSION 3

CONSEQUENCES OF GENE FLOW WITHIN SPECIES

Chairman & Session Organiser:	Mr J Kilpatrick *ADAS Rosemaund, Preston Wynne*

Gene flow in genetically modified herbicide tolerant oilseed rape (*Brassica napus*) in the UK

E C Simpson, C E Norris, J R Law, J E Thomas, J B Sweet
The National Institute of Agricultural Botany, Huntingdon Road, Cambridge, CB3 OLE, UK

ABSTRACT

The environmental and agronomic impact of genetically modified (GM) herbicide tolerant oilseed rape has been studied by NIAB since 1995. This paper reports on results of gene flow frequencies recorded between trial plots at several National List sites and at a large scale release of genetically modified herbicide tolerant rape in the UK. Levels of cross pollination tended to decrease with increasing distance from the pollen source, there was some evidence of varietal differences in receptiveness to foreign pollen. Pollen dispersal was also recorded at distances of up to 400m from a large release of GM herbicide tolerant rape using male sterile 'bait' plants.

INTRODUCTION

In 1995 the Department of the Environment Transport and the Regions commissioned NIAB to monitor some of the first large scale releases of GM herbicide tolerant rape (Sweet *et al.*, 1997, Sweet & Shepperson 1996). The work has recently been extended to include monitoring releases of about 100ha of oilseed rape modified for high lauric acid content. The potential for gene transfer to related *Brassica* species has been examined at a range of sites across the UK in addition to observations made on the incidence and persistence of volunteers and feral populations following GM releases. Further work commissioned by MAFF is investigating agricultural consequences of growing herbicide tolerant oilseed rape and will eventually provide data based on pollen flow between relatively large areas of GM herbicide tolerant and conventional oilseed rape.

Gene flow is an important part of evaluating the risks associated with growing GM herbicide tolerant oilseed rape as the dispersal of transgenes via pollen could potentially contaminate neighbouring crops, feral or volunteer populations or hybridise with related *Brassica* species. Inter and intra-specific transfer of herbicide tolerance transgenes may cause difficulties with weed and volunteer management in agricultural fields.

These studies will provide valuable information required for risk assessment for releases of GM herbicide tolerant oilseed rape and will ultimately contribute to understanding the environmental and economic benefits and costs of the deployment of these crops in farming systems. This paper presents some of the results of initial studies of cross pollination involving herbicide tolerant oilseed rape.

MATERIALS AND METHODS

1. Gene flow In National List Trials - seed and pollen sources

National List (NL) trials of GM herbicide tolerant winter oilseed rape consisted of three replicates containing two GM herbicide tolerant varieties tolerant to the broad spectrum herbicides glufosinate-ammonium and glyphosate. The trials also contained five non-tolerant conventional control varieties (Synergy, Express, Nickel, Falcon and Apex). Each GM trial was isolated from other rape trials by a pollen barrier of a conventional oilseed rape variety which was a minimum of 6m wide. Plots were approximately $40m^2$ and were harvested after swathing by standard small plot combine harvesters. Samples of harvested seed were taken from four NL sites in 1997 by removing the required weight of seed from a bulk of approximately 200g. Samples from non-GM varieties were selected from plots nearest to the GM trial, 50m, and 100m or the furthest point from the GM trial. Seed samples were also tested from the varieties in GM trials to screen for single and double herbicide tolerance.

Screening for herbicide tolerance in National List seed samples

Seed samples collected from NL trials were sown in field plots in a randomised block design replicated three times with negative controls (conventional winter rape variety - Express). Each plot contained an average of 600 plants. In order to allow testing for both glufosinate and glyphosate tolerance the trials were duplicated. Each trial was treated with either 200 g/l glufosinate-ammonium at 3l/ha or glyphosate 360 g/l at 4l/ha when plants were at the 3-5 leaf stage using a tractor mounted sprayer (Sprayranger). The numbers of surviving plants were assessed approximately 7 DAT and 14 DAT for glufosinate and glyphosate treatments respectively. Surviving plants were sprayed on a second occasion with the appropriate herbicide to confirm tolerance.

Seed samples of both GM and conventional varieties from 3 NL GM winter rape trials were also grown in a glasshouse with positive and negative controls. Plants were grown to the two true leaf stage before being treated with either a 1% dilution of 200 g/l glufosinate-ammonium or a 1% dilution of 360 g/l glyphosate using a hand sprayer (Hozelock). The numbers of surviving plants were assessed approximately 7 DAT and 14 DAT after treatment for glufosinate and glyphosate respectively. Surviving plants were treated as appropriate with either glufosinate or glyphosate to detect double tolerance. Leaf tissue samples were taken from plants expressing double tolerance to confirm the presence of multiple transgenes by PCR using specific primers for the PAT and EPSPS GOX genes conferring glufosinate-ammonium and glyphosate tolerance respectively. DNA was isolated from plant tissue using an extraction kit obtained from Qiagen Ltd. UK. PCR reaction conditions were supplied by herbicide manufacturers.

2. Measurement of pollen dispersal from a large area of genetically modified herbicide tolerant winter oilseed rape using male sterile bait plants

Pollen source

The pollen source used for this study was a trial grown at a farm in Cambridgeshire by Plant Genetic Systems (PGS). The trial consisted of an area of approximately 9 ha of winter oilseed rape with about 65% of plants in this area containing the BAR gene conferring tolerance to the herbicide glufosinate ammonium. About 30% of the area contained a mix of 50% male sterile herbicide tolerant and 50% male fertile herbicide susceptible plants. A pollen barrier of non-GM oilseed rape up to 20m wide was grown around the perimeter of the trial.

Male sterile bait plants

The male sterile portion of the spring oilseed rape composite hybrid variety 'Concept' was grown under glasshouse conditions so that flowering coincided with the onset of flowering of the field crop. Six male sterile plants were positioned (approximately 0.5m apart) in linear plots at a range of distances (100m, 200m, 400m) and directions north, south, east and west from the pollen source. Plants were left in position for the duration of the flowering of the crop and were removed after the main flowering period (approximately 4 weeks). Insect activity, weather conditions and the duration of flowering were observed and recorded on each site visit. Detailed weather data was obtained from the meteorological station at the NIAB farm approximately 25km north of the release site. Seeds were screened under glasshouse conditions for glufosinate tolerance as described previously.

RESULTS

1. Gene flow between GM NL and NL trials

Herbicide tolerance was found in the seed samples of non-GM varieties in 3 of the trials tested, Caxton, Cockle Park and Bridgets. No transgenic plants were found in seed samples from Wye Regional Trial Centre. Mean frequencies of glufosinate and glyphosate tolerance detected at a range of distances from the GM trials are shown in Tables 1-3. Frequencies are expressed as the percentage of herbicide tolerant plants per plot. Overall, frequencies of glyphosate and glufosinate tolerance tended to decrease with increasing distance from the GM trial and varied between varieties with the highest levels detected in the composite hybrid Synergy.

Of 38 samples examined from the Caxton site, 23 had no tolerance to either glufosinate or glyphosate. At both Cockle Park and Bridgets 9 samples had no tolerance to glufosinate or glyphosate out of a total of 18 tested. Varieties being tested in NL trials are coded Var. A-I.

Table 1. Frequency of glufosinate and glyphosate tolerance detected at varying distances from the pollen source at Caxton National List trial site

Variety	Distance from source (m)	% glufosinate tolerant	Variety	Distance from source (m)	% glyphosate tolerant
Synergy	4	2	Synergy	4	0.16
Capitol	8	0.1	Lipton	8	0.16
Apex	14	0.05	Var. A	8	0.05
Synergy	34	0.16	Cobra	8	0.33
Synergy	56	0.05	Var. B	8	0.05
			Synergy	20	0.16
			Apex	34	0.05
			Synergy	34	0.05
			Var. C	50	0.05
			Synergy	54	0.11

Table 2. Frequency of glufosinate and glyphosate tolerance detected at varying distances from the pollen source at Cockle Park National List trial site

Variety	Distance from source (m)	% glufosinate tolerant	Variety	Distance from source (m)	% glyphosate tolerant
Var. D	6	0.05	Var. D	6	0.05
Synergy	42	0.05	Var. G	30	0.05
Var. E	50	0.05	Synergy	42	0.33
Var. F	50	0.05	Var. E	50	0.05
			Var. F	50	0.16

Table 3. Frequency of glufosinate and glyphosate tolerance detected at varying distances from the pollen source at Bridgets National List trial site

Variety	Distance from source (m)	% glufosinate tolerant	Variety	Distance from source (m)	% glyphosate tolerant
Var. G	10	0.05	Var. G	10	0.16
Var. H	10	0.05	Var. H	10	0.44
Var. I	10	0.05	Var. I	10	0.05
Synergy	150	0.11	Synergy	20	0.05
			Synergy	150	0.22

Gene flow in GM NL trials

Herbicide tolerance was detected in seed samples from conventional and herbicide tolerant varieties in samples from the three sites tested (Figures 1a-1f). Overall, frequencies of herbicide tolerance decreased with increasing distance from GM plots. Levels of herbicide tolerance observed in the composite hybrid Synergy tended to be higher than in other varieties at the same distance from GM plots. Double tolerance to both glufosinate and glyphosate was detected in seed samples from herbicide tolerant varieties. The presence of the PAT and EPSPS GOX transgenes conferring glufosinate and glyphosate tolerance respectively are currently being confirmed using PCR with specific primers. Glufosinate and glyphosate tolerant varieties are coded GLU1 and GLY1 respectively.

Figures 1a-1f show plot configuration of National List GM winter oilseed rape trials and the levels of glufosinate and glyphosate tolerance detected in seed samples from these plots

Synergy 0	**GLU1**	**GLU1**	**GLY1 0.67**	Express 0	Falcon 1.17
Falcon 0.17	Nickel 9.17	Express 7.5	Falcon 0.33	Synergy 2.17	Nickel 0.33
GLY1 0.17	Express 0.83	Apex 1.0	Nickel 0.5	**GLY1 3.5**	Apex 0
Apex 0	Discard	Synergy 1.17	Discard	**GLU1**	Discard

Figure 1a. % Glufosinate tolerance - Caxton

Synergy 4.33	**GLU1 0.16**	**GLU1 0.83**	**GLY1**	Express 1.17	Falcon 0
Falcon 0.5	Nickel 0.5	Express 0.5	Falcon 7.6	Synergy 9.16	Nickel 0.16
GLY1	Express 0.33	Apex 0	Nickel 0.83	**GLY1**	Apex 0.16
Apex 9.83	Discard	Synergy 1.0	Discard	**GLU1 2.0**	Discard

Figure 1b. % Glyphosate tolerance - Caxton

Falcon 0.17	Synergy1.67	Express 0.33	**GLY1 0.75**	**GLU1**	Nickel 18.33	Apex 0.67
GLU1	Falcon 0.33	Apex 0.5	Synergy 1.5	**GLY1 0.83**	Express1.66	Nickel 0.33
Apex 1.0	Nickel 0.17	**GLY1 0.08**	Falcon 0.33	Express 0.33	**GLU1**	Synergy 21.0

Figure 1c. % Glufosinate tolerance - Cockle Park

Falcon 0.33	Synergy1.83	Express 1.67	**GLY1**	**GLU1 0.58**	Nickel 3.0	Apex 1.5
GLU1 0	Falcon 1.0	Apex 1.83	Synergy 17.5	**GLY1**	Express 2.67	Nickel 0.33
Apex 0	Nickel 0.17	**GLY1**	Falcon 17.7	Express 1.33	**GLU1**	Synergy 1.33

Figure 1d. % Glyphosate tolerance - Cockle Park

Falcon 0	**GLY1 0.17**	Apex 1.17	**GLU1**	Express 10.8	Synergy 4.83	Nickel 1.67
GLY1 0.25	Falcon 0.17	Express 0	Apex 2.5	**GLU1**	Nickel1 4.0	Synergy2.67
Nickel 0.17	Synergy 0.17	**GLY1 1.0**	**GLU1**	Apex 11.5	Falcon 4.17	Express1.83

Figure 1e. % Glufosinate tolerance - Bridgets

Falcon 1.17	**GLY1**	Apex 8.33	**GLU1 0.25**	Express 0	Synergy 0.83	Nickel 0.5
GLY1	Falcon 9.67	Express 2.0	Apex 0.5	**GLU1 0.08**	Nickel 0.17	Synergy0.67
Nickel 1.67	Synergy 5.0	**GLY1**	**GLU1 0.57**	Apex 1.83	Falcon 0.67	Express 0.17

Figure 1f. % Glyphosate tolerance - Bridgets

2. Pollen dispersal from herbicide tolerant winter oilseed rape

Male sterile bait plants were pollinated and set seed in all plots. Numbers of herbicide tolerant seed detected decreased rapidly with distance from the pollen source (Figure 2). Plots to the north of the GM pollen source produced the greatest proportion of herbicide tolerant seed. The highest total level of seed set was recorded from plots to the east of the GM trial. The incidence of pollinating insects particularly honey bees (*Apis mellifera*) and bumble bees (*Bombus* sp.) were notably low at flowering. During the flowering period (01.05.98 - 29.05.98) wind direction was predominantly north and north west, mean high and low air temperatures were 18°C and 6°C respectively, mean rainfall was 0.2mm and mean daily sunshine was 5.9hrs.

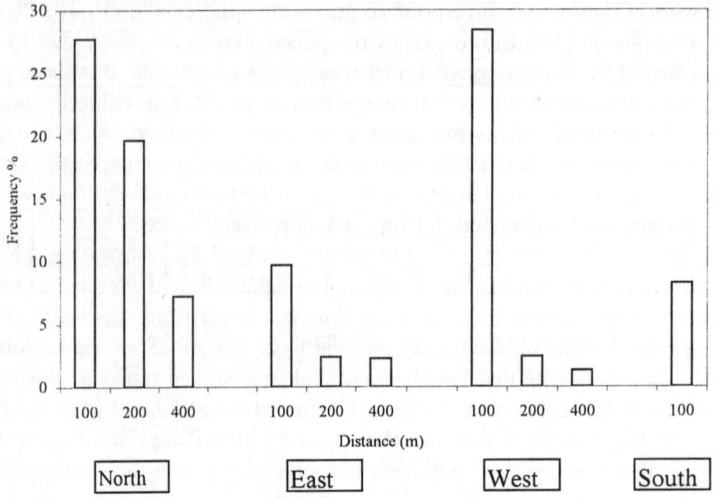

[1]Total numbers of seed set per plot: north 100m=800, 200m=626, 400m=514; east 100m=1187, 200m=1061, 400m=1174; west 100m=88, 200m=967, 400m=317; south 100m=461

Figure 2. Frequency of herbicide tolerant seed harvested from male sterile bait plants at varying distances and directions from genetically modified herbicide tolerant oilseed rape[1]

DISCUSSION

Gene flow levels, indicated by the frequency of herbicide tolerant plants decreased with increasing distance from the pollen source in both tests with NL trials and a large scale release using male sterile bait plants. Differences in experimental design, particularly pollen source size make direct comparisons with other published data difficult (e.g. Timmons *et al.*, 1995, Scheffler *et al.*, 1993). Some of the fluctuations found in the gene flow levels in this study are likely to be due to different positions of GM plots within NL trials, differences in pollen source size, and various agronomic and climatic factors such as carry over of seed in harvesting machinery and prevailing wind direction. High frequencies of herbicide tolerant plants in seed samples of Synergy demonstrates the 'susceptibility' of composite hybrid material to alien pollen. High cross pollination frequencies have been recorded in composite hybrids in other winter and spring rape trials when growing adjacent to GM plots (E. Simpson, unpublished data). Multiple tolerant hybrids tolerant to both glufosinate and glyphosate were detected at all of the NL GM sites (Figures 1a-1f). The highest levels were observed in adjacent plots, 3.5% and 2% at the Caxton site (Figures 1a & 1b) which correspond approximately to rates of multiple tolerance reported by Messean (1997) of 2% at 1m from a much larger pollen source.
Monitoring volunteer incidence at National List sites currently does not indicate that multiple tolerant hybrid plants are more difficult to control than conventional or single tolerant rape varieties. Continued monitoring of these sites will identify any significant future problems with volunteer control.

Detection of herbicide tolerance in seed of male sterile oilseed rape plants at distances of up to 400m show that there is potential for oilseed rape pollen to be dispersed by wind and remain viable over considerable distances. The high levels of herbicide tolerant seed detected in plots to the north and west of the release correspond to prevailing wind conditions. The highest level of total seed set recorded in plots to the east of the pollen source was likely due to dispersal of pollen from a commercial field of conventional winter rape growing near to these plots. Using male sterile plants demonstrates a worst case scenario, as there is reduced competition for foreign pollen, if fully fertile plants had been used, cross pollination levels could have been substantially lower. When seeds (>1000) from feral rape plants flowering at the same time and growing within 120m of the same large scale release of herbicide tolerant rape were tested no herbicide tolerant plants were detected (E. Simpson, unpublished data).

Results from these initial studies of gene flow in oilseed rape show that pollen can be dispersed over considerable distances indicating that gene flow can occur from releases of GM herbicide tolerant oilseed rape. Results of monitoring NL and large scale PGS releases (Norris *et al.*, in press) currently do not indicate that GM herbicide tolerant oilseed rape varieties are any more persistent or invasive than conventional types. Continued monitoring of sites and further gene flow studies using larger areas of GM herbicide tolerant oilseed rape will help to identify any potential ecological and agricultural problems that may occur with the commercial release of these crops.

ACKNOWLEDGEMENTS

The authors would like to thank Ministry of Agriculture, Fisheries and Food and Department of the Environment, Transport and the Regions for financial support, Plant Genetic Systems for their help in carrying out this research, AgrEvo and Monsanto for providing diagnostic primer sequences.

REFERENCES

Messean A (1997) Management of herbicide tolerant crops in Europe. *Proceedings Brighton Crop Protection Conference - Weeds*, **3**, 947-954.

Norris C E; Simpson E C; Sweet J B; Thomas J E (1999) Monitoring weediness and persistence of genetically modified oilseed rape (*Brassica napus*). *BCPC Proceedings Gene Flow and Agriculture Conference, in press.*

Scheffler J A; Parkinson R; Dale P J (1993) Frequency and distance of pollen dispersal from transgenic oilseed rape (*Brassica napus*). *Transgenic Research,* **2**, 356-364.

Sweet J B & Shepperson R (1996) Monitoring commercial releases of genetically modified oilseed rape. *Proceedings 10th International Weed Biology Conference Dijon,* pp 217-222.

Sweet J B; Shepperson R; Thomas J E; Simpson E C (1997) The impact of releases of genetically modified herbicide tolerant oilseed rape in the UK. *Proceedings Brighton Crop Protection Conference - Weeds*, **4**, 291-302

Timmons A M; O'Brien E T; Charters Y M; Dubbels S J; Wilkinson M J (1995) Assessing the risks of wind pollination from fields of genetically modified *Brassica napus* ssp. oleifera. *Euphytica,* **85**, 417-423

Gene flow between sugar beet and weed beet

Y Vigouroux, H Darmency
Laboratoire de Malherbologie, INRA, BV 1540, 21034, Dijon, France.

T Gestat de Garambe, M Richard-Molard
Institut Technique de la Betterave, 45 rue de Naples, 75008, Paris, France

ABSTRACT

Gene flow between transgenic sugar beet and weed beet was investigated at the farm scale. Evidence was given of gene flow between bolters of the crop that occurred at low density in a 1 ha field and weed beet growing spontaneously in an adjacent set-aside field. Up to 0.8 % of the weed progeny was hybrid. A second experiment was specifically designed to check gene flow from a small pollen source of sugar beet to a low density weed beet population. The frequency of hybrids slowed down from 10% three meters away the pollen donor to 1 % at 15 m. The effects of wind, distance, weed aggregation, insects and genotypes on the hybridization rate are discussed.

INTRODUCTION

Risk assessement of transgenic crops on an agricultural scale is currently carried out in France by a group of institutes, seed producers and agrochemical companies. Four different crops are rotating in a farmers' field for several years and are cultivated with usual farmers' practices (see Messean, this volume, for more details). Along with studies on the benefit of growing a given transgenic crop, problems of volunteers in successive crops, pollen flow distances, and gene flow between crops and their wild relatives are monitored and/or investigated. We focus here on the possibility of gene escape from transgenic sugar beet to weed beet in field conditions.

Weed beet are quite common in the sugar beet area in France. They are a real problem in less than 5 % of the fields (Gestat de Garambe et al., 1998). Since weed beet and sugar beet are the same species, no specific herbicide is actually available to solve this problem. Transgenic herbicide-resistant beet have been proposed as a solution for the problem of weed beet infestation (Martin, 1998). However, gene flow between weed beet and sugar beet have been demonstrated to occur in the past: the genetic study of weed beet in the north of France showed they display the cultivated cytoplasm Owen, proving their maternal origin from cultivated beet (Boudry *et al.*, 1993). Although sugar beet is a biennial plant that is not expected to flower the first year, there area few bolting plants occurring per hectare, indicating that pollen escape is possible.

We describe here the pollen flow in a field experiment with transgenic sugar beet and adjacent populations of weed beet in set-aside fields. These results raised questions that were not previously documented in the literature. Especially, pollen flow in beet has been studied in seed production conditions only, not in a root production area (Stewart and Cambell, 1952;

Chamberlain, 1967; Dark, 1971), and the biology of the target weed beet has seldom been considered. Different parameters could interact in cross-pollination as the occurrence of few bolting plants, so that very little transgenic pollen is available, the distance between pollen donor and receiver; and the reproductive behaviour of weed beet and bolting plants (flowering time, incompatibility system, and outcrossing rate). Hence, we designed a specific experiment for studying gene flow using weed beet as pollen receiver in order to assess what gene flow might be when very few pollen donors are present, and how this gene flow is influenced by weed beet as target plants?

MATERIALS AND METHODS

Gene flow between transgenic bolting plants and weed beet in the field

One hectare of sugar beet was sown in 1996 with two transgenic herbicide resistant hybrid lines, one resistant to glyphosate and the other to glufosinate. The plants were heterozygous for the resistance genes. As the main wind direction was South West-North East, ten groups of 20 annual weed beet collected in the Dijon area were planted along the S.W. border of the field. The one ha adjacent field on the N.E. side was left on set aside, and 127 spontaneous weed beet could flower there. Only 94 weed beet were further analyzed for the presence of hybrids because of mortality or no seeds available. In addition, five plots of male sterile beet were planted along the median row of the field, and then five others at 30 m, 90 m and 210 m toward N.E. Every bolting sugar beet (only glufosinate resistant, no glyphosate resistant plants flowered) and all the weed beet were located on a map and their flowering period recorded. Seeds were harvested, weighed and numbered. A sample of 800 seeds from each plant was sown in peat in the greenhouse (22 °C day, 18 °C night). Then 2-4 leaf seedlings were sprayed with 5 l ha^{-1} Liberty (200 g l^{-1} glufosinate ammonium, Agrevo) to check for the presence of resistant plants.

Experimental design to estimate gene flow

Four pollen donor sugar beet were planted out in a small patch (0.2 m^2) in the centre of nine radiating lines. These plants were used as pollen donors and were homozygous for red pigmentation. Red color is inherited as a single dominant allele at one locus (Keller, 1936). One weed beet was planted along the radiating lines at 3, 6, 9, 12 and 15 m. All plants were firstly germinated and grown in the greenhouse before planting. Synchronization of flowering was performed by cutting first flowering ramets. Homogeneous flowering started on July 12 and lasted until the harvest date on August 31. All the beet were harvested, air-dried in bags, and threshed. To prevent contamination during threshing, the more distant plants from the centre of the circle were first processed. Seeds were counted and weighted. A sample of 800 seeds per plant was sown in the greenhouse and the number of red seedlings was recorded.

RESULTS

Transgenic field

They were 58 sugar beet plants bolting in the field, but only 29 were resistant. Hybrids between transgenic sugar beet and male sterile plants were found at a rate of 0.76% (Table 1). Most of the hybrids, 34 out of 36, were found on plants grown within the sugar beet field. The two other hybrids were found at 30 meters and 210 meters from the sugar beet field. As transgenic bolters were diploid and heterozygous, and only half of bolting plants carried the resistance gene, the percentage of pollen that carried the transgene should be 25%. We can deduce that bolters furnished 3% of the pollen cloud that fertilized the male sterile plants.

The progeny of the weed beet planted in the S.W. border contained 0.0065 % of resistant plants (Table 1). That of plants growing naturally in the set aside contained 0.2 %. Using the same correction as above to account for heterozygous resistant and susceptible bolters, that makes 0.026 and 0.8 % hybrids, respectively. The lower frequency of hybrids observed for S.W. planted weeds could be explained by the experimental design. Twenty weed beet were planted together in small plots so that mating inside group within plot was probably enhanced. Moreover, the prevailing wind was coming from S.W. resulting in transport of pollen by air toward the set aside field on the N.E. rather than toward the S.W. border.

Table 1. Hybridization between transgenic bolting sugar beet and other beet as indicated by herbicide resistance

Kind of plant	Number of plants	Number of seedlings	Number of resistant	Number of plants that produced at least one hybrid
Weeds planted in S.W. border	186	30850	2	2
Weeds in N.E. set aside	94	10394	21	4
Male sterile	119	4762	36	19

The 21 resistant hybrids observed in the progeny of the spontaneous weeds in the set-aside were produced by four plants only (Table 1). This could not be expected in the case of random spread of the transgenic pollen, but the fact that three quarters of the hybrids have not been detected is perhaps misleading. From a biological point of view, no data on seed production, weight of seeds and seed viability could indicate that these four plants behaved differently than others (Table 2).

Moreover, this does not seem to be a matter of distance of pollination. These four plants were not closer to bolters, on average, nor closer to the nearest bolters than the other plants which did not produced hybrids (Table 3). However, the number of other weed beet growing in the immediate vicinity of these four plants, within a circle of three meter radius, was slightly lower than for the rest of the weed beet.

Table 2. Comparison of plants that produced hybrids to all other weed beet in the set aside

Type of plant	Number of seeds	Thousand seed weight (g)	Number of seedlings	Number of hybrids	% of hybrids
Plants that produced hybrids (min - max)	658 (330 - 1060)	13,2 (10,8 - 19,1)	148 (52 - 200)	5 (1 - 10)	3,3 (1,8 - 5,8)
All other plants (se)	665 (54)	11,0 (0,36)	110 (11)	0	0

Table 3. Localization of weed beet that produced hybrids

Type of plant	Number of weed plants within a 3 m radius	Average distance from all transgenic bolters (m)	Distance from the neare transgenic bolter (m)
Plants that produced hybrids (se) (min - max)	1,25 (0,25) (1 - 2)	87 (6,0) (77 - 96)	59 (6,6) (47 - 73)
All other plants (se) (min - max)	1,73 (0,16) (0 - 7)	94 (1,7) (61 - 139)	63 (1,4) (34 - 90)

Experimental design

The number of seedlings germinated from samples of 800 seeds was 571 (se 49), on average, and at least 176 seedlings. Only five plants presented no red pigmented hybrids, at 6, 9 and 15 m. Pollen flow from red beet to weed resulted in 10.7 % hybrids found at 3 metres (Table 4). At a distance of 9, 12 and 15 m, the rate of hybrids decreased to values around 1 %. Quite homogeneous percentages of hybrids were observed among plants at a same distance, except for the six meters distance where values ranged from 0% to 16.4%. There was no evidence of marked effect of wind direction.

Table 4. Average rate of hybridization between red beet as pollen donor and weed beet as pollen receiver (standard error in brackets).

Distance (m)	3	6	9	12	15
Hybrid %	10.7 (2.2)	7.1 (3.4)	1.0 (0.3)	0.8 (0.2)	1.3 (0.3)

DISCUSSION

The obvious result is that hybridisation between bolting transgenic sugar beet and weed beet occurred in this study under field conditions and should occur when transgenic material is used in farmers' fields. No genetic barrier separates weed from cultivated beet, as already reported in other studies (Boudry *et al.* , 1993; Bartsch and Pohl-Orf, 1996). The rate of hybrids found in the progeny of weed beet in the set aside was around 0.8 %. That means that the pollen cloud that fertilized the weed beet contained, on average, more than 99 % of pollen produced by the 127 weed beet, and that the 58 crop bolters contributed for 0.8 % only. This is not surprising as the smallest distance between a weed beet and its closest bolter ranged from 34 to 90 m, a distance at which gene flow is expected to be equal to, or lower than, 1 % according to the result of our specific pollen flow experiment. However, hybrid production was not at random. Only four plants out of 94 produced the 21 hybrids. These plants were not placed in particular conditions and had no peculiar reproductive output, so that random pollination by the wind is questionable. Even if we imagine that the rate of 1 % hybridization, that was observed at 9, 12 and 15 m in the second experiment, keeps stable at farther distances, it cannot account for the higher rate of hybrids produced by these four plants (2 to 6 %). This level in our experiment of gene flow could only be observed for weeds at a distance close to 6 m, while plant mapping showed that these four plants were 47 to 73 m from the closest bolter.

To explain why all the hybrids were restricted to four families only, two hypothesis could be proposed. The one is that a complex relationship between self-fertility and incompatibility among weed and cultivated beet leads to this pattern of gene flow. But little is known about the reproductive behavior of weed beet, which is currently approached by the study of half-sib progeny. The alternate hypothesis is that insects could have been pollen carriers, which could explain the high level of hybrids for a few weed beet that should have been visited by a single insect bearing pollen from a transgenic bolter, while other weeds have not been visited under similar conditions. Work on pollination of sugar beet carried out in seed production fields where high density of flowering plants occurs has concluded that wind is the main vector of pollen. A few studies also showed that some insects could carry beet pollen and contribute to pollination (Free *et al.* 1975). Perhaps, the effect of insects is more important in root production fields where there is a low density of pollen donors. In order to gain improved understanding on transgene flow in production fields of transgenic sugar beet, therefore, more studies on the biology of the pollen recipient plants, the weed beet, are needed.

ACKNOWLEDGEMENTS

We wish to thank the Burgundy Region Council (France) for a grant to Y. V.

REFERENCES

Bartsch D; Pohl-Orf M (1996). Ecological aspect of transgenic sugar beet : transfert and expression of herbicide resistance in hybrids with wild beet. *Euphytica* **91**, 55-58

Boudry P; Mörchen M; Saumitou-Laprade P; Vernet P; Van Dijk H (1993). The origin and evolution of weed beet : consequences fore the breeding and release of herbicide resistant transgenic sugar beet. *Theoritical and applied genetics* **87**, 471-478.

Chamberlain AC (1967). Cross pollinisation between fields of sugar beet. Quarterly Journal of the Royal Meteorology Society **93**, 509-515.

Dark SOS (1971). Experiments on the cross-pollination of sugar beet in the field. Journal of National Institut of Agricultural Botany **12**, 242-266.

Free JB; Williams IH; Longden PC; Johnson MG (1975). Insect pollination of sugar-beet (Beta vulgaris) seed crop. Annals of Applied Biology **81**, 127-134.

Gestat de Garambe T; C Muchembled; Richard-Mollard M (1998). Utilisation de betteraves tolérantes à un herbicide non sélectifs-conséquences sur les pratiques agricoles. In: *Seventeenth COLUMA conference*, pp951-958.

Keller W (1936). Inheritance of some major color types in beet. Journal of Agricultural Research 52, 27-38.

Martin J L (1998). Betteraves Roundup Ready™ : intêrets et utilisations. In: *Seventeenth COLUMA conference*, pp975-982.

Stewart D; Campbell SC (1952). The dispersion of pollen in sugar beet seed plots. Proceeding of the America Society of Sugar Beet Technologists 7, 459-469.

GENESYS: A model of the effects of cropping system on gene flow from transgenic rapeseed

N Colbach
Unité d'Agronomie, INRA; 17 rue Sully, BV 1540, 21034 DIJON Cedex, France

J M Meynard, C Clermont-Dauphin
Laboratoire d'Agronomie, INRA-INAPG, 78850 THIVERVAL-GRIGNON, France

A Messéan
CETIOM, 174 avenue Victor Hugo, 75116 PARIS, France

ABSTRACT

The aim of the model is to evaluate the influence of cropping systems on transgene escape from rapeseed crops to rapeseed volunteers in time and space. The model input variables are the regional field pattern, crop succession and cultivation techniques. The main output variables are, for each year and plot, the number of individuals per m² and the genotype proportions of the adult rapeseed plants, the newly produced seeds and the seedbank. The model comprises an annual life-cycle for volunteer and cropped rapeseed plants simulated for each plot and year; the relationships between the various life-stages depend on crop type and management. Pollen and grain exchanges between plots depend on distance between plots and cropping system. With the help of the simulations performed with the model, it is possible to identify low-gene-flow cropping systems or the minimum distance between rape plots to avoid contamination of the harvest product, to estimate the consequences if one farmer does not manage his rape volunteers correctly etc.

INTRODUCTION

In the current discussion on transgenic herbicide-tolerant rapeseed varieties, the risk of gene transfer to rape volunteers is often brought up (Messéan, 1996). It is therefore essential to rapidly evaluate this risk and identify means to reduce it. Long-term field trials were set up by French technical institutes on several locations to evaluate this risk (Champolivier *et al.*, 1997), but besides being too slow, these trials only comprise a limited number of cropping systems. Therefore, we built a model of the effects of cropping systems on gene flow from transgenic rapeseed varieties to rapeseed volunteers in neighbour plots and following crops; the first version of the model concerns herbicide tolerance transgenes. Hybridisation between rapeseed and *Brassica* weeds, though possible (Lefol *et al.*, 1996; Chèvre *et al.*, 1997), were not integrated. In this paper, only the general structure of the model is presented; details are given by Colbach and Meynard (1996), Clermont-Dauphin *et al.* and Colbach *et al.* Simulations performed with the model contribute to determine the effect of field plan, crop rotation and management on gene flow, to identify low- and high-risk situations, to propose agricultural practices minimising gene flow etc.

MODEL STRUCTURE

The input variables of the model are: (i) the field plan of the region, comprising cultivated fields and field-edges or waysides (henceforth "borders") consisting of spontaneous vegetation; (ii) the crop rotation of each field (Table 1) and (iii) the cultivation techniques applied to each crop (Table 2). The type of the transgene, either a dominant allele *A* or a recessive *a*, and of the rapeseed varieties must also be chosen. All examples presented in this paper are based on a dominant allele *A*, genotype *AA* for the transgenic and genotype *aa* for the classic variety. The major output variables are, for each field and year, the number per m² and the genotype proportions of adult rapeseed plants, of seeds produced and of the seedbank containing the viable rape seeds in soil.

Table 1. Example of crop successions of a simulated agricultural region.

Spatial unit	Type	Crop succession				
		year 1	year 2	year 3	year 4	year 5
1	border	border	border	border	border	border
2	field	transgenic rape	winter crop	spring crop	unsown set-aside	classic rape
3	field	winter crop	sown set-aside	classic rape	winter crop	spring crop

Table 2. Example of cultural practices used to manage the simulated crops

Crop	Cultural techniques: stubble breaking	soil tillage	sowing date	sowing density (seeds/m²)	herbicide[a] sowing C	herbicide[a] sowing T	herbicide[a] spring C	herbicide[a] spring T	cutting[b] 1st	cutting[b] 2nd	harvest seed loss rate
transgenic rape	yes	plough	31/08	70	90	0	0	0	[c]		10%
winter crop	yes	chisel	4/10	350	90	90	90	90			100%
unsown set-aside							0	0	3	no	
border							0	0	3	no	

[a] Mortality rates (%) for classic (C, of genotype *aa*) and transgenic plants (T, of genotype *Aa* or *AA*).
[b] Date of cutting in number of days after rape flowering onset.
[c] This technique cannot be used for this crop.

Description of the temporal sub-model

Figure 2 shows the general organisation of the annual rape life-cycle simulated for each plot and year. For each life-stage such as *seedlings, adults* etc, the model calculates the number of individuals per m² and the proportion of the three possible genotypes *AA, Aa* and *aa.* The life-cycle is modulated slightly according to the crop grown on the simulated field; for instance, the compartment *sown rape seeds* only exists in rape fields and simulates the arrival of rape seeds at the sowing of the crop. The relationships between the various life-stages depend on (further details are given in Table 3):

- on the crop grown in the plot. For instance, rape volunteers growing in a spring crop cannot complete their cycle because of insufficient vernalization and/or time and produce neither flowers nor seeds (relations 8 and 9).

- on the cultural techniques applied to the crops. For instance, the vertical seed distribution in the pre-sowing seedbank depends on soil tillage (relation 3); ploughing mixes seeds of various layers whereas chiselling little disturbs the initial seed distribution.
- on rapeseed characteristics. For instance, plant mortality after herbicide treatments strongly depends on plant genotype (relations 5 and 7); if the applied herbicide is the one against that the transgene confers tolerance, then the plants of genotypes *Aa* and *AA* survive whereas most of those of genotype *aa* die; for any other herbicides however, mortality is independent of genotype.

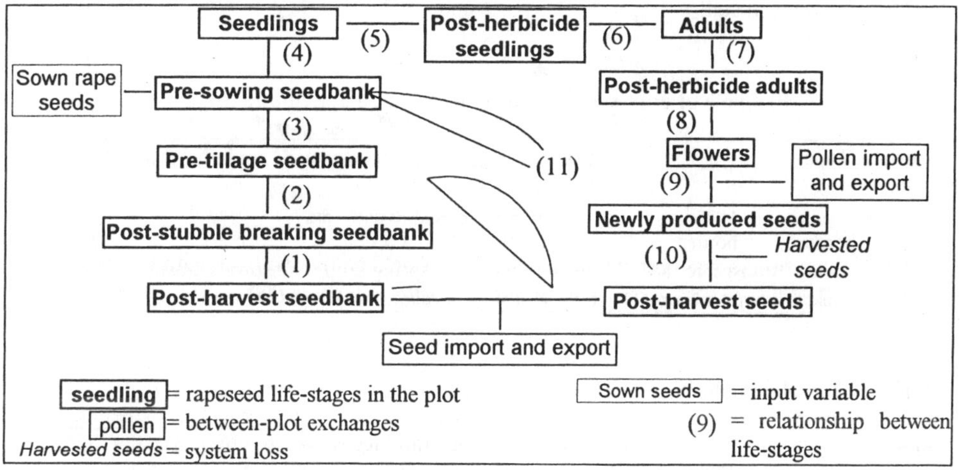

Figure 2. Annual rape life-cycle (for cropped and volunteer plants) simulated for each plot and year

Description of the spatial sub-model

The annual life-cycle of Figure 2 is simulated independently on each plot, but on two occasions, the various spatial units connect by exchanging pollen and seeds that are illustrated by the "pollen or seed import and exports" compartments. These exchanges are mostly caused by wind. The amount of pollen or seeds dispersed from plot *i* to plot *j* results from multiplying the amount produced (and not exported by harvest-combines in the case of seeds) in plot *i* with the proportion of pollen or seeds dispersed from *i* to *j*. This proportion is deduced by quadruple integration from equations giving the proportion of pollen or seeds as a distance from the parent-plant; it increases with plot areas and decreases with the distance between the two units. The determination of pollen exchanges is a little more complicated as the pollen must not only arrive in plot *j*, but plot *j* must also present flowering rapeseed plants ready for pollination. Therefore, besides calculating the number of pollen-producing or receiving flowers, it is also necessary to determine when these flowers are open. The flowering dates depend on the emergence date of the seedlings which depend, in the case of cropped rape plants, on the sowing date of the rape crops or, in the case of rape volunteers, on the sowing date of the crops sown in the plot where these volunteers emerge.

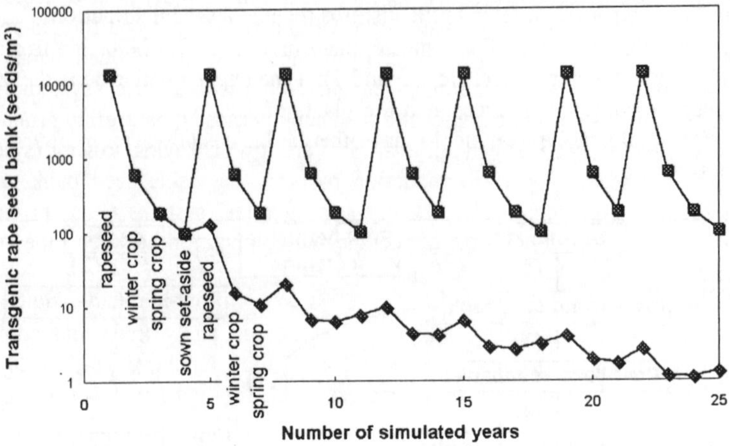

Figure 3. Evolution of transgenic rape seedbank of a field cultivated once with the transgenic, herbicide-tolerant rapeseed variety (◆) or every four to five years (■).

MODEL EVALUATION

GENESYS is currently evaluated, using surveys of farmers' plots for the dynamic part and observations from the long-term experiments of the technical institutes for the genetic and dispersal part. Preliminary results (Couturaud, 1998) show that the model correctly ranks cropping systems for rape volunteer infestation, but underestimates pollen and seed dispersal, probably because dispersal by insects and animals is insufficiently integrated into the model.

SIMULATIONS

Effect of cropping system in a single field

Figure 3 illustrates the effect of crop rotation on the evolution of the post-harvest transgenic seed bank. The basic rotation is the same for both simulations: rape/winter crop/spring crop/ set-aside/rape/winter crop/spring crop. The systematic growing of transgenic herbicide-tolerant varieties (of genotype *AA* with *A* conferring tolerance) results, of course, in the largest transgenic seed bank, with substantial increases after rape crops, because of seed loss before or during harvest, and reductions after winter and spring crops. If, except for the first year, only classic varieties are grown, then the seed bank decreases with time, with again slight increases after rape crops, despite these being classic varieties. Set-aside also increases seed bank because of conditions favouring seed production. After 20 years of simulation, there are still a few transgenic seeds left, despite there having been no transgenic variety since the first year.

The transgenic seed bank after four years is, of course, more important. If a classic variety is to be grown at the fifth year, minimising this seedbank is essential to limit the emergence of transgenic volunteers in the classic crop and thus avoid the pollution of the classic harvest by the unwanted gene which could lead to a refusal of a "non-transgenic" quality label for the harvest or to a decrease in price if certain fatty acids were required. The simulations show that set-aside management is crucial: sowing a set-aside field in spring (system 2 of Table 4) instead of leaving it unsown (system 1) considerably decreases the transgenic seed bank as rape volunteers cannot produce seeds in spring crops. Frequent cutting of set-aside plots is also necessary: a single cutting (system 3) only delays seed production without reducing it and thus increases the seed bank 40 times compared to two successive cuttings (system 1). Seed banks can also be regulated via winter and spring crop management. Tilling with a chisel (system 4) instead of a mouldboard plough (system 1) decreases the seed bank because most freshly produced seeds are left close to soil surface where seed survival rates are low. Delaying sowing (systems 5 vs. 1) also decreases the seed bank via a reduction of both the viable seed bank at sowing and subsequent seedling emergence rates. Larger sowing rates (systems 6 vs. 1) limit rape seed production in winter crops by favouring interspecific competition between rape volunteers and cropped plants.

Table 4. Effect of cultural techniques for set-aside and winter and spring crops on transgenic post-harvest seedbank after four simulated years with a transgenic rapeseed/winter crop/spring crop/set-aside rotation.

System	Set-aside management		Winter and spring crop management			Post-harvest transgenic seed bank	
			soil tillage	sowing date	sowing density (seeds/m²)	seeds/m²	relative evolution
	sowing	cutting					
1	none	2	plough	Oct./Feb.[a]	350	14150	100
2	spring	2	plough	Oct./Feb.	350	69	0.49
3	none	1	plough	Oct./Feb.	350	570435	4031
4	none	2	chisel	Oct./Feb.	350	2296	16
5	none	2	plough	Nov./March	350	3661	25
6	none	2	plough	Oct./Feb.	400	4405	31

[a] The first sowing date concerns winter crops, the second spring crops.

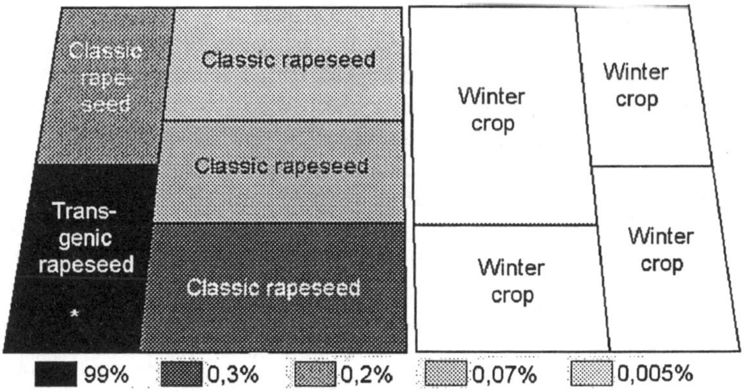

Figure 4. Proportion of newly produced seeds in rape crops carrying one or two copies of the transgene *A* coding for a fatty acid after five simulated years in a region cropped with a rape/winter crop/spring crop/set-aside/rape/winter crop/spring crop rotation, transgenic varieties (of genotype *AA*) in field * and classic varieties in all other fields.

Effect of regional cropping system

Figure 4 gives the proportion of seeds produced by rape crops and carrying one or two copies of a transgene *A* coding for a fatty acid. Basic rotation is the same as in the previous simulations and again identical for all plots. In this example, all rape crops of a given year were concentrated in the left-hand part of the simulated region; in the second sub-region, rape crops were grown a year later. Only one field was cultivated with the transgenic variety *AA*, but the classic varieties of the neighbour fields also produced seeds carrying the transgene. Furthermore, less than 100% of the seeds produced by the transgenic variety carried the gene. Removing the classic varieties to the right-hand sub-region to increase the distance to the transgenic variety divides the pollution of the classic harvest by at least three.

CONCLUSION

Despite the consistent results of the first evaluation and simulation steps, GENESYS is not yet ready to be used as a decision aid tool. The major sub-models needing improvement concern pollen dispersal processes and rape volunteer demographics in borders. Furthermore, the effect of rape genotype which is presently restricted to herbicide tolerance and selfpollination rates, should be extended to other rape characteristics (*e.g.* pod shattering, pollen production) to identify those features able to limit gene flow. Moreover, even though GENESYS was developed for herbicide tolerance dispersal, the model can be used for other genes, such as those coding for fatty acid content. The model therefore constitutes a reflection tool for those concerned with production, transformation and marketing of rapeseed, thus permitting a prospective analysis of the impact of changes in agricultural practices.

REFERENCES

Champolivier J; Messéan A; Gasquez J (1997). Crop management of transgenic rapeseed: risk assessment of gene flow. *Bulletin GCIRC* **14**, 63-66.

Chèvre A M; Eber F; Baranger A; Renard M (1997). Gene flow from transgenic crops. *Nature* **389**, 924.

Clermont-Dauphin C; Colbach N; Meynard J M. Modelling the influence of cropping system on gene escape from herbicide resistant rapeseed crops to rape volunteers. I. Temporal dynamic and genetic structure of the rape volunteer population in a field.

Colbach N; Meynard J M (1996). Modelling the influence of cropping system on gene flow for herbicide resistant rapeseed. Presentation of model structure. *10ᵉ Colloque International sur la Biologie des Mauvaises Herbes, Dijon 11-13 septembre 1996*, 223-230.

Colbach N; Clermont-Dauphin C; Meynard J M. Modelling the influence of cropping system on gene escape from herbicide resistant rapeseed crops to rape volunteers. II. Spatial dynamic and genetic structure of the rape volunteer population in a small region.

Couturaud M C (1998). Effet des systèmes de culture sur les risques de dissémination d'un transgène de colza dans l'environnement: validation et utilisation du modèle GENESYS. *Rapport d'ingénieur agronome INA-PG*, 50 p.

Lefol E; Danielou V; Darmency H (1996). Predicting hybridization between transgenic oilseed rape and wild mustard. *Field Crops Research* **45,** 153-161.

Messéan A (1996). Impact des plantes résistantes à un herbicide. *OCL* **3**, 5-9.

Regional patterns of gene flow and its consequence for GM oilseed rape

C E Thompson, G Squire, G R Mackay, J E Bradshaw, J Crawford and G Ramsay.
Scottish Crop Research Institute, Invergowrie, Dundee, DD2 5DA, UK.

ABSTRACT

The imminent commercialisation of genetically modified oilseed rape varieties requires accurate quantification of transgene movement via pollen within realistic agricultural contexts. This study measured gene flow on a regional and local scale. A 70 square km area west of Dundee was used, in which oilseed rape crops were common, but no GM OSR was grown. Fifty-two sites were selected at distances between zero m and 4000 m from the nearest non-GM spring oilseed rape crop. At each site male-sterile bait plants were placed for a 14 day period and airborne pollen density measured. Pollination occurred at all sites, even those at the greatest distance from flowering oilseed rape fields. Although airborne OSR pollen was recorded at all sites, the density declined rapidly with distance from source. There was also evidence that bees may be important pollen vectors over a range of distances. Additional sites 100 to 900 m from a GM spring oilseed rape crop in a second study area were also investigated. Despite the predominance of non-GM OSR crops in the immediate locality, all sites were pollinated by a mixture of GM and non-GM sources. The results suggest that the farm to farm spread of OSR transgenes will be widespread.

INTRODUCTION

Accurate quantification of transgene escape from and movement within realistic agricultural contexts is required now that transgenic oilseed rape (*Brassica napus* L.) is being grown experimentally. Oilseed rape (OSR) is notable among GM crops as having a widespread occurrence of feral and volunteer weedy populations. Gene flow via pollen movement is particularly hard to quantify as it depends on several elements outwith the control of the experimenter. For example as well as the size of pollen source, the competitiveness of the GM pollen, and meteorological and environmental factors, gene flow will depend on the regional pattern of OSR crops and weedy populations.

Early trials with GM OSR imposed artificial experimental designs to reduce the possibility of gene flow, and many used relatively small plots of GM OSR as pollen sources. Both may lead to an underestimation of the frequency of gene flow. The aim of this study is to measure gene flow through pollen dispersal in a realistic agricultural context. On a regional scale this was done by looking at pollination of male sterile plants in an area with a heterogeneous range of pollen sources. By taking a similar approach in a second study area including a transgenic crop, it was possible to measure gene flow from a single source on a more local scale.

MATERIALS AND METHODS

An area west of Dundee, approximately 70 square km, was selected for study. The acreage of OSR has increased locally over the past decade and volunteer OSR populations are widespread. The cytoplasmic male-sterile component of the cv. Triolo (kindly donated by CPB, Twyford, Cambridge, UK) were used as bait plants to measure gene flow via pollen dispersal. Ten such plants (area = $0.15m^2$) were used to form a bait colony *in situ* for 14 days during the main spring OSR flowering period in the region. Passive pollen traps (double-sided adhesive tape on microscope slides) were used at each site for seven days to measure the density of airborne pollen. Motorised Burkard pollen traps (Burkard Manufacturing Co Ltd, Rickmansworth, UK) were operated at five of the sites, measuring pollen density continuously throughout the two week period.

The 52 bait colonies were sited from zero to approximately 4000 m from the nearest identifiable pollen source. After two weeks the plants were removed to a glasshouse, subsequent flower buds removed, and the seeds grown to maturity. The percentage level of pollination (i.e. the percentage of flower buds forming siliquae during the experimental period) was counted for each bait colony. Also the average number of seeds set per siliqua was calculated for each plant, based on counts from five siliquae evenly distributed along the primary stem. Mature seed was harvested from a number of the bait plants, germinated and DNA extracted from green leaf samples. DNA profiling using anchored microsatellites (Charters *et al.*, 1995) was used to identify the source of pollen where the genotypes of possible sources were known.

As a special case, a similar experimental approach was taken to measure gene flow from an 11 ha field of high laurate transgenic OSR in Perthshire. The OSR carried a thioesterase gene from *Umbellularia californica* and a *npt*II marker gene conferring kanamycin resistance. Bait colonies were placed from 100 m to 900 m from the field. Several other OSR crops were being grown in the same area, the nearest to the field being some 320 m away. To detect trangenic progeny among the bait plant progeny, seeds were germinated on selection media containing kanamycin ($300mgl^{-1}$). To confirm the scoring of bait plant progeny, a subsample were grown to seedling stage and DNA extracted. PCR was used to detect the *npt*II gene (Hamill *et al.*, 1991), and further primers were designed to amplify the thioesterase transgene (EMBL M94159). For the analysis of the general relationships between pollination and source, results from these bait colonies were pooled with the rest.

RESULTS

Pollen density profile

Four of the Burkard traps at sites in a transect at zero, five, 170 and 400 m from edge of a 55 ha field were used to measure daily pollen counts. Average pollen counts over the two week period show a steep decline with distance with levels at 170 m from the field being less than 25% of that at the edge of the field. The three nearest pollen traps show similar peaks during the time period, while levels at the most distant site are relatively constant. Although 'clumps' of OSR pollen grains (up to 50 grains) were infrequently detected up 700 m from source, over 82% of pollen recorded more than 100 m from source was present as single grains.

Regional scale gene flow

Pollination occurred at all sites, ranging from 88.4% of flowers buds being pollinated at 1 m from a field edge, to 5% at extreme distances from source. Pollination declined with distance from source ($r^2 = 0.5178$, $P < 0.0001$), under a Gaussian distribution of values (Figure 1). The density of airborne pollen, as measured on microscope slides, also declined but with a pronounced leptokurtic distribution and values in the second phase of the graph (distance greater than 150 m from source) were not correlated with distance (Spearman $r = -0.2956$, $P = 0.1427$).

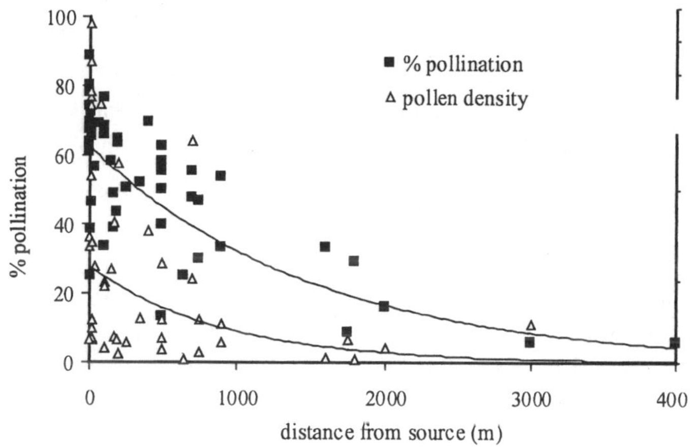

Figure 1. Decay curves of % pollination and airborne pollen density with distance from nearest pollen source.

At one of the bait sites, with a pollination rate of 33%, a sample of bait plant progeny were used for DNA profiling. The majority of the sample (>80%) were shown to have been fertilised by pollen from the nearest crop, 900 m distant. However, seeds from one siliqua sampled were fertilised by pollen of a second genotype whose nearest *known* source is some 4 km away. There was little pollen recorded at the Burkard trap at this site (<1 grain m^{-3} per day) suggesting that such longer range pollination events could have been mediated by insects.

The numbers of seed set per siliqua in bait plants close to a pollen source were identical to typical values within the actual crop (Figure 2). Values then declined over the first hundred metres, but changed relatively little thereafter. At 400 m from source, a mean of 9.8 seeds (SD ± 8.1) seeds per siliqua was recorded. At such extreme distances the large range of values (1 to 24 seeds per siliqua), despite the predominance of isolated airborne pollen grains, suggest that pollination is not wholly due to wind dispersal, and may therefore be due to (occasional) visits by pollinating insects.

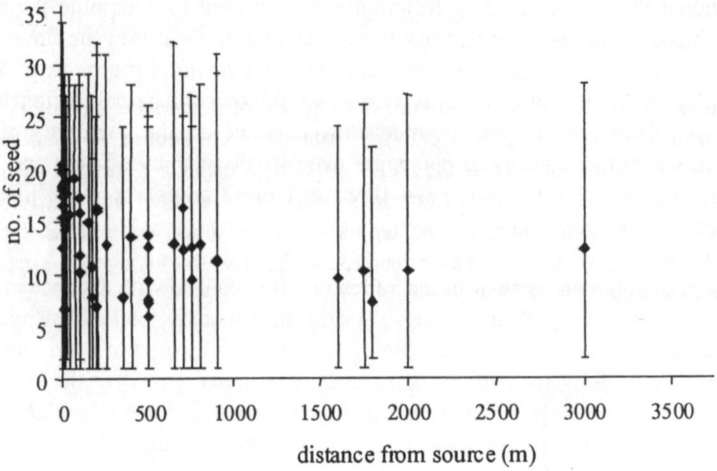

Figure 2. Mean number of seed set per siliqua and distance from source. Bars indicate maximum and minimum values

Measurement of gene flow from a single GM source

Fertilisation by GM and non-GM pollen was detected at all bait colony sites around the transgenic field (Table 1). While the level of total pollination remained relatively constant, the percentage of seeds set that proved to be transgenic was highest at the site closest to the GM field, and lower at more distant sites or those nearer to a non-GM crop.

Table 1. Pollination profiles of GM and non-GM pollen sources

	Distance from nearest crop (m) GM	Distance from nearest crop (m) non-GM	% total pollination	% of GM seed set
site 1	100	1160	65.6	74.7
site 2	580	220	61.2	12.4
site 3	670	900	58.0	30.3
site 4	800	500	49.3	31.1
site 5	930	1200	54.8	10.3

DISCUSSION

The experimental approach described here is highly realistic in that it uses existing field crops of OSR as sources to measure the potential transfer of transgenes. We have also demonstrated that DNA profiling techniques can be used to determine the source of pollen. The use of male-sterile bait plants was essential to measure gene flow on such a large scale, but the results obtained represent only the *potential* for gene flow. While we may anticipate that pollination decay curves would be similar, pollination levels for male-fertile plants are likely to be much lower. However, the popularity of varietal hybrids in which the majority of plants are male-sterile (e.g. cvs. Triolo and Synergy) has important consequences for gene flow, and these crops are more likely to be pollinated by external sources.

Both levels of pollination and airborne pollen density declined with distance in a similar manner to that found by Timmons *et al.* (1995). In that study, pollen density at 360 m was found to be 10% of that found at the field margin, compared to approximately 5% at 400 m in this study (at both Burkard and slide traps). However the frequency of pollination events (0.8% at 2500 m from source) reported in Timmons *et al.* (1995) has been easily exceeded here. Notably, petals were removed from the bait plants used by Timmons *et al.* (1995), in order to make them less attractive to honeybees which maybe in part responsible for the lower levels found in that study. OSR pollen dispersal by bees has been previously reported (e.g. Mesquida *et al.*, 1988; Bilsborrow *et al.*, 1998), and the behaviour of honeybees in this respect is reported elsewhere (G. Ramsay *et al.*, these proceedings). The relative importance of wind and insect pollination is difficult to examine in field conditions, however indirect evidence suggests bee involvement. For example high numbers of seed set per siliqua occurred at distant sites, despite an overall low frequency of pollination events. This, and the occurrence of pollination events in the absence of high levels of airborne pollen, suggest that insects play an important role and could mediate gene flow over a range of distances.

This study also shows bait plants pollinated over both short- and long-range distances by a mixture of pollen sources indicating that their regional distribution is an important factor in contributing to overall pollination levels. This is also apparent from the variation in pollination levels at any one particular distance from source. For example, pollination in bait colonies 500 m from the nearest source ranged from 13.0% to 57.9% of flowers. This may be due in part to the microgeographical habitat, (the lowest value came from a sheltered site), but may also be due to the clustering of crop sources irrespective of the distance to simply the nearest source.

Reported gene flow levels from transgenic OSR plots in the presence of normal pollen competition (e.g. Scheffler *et al.*, 1993; Paul *et al.*, 1995) conflict with results reported by Timmons *et al.* (1996), and by implication with the results from this study. Since 1993 over 60% of winter OSR crops in the wider Tayside area were grown within 100 m of another crop simultaneously in flower, and at least 20% within 100 m of a feral population (DETR, 99). In the current study, an average of 61% of flowers on male-sterile bait plants were pollinated at 100 m from source with 9.5 seed set per siliqua, and 49% of seed set on male-sterile plants at 100 m from a GM crop were transgenic. Despite the over-estimation of gene flow levels detected by the approach used here, the results have implications for growing non-traditional OSR types. The slow decline in pollination levels with increasing distance from source illustrates the potential distance OSR transgenes may move, and that background gene flow occurs across substantial distances from the more intensely farmed areas. Thus potential exists for a continuous network of cross-pollination of OSR across any region growing the crop. This far-reaching gene flow, even if in reality less frequent than these results suggest, will lead to volunteers and feral populations acting as gene pools for introduced genes coming from adjacent and even distant farms.

ACKNOWLEDGMENTS

This work was undertaken as part of a three year MAFF Open Contract.

REFERENCES

Bilsborrow P E; Evans E J; Bowman J; Bland B F (1998). Contamination of edible double-low oilseed rape crops via pollen transfer from high erucic cultivars. *Journal of the Science of Food and Agriculture*, **76**, 17-22.

Charters Y M; Robertson A; Wilkinson M J; Ramsay G (1996). PCR analysis of oilseed rape cultivars (*Brassica napus* L. spp. *oleifera*) using 5' anchored simple sequence repeat (SSR) primers. *Theoretical and Applied Genetics* **92**, 442-447.

DETR (1999). *Investigations of feral oilseed rape populations. Research Report No. 12.* Department of the Environment, Transport and the Regions: London.

Hamill J D; Rounsley S; Spencer A; Todd G: Rhodes M J C (1991). The use of polymerase chain reaction in plant transformation studies. *Plant Cell Reports* **10**, 221-224.

Mesquida J; Renard M; Pierre J S (1988). Rapeseed (*Brassica napus* L.) productivity: the effect of honeybees (*Apis mellifera* L.) and different pollination conditions in cage and field tests. *Apidologie* **19**, 51-72.

Paul E M; Thompson C; Dunwell J M (1995). Gene dispersal from genetically modified oil seed rape in the field. *Euphytica* **81**, 283-289.

Scheffler J A; Parkinson R; Dale P J (1993). Frequency and distance of pollen dispersal from transgenic oilseed rape (*Brassica napus*). *Transgenic Research* **2**, 356-364.

Timmons A M; O'Brien E T; Charters Y M; Dubbels S J: Wilkinson M J (1995). Assessing the risks of wind pollination from fields of genetically modified *Brassica napus* ssp. *oleifera*. *Euphytica* **85**, 417-423.

Timmons A M; Charters Y M; Crawford J W; Burn D; Scott S E; Dubbels S J; Wilson N J; Roberston A; O'Brien E T; Squire G R; Wilkinson M J (1996) Risks from transgenic crops. *Nature* **380**, 487.

Modelling the potential for gene escape in oilseed rape via the soil seedbank: its relevance for genetically modified cultivars

C Pekrun
Institute Agronomy Grassland (340), University Hohenheim, 70593 Stuttgart, Germany

P W Lane, P J W Lutman
IACR-Rothamsted, Harpenden, Herts, AL5 2JQ, UK

ABSTRACT

A simple mechanistic model is presented which describes population dynamics of volunteer oilseed rape within a field. The model calculates the number of volunteers appearing in each crop and the seedbank after each crop. The main input variables are harvesting losses when the crop is oilseed rape, crop rotation, soil cultivation, moisture distribution in soil and the level of volunteer control in each crop in the rotation. The impact of the amount of harvesting losses and of the level of volunteer control are tested and presented in this paper. Simulations of the effects of replacing oilseed rape with linseed are shown. Sensitivity analysis revealed that the relationship between the number of volunteer seedlings and the size of the seedbank in the absence of direct control measures was of particular importance and further research is necessary to improve the definition of this relationship.

INTRODUCTION

From an ecological point of view the soil seedbank of a species ensures genes for the future. Oilseed rape, more than many other crops, has the potential to develop large and persistent seedbanks (Schlink, 1994, 1998). At the same time oilseed rape is very amenable to genetic engineering and is the subject of work relating to herbicide resistance, oil quality and to resistance against pests and diseases. The ability of oilseed rape to persist in soil may cause agronomic problems and is of concern for ecological reasons. If volunteers confer a different oil quality to that of the crop sown, impurities and thus quality reductions may be the result. If volunteers are resistant to glyphosate or glufisonate, herbicide programmes based on either of these herbicides will not be effective in their control. Oilseed rape can hybridise with several wild crucifers and hence genes can be transferred into the genome of wild plants of the agro-ecosystem (Mikkelsen *et al.*, 1996; Chèvre *et al.*, 1997). This again may result in enhanced weed problems both within and outside fields.

There are no data available on the long-term behaviour of seedbanks of oilseed rape under arable conditions. To date, in the limited number of field experiments which have studied seedbank dynamics, data are only available for a maximum of three and a half years (Lutman *et al.*, 1998; Chadoeuf *et al.*, 1998). So, the only way to study seedbank dynamics of volunteer oilseed rape over a longer period of time is by modelling. There are data available on most steps of the life-cycle of volunteer rape and so a synthesis of the different pieces of information into a simulation model provides a mechanism for further insight into the problem. In this paper we present a simple deterministic model of the life-cycle of volunteer oilseed rape along with the output of some simulation runs.

PRESENTATION OF THE MODEL

The basic structure of the model is presented in Fig. 1. The boxes represent the state of some aspect of the seeds or plants at some time during the year and the valves represent factors that influence the transition from one state to the next. Boxes indicating the seedbank in the soil are divided into four sections representing layers of the soil profile: 0 – 5 cm, 5 – 10 cm, 10 – 15 cm and 15 – 20 cm.

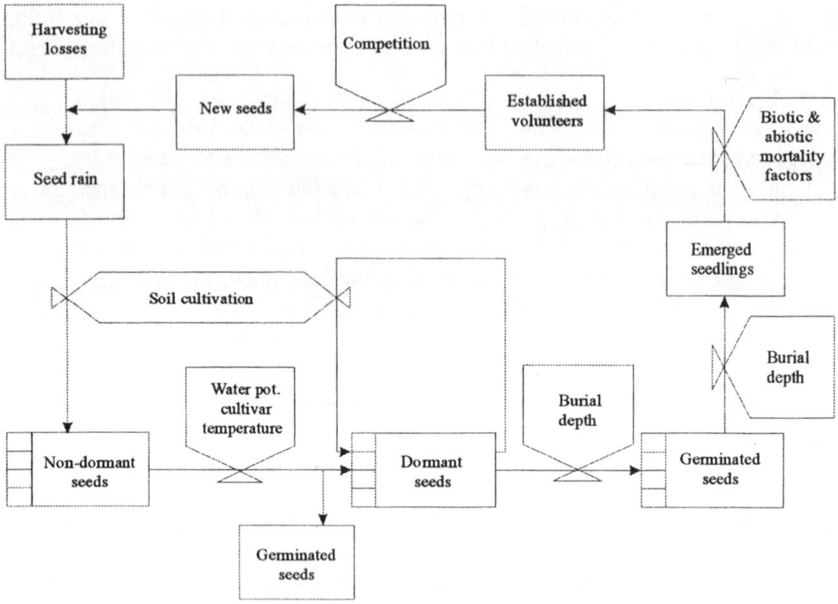

Fig. 1: General structure of the model. Boxes represent individuals (seeds or volunteer plants). Valves represent factors that influence the transition from one step to the next.

The formula for the model is:

$$s_{t+1} = D_j\,(C_j\,(I\text{-}G)\,s_t + d\,C_j\,\{l_t + p\,e\,G\,s_t\,(b_0 + b_c\,(p\,e\,G\,s_t\; + \;n_x\,e_x))^{-1}\}\,),\;\; j = 1 \ldots m$$

s_t matrix for seedbank in year t.

C_j, D_j matrices for seed movement within the soil profile caused by soil cultivation j ($j = 1 \ldots m$). C_j represents the first cultivation after harvest, D_j all the following cultivations. Matrices for ploughing and spring-tining were taken from Cousens & Moss (1990). For discing an additional matrix was constructed by extending the matrix for spring-tining.

I identity matrix when no redistribution between layers occurs.

d proportion of seeds developing dormancy in year t after first soil cultivation. A logistic curve was derived from an experiment by Pekrun *et al.* (1998a) with water potential as the explanatory variable and the proportion of dormant seeds as response variable.

l_t harvesting losses in year t.

G yearly germination from the soil seedbank. Based on findings by Schlink (1994) and Pekrun *et al.* (1997, 1998a) germination is assumed to be 100% in the 0–5 cm layer. For the 5–20 cm layer, seedbank data published by Lutman *et al.* (1998) were used. From this a yearly rate of decline of 0.55 was calculated, with a 95% confidence interval of 0.37 and 0.82. As this value was calculated for 0–20 cm and as the confidence interval for it was very large, 0.50 was assumed to be a reasonable value for the soil seedbank in 5–20 cm.

e emergence of seedlings in relation to burial depth derived from data by Lutman (1993).

p survival of emerged seedlings. There are no data available on this parameter in the absence of direct volunteer control. With emergence being depth dependent and $p = 0.1$ a maximum of 5 % of the seedbank was found to give rise to volunteer seedlings. This approximately reflects the situation in our own experiments (Lutman *et al.*, 1998; unpublished observations) and therefore was used in simulation runs without direct volunteer control.

e_x density equivalent for crop x - peas, beans or linseed - calculated on the basis of Lutman *et al.* (1996).

n_x density of crop x.

b_0, b_c 1/maximum seed production of a plant in isolation and 1/maximum seed production per unit area were based on data by Leach *et al.* (1994).

Sensitivity analysis on each of the parameters suggests that the proportion of seeds germinating from the seedbank each year has a particularly large impact on the output of the simulations.

SIMULATION RUNS

A number of simulations have been run with a range of different input parameters such as seed losses, rotations and levels of control of volunteers. The basic rotation tested was oilseed rape, wheat, field beans, wheat, as this is quite typical of UK rotations where farmers are trying to maximise the proportion of first wheat crops. The first simulation shown in this paper compares the impact of different levels of harvesting losses. Seed losses from oilseed rape have been found to vary between 20 and 650 kg/ha (Pekrun *et al.*, 1998b). Fig. 2 shows the effect of seed losses varying between 20 and 400 kg/ha. Assuming seed losses of 20 kg/ha and average agronomic practices the model suggests that after three rotations no volunteers appear in beans and 2 volunteers m $^{-2}$ in oilseed rape. In a simulation which reflects a more average situation with seed losses of 200 kg/ha the model predicts 2 volunteers m $^{-2}$ in beans and 15 m $^{-2}$ in oilseed rape, after three rotations. Where seed losses are 400 kg/ha the model predicts 4 and 30 volunteers m $^{-2}$, respectively. In Fig. 3 the impact of varying levels of volunteer control in beans is shown. Seed losses are 200 kg/ha each year if oilseed rape is grown. In beans, control efficiency of volunteer oilseed rape can vary between 85 and 100 % with an average of 90 % (Knott, pers. commun.). At 100 % control no volunteers will appear in beans and approximately 2 volunteers m $^{-2}$ in oilseed rape. When assuming 90 % volunteer control in beans about 2 volunteers m $^{-2}$ will appear in beans and about 15 m $^{-2}$ in oilseed rape. This will be increased again if control is decreased to 85 %, resulting in 3 and 22 plants m $^{-2}$ in beans and rape, respectively. In Fig. 4 the effect of growing linseed instead of oilseed rape is shown. The starting seedbank is a seedbank reached after growing 4 rotations of oilseed rape, wheat, beans, wheat at average input variables and agronomic practices. After changing to a rotation with linseed, wheat, beans, wheat, the seedbank declines such that after only two rotations seedbanks are so small that less than 1 volunteer plant m $^{-2}$ sets seed.

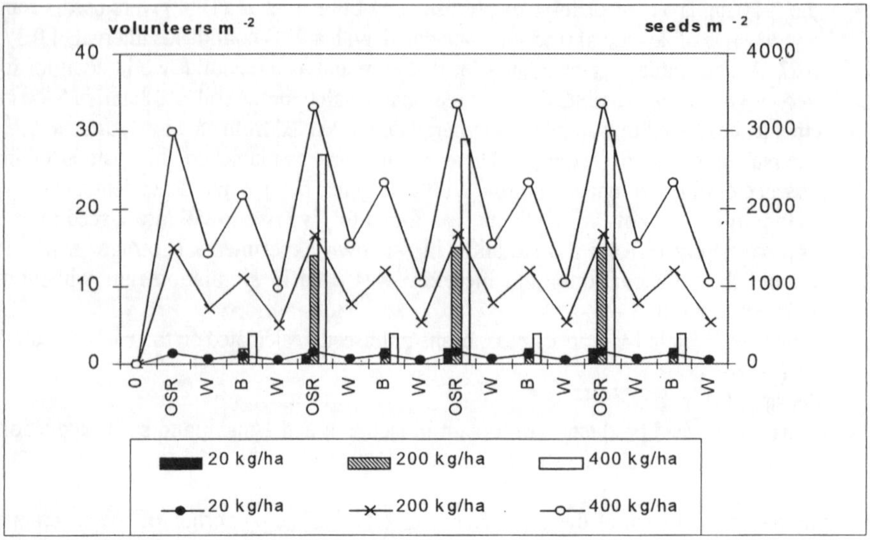

Fig. 2: Effect of varying seed losses from oilseed rape in a rotation with oilseed rape (OSR), wheat (W), beans (B), wheat (W). Seed losses are varied between 20, 200 and 400 kg/ha. Bars represent volunteers m $^{-2}$ in each crop, lines represent seeds m $^{-2}$ at 0-20 cm after each crop.

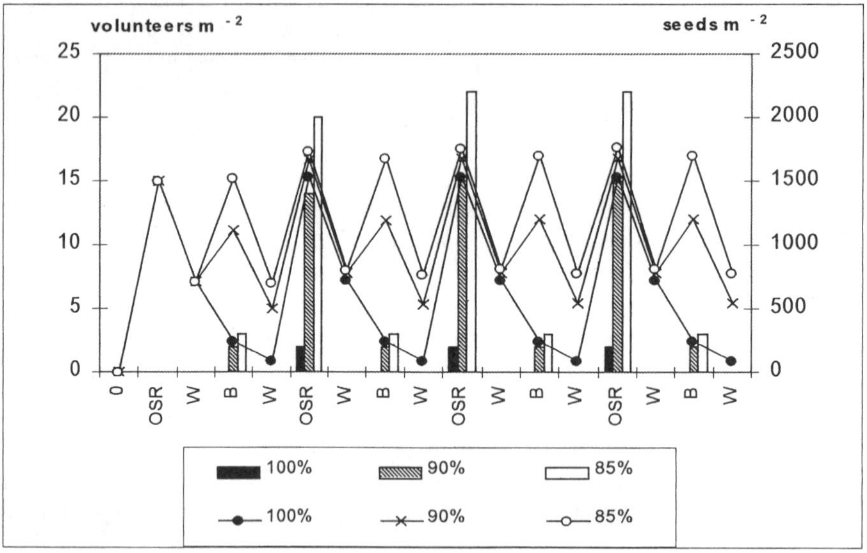

Fig. 3: Effect of varying levels of volunteer control in beans in a rotation with oilseed rape (OSR), wheat (W), beans (B), wheat (W). Volunteer control is varied between 85, 90 and 100%. Bars represent volunteers m $^{-2}$ in each crop, lines represent seeds m $^{-2}$ at 0-20 cm after each crop.

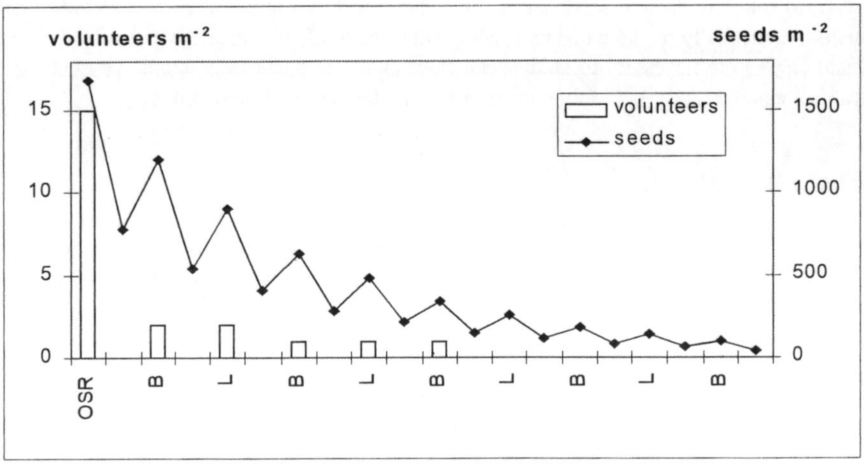

Fig. 4: Reducing the seedbank by growing linseed instead of oilseed rape. Simulations starts from an average seedbank reached after growing 4 x oilseed rape (OSR), wheat (W), beans (B), wheat (W). Bars represent volunteers m $^{-2}$ in each crop, lines represent seeds m $^{-2}$ at 0-20 cm after each crop

DISCUSSION

At present the model presented in this paper cannot be fully validated or improved as there is no independent data set to use for comparison. So, the model cannot be used for prediction. However, it can be used to identify major gaps in knowledge and to study the relative impact of various agronomic practices. Additionally, it helps to understand the processes underlying population dynamics of volunteer oilseed rape. Sensitivity analysis showed that the proportion of seeds germinating from the seedbank each year is a critical factor. As this parameter is based on very limited data there is a clear need for future research to define it more precisely. Also, there is a need for research to quantify survival of emerged seedlings in the absence of direct control measures. We have been unable to find any data on this parameter and therefore derived it on the basis of generating outputs that realistically describe the situation in the field.

The simulation runs presented in this paper show that the amount of seed losses and the level of volunteer control are important input variables. The former is not surprising but the latter is particularly interesting. Volunteer control in beans was varied only marginally but still had a significant impact. It has an even bigger impact when volunteer control in beans is assumed to be 0 % (not shown), a situation that could occur when herbicide programmes are based solely on glyphosate or glufosinate and the oilseed rape volunteers arise from crops resistant to these herbicides, a result also supported by the model of Squire *et al.* (1997). In all runs the model suggests relatively large volunteer densities in oilseed rape. So, there is considerable potential for contamination when the quality of volunteers differs from the sown crop. Also, it shows that if a farmer decides not to grow gm-rape any more but to return to conventional oilseed rape a relatively large hidden population of volunteer rape would still appear for a number of years, hence potentially spreading unwanted genes. In contrast, replacing oilseed rape by a crop in which reasonable volunteer control is possible (eg linseed) would result in a rapid decrease in seedbanks.

Gene escape in time via the soil seedbank is only one aspect that needs to be considered in risk assessments of genetically manipulated rape. More complex models such as the model by Colbach & Meynard (1996) are necessary to study gene flow as a function of agronomic practices and ecological factors. Our model could function as a sub-model in one of these models.

ACKNOWLEDGEMENTS

We thank John Cussans for providing the competition part of the model and for giving valuable advice in the process of producing the model. IACR-Rothamsted receives grant aided support from the Biotechnology and Biological Sciences Research Council of the UK.

REFERENCES

Chadoeuf R; Darmency H; Maillet J; Renard M (1998). Survival of buried seeds of interspecific hybrids between oilseed rape, hoary mustard and wild radish. *Field Crops Research* **58**, 197-204.

Chèvre, A-M; Eber F; Baranger A.; Renard M (1997). Gene flow from transgenic crops. *Nature* **389**, 924.

Colbach N; Meynard J M (1996). Modelling the influence of cropping system on gene flow for herbicide resistant rapeseed. Presentation of the model structure. *Proceedings of the Xe Colloque international sur la biologie des mauvaises herbes, Dijon*, 223-230.

Cousens R; Moss S R (1990). A model of the effects of cultivation on the vertical distribution of weed seeds within the soil. *Weed Research* **30**, 61-70.

Leach J E; Darby R J; Williams I H; Fitt B D; Rawlinson C J (1994). Factors affecting growth and yield of winter oilseed rape (*Brassica napus*), 1985-89. *Journal Agricultural Science, Cambridge* **122**, 405-413.

Lutman P J W (1993). The occurrence and persistence of volunteer oilseed rape (*Brassica napus*). *Aspects of Applied Biology* **35**, *Volunteer crops as weeds*, 29-36.

Lutman P J W; Risiott R; Ostermann H P (1996). Investigations into alternative methods to predict the competitive effects of weeds on crop yields. *Weed Science* **44**, 290-297.

Lutman P J W; Pekrun C; Albertini A (1998). Dormancy and persistence of volunteer oilseed rape. *HGCA-report No*. OS32.

Mikkelsen T R; Andersen B; Jorgensen R B (1996). The risk of crop transgene spread. *Nature* **380**, 31.

Pekrun C; Lutman P J W; Baeumer K (1997). Germination behaviour of persisting oilseed rape seeds in relation to temperature. *Weed Research* **37**, 419-431.

Pekrun C; Hewitt J D J; Lutman P J W (1998a). Cultural control of volunteer oilseed rape (*Brassica napus* L.). *Journal Agricultural Science, Cambridge* **130**, 155-163.

Pekrun C; Lutman P J W; Baeumer K (1998b). Research on volunteer rape: a review. *Pflanzenbauwissenschaften* **2**, 84-90.

Schlink S (1994). Ökologie der Keimung und Dormanz von Rapssamen (*Brassica napus* L.) und ihre Bedeutung für eine Überdauerung der Samen im Boden. *Dissertationes Botanicae* 222, Cramer, Berlin, Stuttgart.

Schlink S (1998). 10 years survival of rape seed (*Brassica napus* L.) in soil. *Zeitschrift Pflanzenkrankheiten Pflanzenschutz Sonderheft* **XVI**, 169-172.

Squire G R; Burn D; Crawford J W (1997). A model for the impact of herbicide tolerance on the performance of oilseed rape as a volunteer weed. *Annals Applied Biology* **131**, 315-338.

SESSION 4

CONSEQUENCES OF GENE FLOW BETWEEN SPECIES – I

Chairman & Session Organiser: Mr J Orson
Morley Research Centre, Wymondham

Gene flow and rape - the Canadian experience

R K Downey
Agriculture and Agri-Food Canada, Saskatoon Research Centre, 107 Science Place, Saskatoon, SK S7N 0X2, Canada

ABSTRACT

Herbicide tolerant cultivars developed by both induced mutation and gene transfer occupied 50% of Canada's 5.2 million hectares of canola (*Brassica napus* plus *Brassica rapa*) in 1998 and an expansion to 70 or 75% of the canola area is expected in 1999. Cultivars tolerant to S-triazine herbicides have been successfully grown in Canada since 1981. Early isolation studies in western Canada indicated an average outcrossing rate from large fields to small plots of 0.6 and 3.7% at 366 m for *B. napus* and *B. rapa*, respectively. However, 1998 data indicate a much lower outcrossing rate between large fields of herbicide tolerant and standard *B. rapa* and *B. napus*. Gene flow between herbicide tolerant *B. napus* cultivars and stacking of resistance genes in volunteer *B. napus* plants was observed in a commercial field when a producer did not follow agronomic and herbicide recommendations. The weeds *Raphanus raphanistrum, Hirschfeldia incana* and *Brassica nigra* are absent or uncommon in western Canada. However, the weed dog mustard (*Erucastrum gallicum*) is abundant and inter-generic hybrids, using *E. gallicum* as the male, have been obtained in crosses with *B. napus* and *B. rapa*. Selfed and backcross progeny are under investigation.

INTRODUCTION

The area sown to herbicide tolerant canola cultivars, resulting from both induced mutation and gene transfer has grown rapidly since their introduction and release in 1995 (Table 1). Such rapid expansion is indicative of the widespread acceptance by producers as well as consumers in North and Central America and Asia. It is predicted that some 70% or perhaps 75% of the six million ha of canola expected to be sown in western Canada in 1999 will be tolerant to one or the other of the three broad spectrum herbicides, glyphosate (Roundup), glufosinate (Liberty) or the imidazolinones, imazethapyr (Pursuit) and imazomox plus imazethapyr 1:1 (Odyssey).

Herbicide tolerant *B. napus* cultivars have been used in Canada since 1981 when OAO Triton, the first canola cultivar with tolerance to the S-triazine family of herbicides, was introduced. This herbicide tolerance results from a natural mutation (chloroplast gene *psbA)* originally identified in *B. rapa* plants growing in a field where a continuous atrazine/corn management system had been practised. The usage peaked at 145,000 ha in 1988, some 4.0% of the total canola acreage, and diminished to less than 1% by 1996. No environmental or consumer problems were encountered and since the herbicide tolerance characteristic was cytoplasmically inherited, gene flow via pollen was not a concern. The usage of triazine tolerant cultivars in Canada had limited grower acceptance because the mutated chloroplast gene, that made the plants tolerant to the herbicide, also reduced the photosynthetic efficiency of such plants, resulting in slower growth, lower seed yield and a reduced oil content. In addition, the slow breakdown of the herbicide under western Canada conditions placed serious limitations on other crops that could be included in the rotation. On the other hand, Australian canola growers have found that triazine tolerant cultivars solved their weed problems and provided an economic return even though the triazine tolerant cultivars yielded some 20-30% less seed and produced two to three percent less actual seed oil than the corresponding non-triazine tolerant cultivars. Breeding for superior performing triazine cultivars in Canada was discontinued in the 1990's because cultivars with inserted or induced mutated genes that imparted tolerance to more environmentally friendly broad spectrum herbicides, which do not impose a physiological cost to yield and oil content, would soon be available.

Table 1. Estimated area sown ('000 ha) to herbicide tolerant canola *(B. napus)* cultivars by herbicide type and total canola production in western Canada, 1995-1999.

Year	Glyphosate	Glufosinate	Imidazolinone	Total canola area sown	% herbicide tolerance
1995	0	14	10	5,273	<1%
1996	18	103	227	3,451	10%
1997	182	404	688	4,869	26%
1998	1,214	610	850*	5,260	51%
1999**	2,000	1,226	1,226*	6,070	73%

* Significant usage of non-certified seed included
** Projected production and usage

WITHIN SPECIES GENE FLOW

Prior to the introduction of the new herbicide tolerant forms of *B. napus,* background information on outcrossing within and among species, regardless of whether the tolerance resulted from mutation breeding or gene transfer, was a primary point of interest. Data on gene flow from large commercial fields (approximately 65 ha) to small (46 m^2) plots of *B. napus* and *B. rapa* using a chlorophyll deficient marker were obtained in 1974 and 1977 (Stringam & Downey, 1978 & 1982). These data indicated that within species, pollen outflow resulted in an average of 0.6% and 3.7% outcrossing in *B. napus* and *B. rapa*, respectively, even at 366 m isolation (Figure 1). More recently, Raney and Falk (1998) found that the amount of outcrossing from pollen outflow from a small (0.4 ha) high erucic acid contaminant *B. rapa* source into a large (64 ha) low erucic acid recipient field was observed up to 260 m and that a 100 m buffer would be sufficient to contain more than 99% of the pollen flowing from the contaminating source.

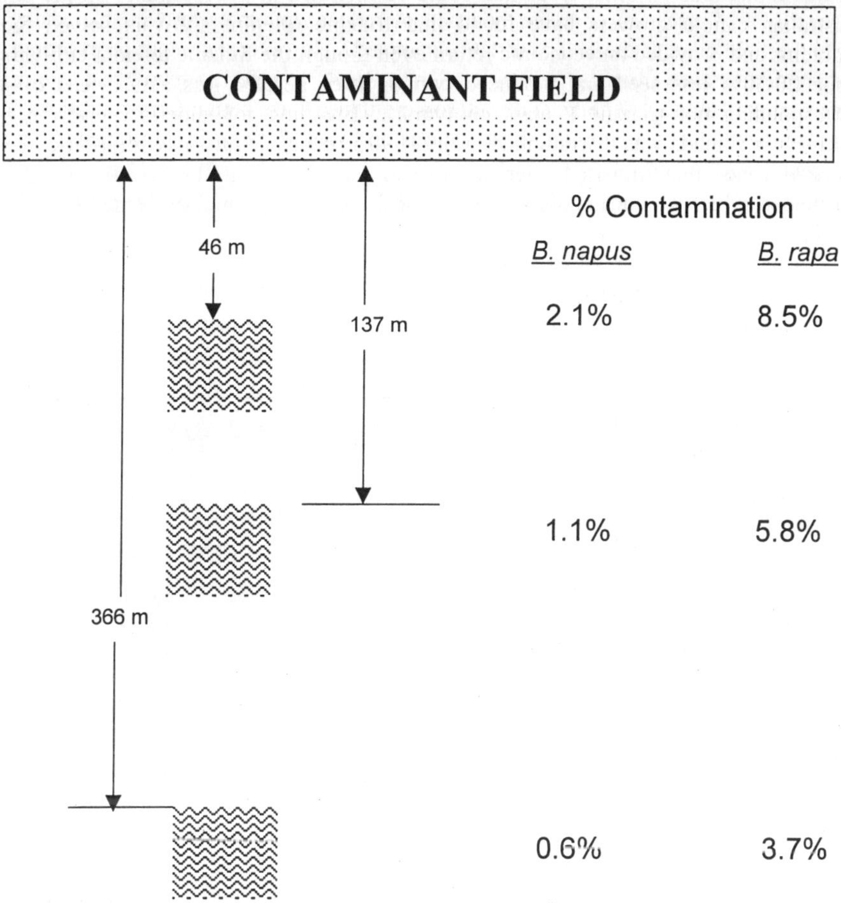

Figure. 1. Typical arrangement of 46m² isolation blocks and the average level of outcrossing from six fields of *Brassica napus* and of *B. rapa* (after Stringam & Downey 1978 & 1982).

In 1998, it was possible for the first time to investigate large field to large field (>16 ha) pollen movement with the first commercial production of the Roundup Ready (RR) *B. rapa* cultivar 41A51 from Pioneer Hi-Bred. Seed samples were harvested from herbicide susceptible canola in three fields that were paired with fields sown to RR 41A51. These paired fields were located at widely different locations in the province of Alberta and the distance between the paired fields differed with location (Table 2). Seed samples were harvested from the herbicide susceptible fields at 0, 33, 66 and 100 m from the edge of the fields closest to their corresponding RR field. Random samples of about 900 seeds from each distance and each susceptible field were sown in the greenhouse and the seedlings sprayed twice with a 2x concentration of glyphosate herbicide. Seedlings of a susceptible check cultivar, Tobin were all killed while 11% of the RR 41A51 seedlings were also herbicide susceptible. There was no clear evidence of reduced outcrossing with increasing distance from the field edge to the 100 m distance sampled. However, as might be expected, fields at the greater distance from the herbicide tolerant RR 41A51 pollen source had a lower level of outcrossing (Table 2). The level of outcrossing under these actual farm conditions was lower than anticipated given previous observations and that *B. rapa* is an obligatory outcrossing species.

Table 2. Percent outcrossing between large fields* of herbicide susceptible and glyphosate tolerant *B. rapa* grown at three Alberta, Canada locations in 1998.

Location	Distance (metres) Between fields	Distance (metres) From edge of susceptible crop	Percent outcrossing**
A	4	0	0.2
		33	0.4
		66	0.3
		100	0.2
		Average	0.28
B	250	0	0.0
		33	0.1
		66	0.0
		100	0.0
		Average	0.02
C	600	0	0.0
		33	0.0
		66	0.0
		100	0.1
		Average	0.02

* Field size ranged from16 to 64 ha

** Based on herbicide reaction of 900 seedlings per sampled distance

Seed from two herbicide susceptible *B. napus* fields were sampled by Monsanto at 20, 50 and 100 m from adjacent RR *B. napus* fields. Greenhouse grow outs of between 991 and 1009 seedlings per sampling site confirmed 1.5, 0.4, and 0.1% were RR survivors from Field 1 at 20, 50, and 100 m, respectively and 0.01, 0.00 and 0.40% from Field 2, for an average outcrossing rate of 0.6 and 0.16% for Field 1 and 2, respectively. Thus in 1998 the degree of cultivar outcrossing on a large field to large field basis was substantially lower for both *B. rapa* and *B. napus* than that recorded for large field to small plots, as reported by Stringam and Downey (1978 & 1982).

It was anticipated that useful information on *B. napus* gene movement could be obtained from an unusual agronomic situation that occurred on a farm in the Peace River region of northern Alberta. In 1997 a farmer, apparently wishing to compare the performance and weed control benefits of the three different herbicide tolerant *B. napus* cultivars, sowed a Roundup Ready (RR) cultivar in one field and 20 m to the east, across a road and parallel to the RR field sowed 8 ha of a Liberty Link (LL) cultivar. In the remainder of the 40 ha LL field he sowed a Pursuit Smart (PS) cultivar. The RR field was sown, swathed and combined two weeks earlier than the LL-PS field. In 1998 severe drought prevented the farmer from sowing a crop into his canola stubble in the LL-PS field. He therefore decided to chemically fallow the field, using glyphosate herbicide and noted that many of the volunteer canola plants did not die. Later the same season he twice more sprayed the same field with glyphosate with little or no control of the canola volunteers. The press then learned of the weeds that would not die resulting in much publicity and wrong conclusions. A field plot of the number and location of the volunteer plants was made by Alberta Agriculture. However, no clear or meaningful distribution pattern was evident, probably due to the swathing and combining process which almost certainly distributed seed from the area adjacent to the RR field to the far side of the LL-PS field. Nevertheless there was evidence of a significant pollen flow from the RR field to plants in the LL-PS field 20 m away. Seed harvested from the surviving volunteer plants was grown in the greenhouse by Alberta Agriculture and sprayed with glyphosate. Those plants that survived were also sprayed either with glufosinate or imidazolinone that resulted in clear evidence of multiple herbicide tolerance or gene stacking. However, it is important to note that when the 1998 field of barley, sown into the stubble of the RR canola field, was surveyed, no canola plants were found, except a few on the field edge where the spray boom had missed. From these observations we can conclude:

- That a significant amount of pollen flowed across the 20 m roadway and effected fertilization, resulting in gene stacking.
- The volunteer plants that survived the glyphosate treatment were not "super weeds" as the press and Greenpeace are wont to call them for had the producer followed recommended practices for chemical fallow and tank-mixed glyphosate with 2, 4-D or MCPA, he would have controlled the volunteers just as he controlled them in the barley that was grown on RR canola stubble.
- Growers should avoid sowing cultivars with different herbicide tolerances in the same or adjacent fields.
- All segments of the industry and particularly Canadian seed and chemical companies need to take a much greater effort to inform producers as to the management do's and don'ts of growing herbicide tolerant canola.

BETWEEN SPECIES GENE FLOW

Between crop species gene flow

Gene flow among related *Brassica* oilseed species is of minor concern. Canadian oilseed *Brassica* breeders have long known that inter-specific crossing among *B. napus, B. rapa* and *Brassica juncea* can and does occur in nature as recently documented by Bing *et al.* (1991). However, since these species are all major commercial crops one or another of the herbicide tolerant genes will almost certainly be deliberately introduced. Indeed limited commercial production of RR and LL *B. rapa* cultivars occurred in western Canada in 1998. If gene transfer of the herbicide tolerant trait were to occur among these species, the agronomic and environmental consequences would be essentially nil since the presence of such genes do not make the recipient oilseed *Brassica* plants more competitive or invasive of cultivated or natural ecosystems in the absence of the specific herbicide (MacDonald, 1994; Belyk & MacDonald, 1994 & 1995 a, b, c).

Gene flow to weed species

Of greater importance is the possibility of herbicide tolerance genes escaping to closely related weedy species (Table 3). From the data available to date there appears to be general agreement that no natural gene flow would occur between the rapeseed species *B. napus* and *B. rapa* to *Sinapis arvensis*. French researchers have also found significant barriers to the introgression of *B. napus* marker genes as well as the herbicide tolerance gene into the genomes of *Raphanus raphanistrum* and *Hirscheldia incana* (Darmency *et al.* 1998; Darmency, personal communication; Chèvre *et al.* 1997; Chèvre, personal communication). Fortunately these two weeds and *B. nigra* are either absent or uncommon in western Canada. Thus of greatest interest is the possibility of natural outcrossing of herbicide tolerant *B. napus* and *B. rapa* with *Erucastrum gallicum*. Studies by Lefol *et al.* (1997) produced no hybrids when *E. gallicum* was used as the female in crosses with *B. rapa, B. napus* and *B. juncea*. However, one hybrid was obtained from each of the crosses, *B. napus* x *E. gallicum* and *B. rapa* x *E. gallicum*.

Table 3. Closely related weeds to the *Brassica* oilseed species, *B. napus, B. rapa* and *B. juncea.*

Common name	Scientific name
Wild mustard	*Sinapis arvensis*
Black mustard	*Brassica nigra*
Wild radish	*Raphanus raphanistrum*
Hoary mustard	*Hirschfeldia incana*
Dog mustard	*Erucastrum gallicum*

Observations on the *B. napus* x *E. gallicum* hybrid strongly suggested that such a plant would not compete with *B. napus* plants and would have a very limited opportunity to backcross to *B. napus*. On the other hand, the hybrid readily backcrossed when *E. gallicum* was the male parent. The progeny of the backcross were poor competitors and all their progeny appeared to have completely reverted to *E. gallicum*, suggesting that the *B. napus* chromosomes were lost. The 13 seeds obtained from the backcross *B. napus* x [*B. napus* x *E. gallicum*] are being grown out to determine if they are true hybrids. In any case, a backcross to *B. napus* would result in retaining the gene of interest in the putative parent. These data indicated that the possibility of gene transfer from *B. napus* to *E. gallicum* exists but the probability for it to occur is very low.

The hybrid *B. rapa* x *E. gallicum*, when grown in a greenhouse pot, produced a strong, fertile plant which would probably have survived under field competition. This intergeneric hybrid produced 67 selfed seeds and on backcrossing 690 seeds from the cross *B. rapa x* [*B. rapa x E. gallicum*] as well as 14 seeds from the reciprocal backcross. No seed was produced when *E. gallicum* was the female parent. Backcrossing to the *B. rapa* parent would retain the gene(s) of interest in the putative parent and would not increase the risk of gene transfer to *E. gallicum* to any degree. However, the production of 14 backcross seeds on the *B. rapa* x *E. gallicum* hybrid, when pollinated by *E. gallicum*, indicates a strong possibility that gene flow could occur from *B. rapa* to *E. gallicum*. The observations on the 14 backcross and 67 selfed seed obtained on the *B. rapa* x *E. gallicum* hybrid are underway.

DISCUSSION

The degree of large field to large field gene flow, recorded for both *B. napus* and *B. rapa* in 1998, was substantially lower than suggested in previous studies which measured gene flow from large fields to small plots. Although field to field crossing levels were low, a considerable number of cultivar x cultivar hybrid seed could remain in a field after harvest. For example, a gene flow of 0.4% into a field yielding the Canadian average of 1400 kg/ha, with a harvest loss of 5% could result in some 70,000 outcrossed seed per ha remaining (7 seeds/m^2) in the recipient field. The vast majority of such seed would germinate and be killed by frost and/or cultivation. However, the presence in a producer's field of a significant number of volunteer *B. napus* plants, tolerant to too different herbicides confirms that field to field gene flow can result in gene stacking if careful rotation and herbicide management practices are not followed. The absence of volunteer canola plants in the barley crop sown on RR canola stubble is clear evidence that such volunteers can be readily controlled with standard practices. Although distance isolation was found to be effective in reducing gene outflow between *B. rapa* fields, extremely low outcrossing levels were detected even at 250 and 600 m isolation. Thus, under commercial production conditions, complete containment using distance isolation alone becomes impractical. The possible escape of herbicide tolerant genes to other *Brassica* oilseed species or to weedy species, other than *E. gallicum*, is not a primary concern in Canada. Studies to determine the possibility of natural field crosses between *B. rapa* and *E. gallicum* were undertaken in 1998 and the self and backcross progenies of the hand crossed intergeneric hybrid obtained by Lefol *et al.* (1997) are under evaluation. *E. gallicum* is highly self-fertile and under field conditions *B. rapa* and *B. napus* pollen would have little or no opportunity to fertilise *E. gallicum* plants. However, the reciprocal cross could occur. The question then becomes what would be the consequences of herbicide tolerance genes occurring in *E. gallicum*? From what is known about the biology of this weed it is more of a nuisance than an ecological and agronomic threat. The species may continue to build up populations in infested canola fields, as it does now, but it is easily controlled in cereal and other crops with Group 4 and 6 herbicides.

In all such interspecific studies it is important to keep in mind that if a herbicide tolerant gene were to escape into a weedy relative that weed will be no more difficult to control than before the broad spectrum herbicide became available. Further, all data to date has shown that the trait does not confer an advantage to the plant unless the specific herbicide is applied and normal competition is eliminated.

ACKNOWLEDGEMENTS

Thanks are due to United Grain Growers personnel for obtaining the 1998 *B. rapa* seed samples, Monsanto Canada for the 1998 *B. napus* data and funding for growouts and to Dr. D J Bing for supervising the growing and rating the *B. rapa* seedlings.

REFERENCES

Belyk M; MacDonald R (1994). Invasive potential of transgenic *B. napus* (HCN92) under agronomic conditions. In: *Report No. AC194-06,* pp. 20. AgrEvo Canada Inc.: Regina SK.

Belyk M; MacDonald R (1995a). Invasive potential of transgenic pHOE4 *B. napus* (HCN28) under agronomic conditions. In: *Report No. AC 195-18,* pp. 21. AgrEvo Canada Inc.: Regina SK.

Belyk M; MacDonald R (1995b). Invasive potential of transgenic *B. rapa* (pHOE4) under disturbed and undisturbed field conditions. In: *Report No. AC195-17,* pp.18. AgrEvo Canada Inc.: Regina SK.

Belyk M; MacDonald R (1995c). Competition potential of transgenic pHOE4 *B. rapa* (HCR1) with commercial varieties under agronomic conditions. In: *Report AC195-19,* pp. 23. AgrEvo Canada Inc.: Regina SK.

Bing D J; Downey R K; Rakow G F W (1991). Potential of gene transfer among oilseed *Brassica* and their weedy relatives. *Proceedings 8th International Rapeseed Congress - Saskatoon,* pp. 1022-1027.

Chèvre A-M; Eber F; Baranger A; Renard M (1997). Gene flow from transgenic crops. *Nature* **389**, 924.

Darmency H A; Lefol E; Fleury A (1998). Spontaneous hybridization between oilseed rape and wild radish. *Molecular Ecology* 7, 1467-1473.

Lefol E; Séguin-Swartz G; Downey R K (1997). Sexual hybridization in crosses of cultivated *Brassica* species with *Erucastrum gallicum* and *Raphanus raphanistrum*: potential for gene introgression. *Euphytica* **95**, 127-139.

MacDonald R (1994). Invasive potential of transgenic *B. napus* (HCN92) under disturbed and undisturbed field conditions. In: *Report No. AC194-05*, pp14. AgrEvo Canada Inc.: Regina SK.

Raney J P; Falk K C (1998). Outcrossing in turnip rape *(Brassica rapa* L.). *11th International Crucifer Genetics Workshop*, Montreal PQ. *Book of Abstracts,* p. 36.

Stringam G R; Downey R K (1978). Effectiveness of isolation distance in turnip rape. *Canadian Journal Plant Science* **58,** 427-434.

Stringam G R; Downey R K (1982). Effectiveness of isolation distance in seed production of rapeseed *(Brassica napus). Agronomy Abstracts,* 136-137.

Gene flow from oilseed rape (*Brassica napus*) to related species

R B Jørgensen
Plant Biology and Biogeochemistry Department, Risø National Laboratory, DK-4000, Roskilde, Denmark

ABSTRACT

The spontaneous gene transfer from oilseed rape (*Brassica napus*, 2n=38, genomic constitution AACC) to the weedy relatives *B. rapa* (2n = 20; AA), *B. juncea* (2n=36, AABB), *Hirschfeldia incana* (*B. adpressa,* 2n=14, AdAd), *Raphanus raphanistrum* (2n=18, RrRr) and *Sinapis arvensis* (2n=18, Sar Sar) has been studied in controlled experiments or in natural populations. The extent and possible consequences of this interspecific and intergeneric gene flow from oilseed rape are evaluated in a review of relevant European research.

INTRODUCTION

For many common traditionally bred crops the gene flow to their wild relatives has resulted in increased weediness as well as crop gene flow has been a factor responsible for extinction of some wild crop relatives (Ellstrand et al., 1999). Recently hybridisation as a way for the escape of crop transgenes has put extra focus on the spontaneous gene flow from cultivated plants to their wild relatives. As genetic modification differs from traditional breeding in that totally new characteristics can be added to crop plants at a high rate - i.e. many at a time as precisely designed gene pyramids with regulating sequences – the spread of transgenes by introgression represent a new potential of increasing weediness. Oilseed rape, *Brassica napus*, is easily genetically modified and related *Brassica* species or species from related genera are known to be potential recipients of new oilseed rape genes. The extent and consequences of spontaneous introgression of *B. napus* genes are evaluated.

Spontaneous hybridisation between oilseed rape and related species demands that some basic conditions are met: Oilseed rape and the wild recipient must have overlapping flowering periods and the distance between the crop and recipient must allow transfer of pollen between them by the aid of appropriate vectors. There must be genetic cross compatibility so that the heterospecific pollen can germinate and fertilise the egg cell. Subsequently a hybrid embryo must develop from the germinating seed and produce a viable plant. If the hybrid survives to the adult stage, and shows at least some fertility the introgression process may continue with backcrossing, or a combination of backcrossing and selfing. Many other factors have an impact on hybridisation, i.e. the agricultural management, the environmental conditions, the direction of the cross and the breeding system.

In 1994 Scheffler & Dale reviewed the literature to access the ease with which *Brassica napus* crossed with related species and formed hybrids, F_2 and backcross progeny. They ranked the different wild recipients as to their cross compatibility with oilseed rape in open pollinations, hand pollinations and by embryo rescue. Four species - *B. rapa, B.*

juncea, Hirschfeldia incana (*B. adpressa*) and *Raphanus raphanistrum* - were reported to hybridise spontaneously with *B. napus*. In the light of new experimental data on spontaneous hybridisation, also *Sinapis arvensis* is now added to this list of species (Lefol et al., 1996a), however, hybridisation with this species is rare.

INTERSPECIFIC GENE FLOW

Brassica rapa

B. rapa (2n=20, genomes AA) is one of the parental species of oilseed rape. *B. rapa* is a common annual weed in agricultural fields worldwide (in the temperate zone) particulary in oilseed rape fields with a potential for introgression of oilseed rape genes from the crop to the weed. Outside the field *B. rapa* populations are ephemeral, as seeds will only germinate when the soil is turned. Harberd (1975) reported the spontaneous occurrence of the *B. napus* x *B. rapa* hybrid (*B. x harmsiana*) in oilseed rape fields. Frequencies of hybridisation between oilseed rape and the wild *B. rapa* have been reported from field experiments and survey of natural populations of the wild species. Jørgensen & Andersen (1994), Jørgensen et al. (1996, 1998), Landbo et al. (1996) and Scott & Wilkinson (1998) found hybrid frequencies between 0-69 % of the seeds depending of i.e. parental genotypes, experimental design, agricultural practise, population size and site. Generally *B. rapa* produces more hybrids than oilseed rape (Jørgensen & Andersen, 1994; Jørgensen et al., 1998; Hauser et al., 1997).

The interspecific hybrids have reduced fertility (Jørgensen & Andersen,1994), but spontaneous backcrossing can take place. *B. rapa* and interspecific hybrids with a gene providing Basta resistance were sown together in field experiments to assess the extent of backcrossing (Mikkelsen 1996, Mikkelsen et al., 1996a). Seed set per pod on interspecific hybrids was low (approximately 2.5) compared to seed set on the parental species (normally16-23). The reciprocal cross *B. rapa* x hybrid was not observed among more than 2000 offspring from 30 *B. rapa* plants. A large number of herbicide resistant plants developed from seeds harvested on the interspecific hybrids, and herbicide resistant individuals with a *B. rapa*-like morphology were selected for further analysis. Most of the selected plants were clearly backcross plants and a few (0.5%) were almost identical to *B. rapa* (chromosome number 2n=20, high pollen fertility) and set a normal number of seeds in crosses with genuine *B. rapa* (Mikkelsen et al., 1996a).
The field experiments quantifying introgression are supplemented by descriptive analysis of introgression in natural populations. A natural Danish population of *B. rapa* showing introgression beyond the stage of F_1 hybrids (more advanced generations) is presently being characterised by AFLP.

A successful introgression of crop genes to the wild relatives is very dependent on the fitness of the hybrids and subsequent backcross generations. In this respect the reproductive fitness is very important but also other stages of the life cycle, i.e. establishment of introgressed plants in the field and longevity of the seeds in the soil. As all known relatives capable of receiving genes from oilseed rape are weedy species or crops, the fitness of the introgressed plants should undergo long time monitoring in the agro-ecosystem with the appropriate management strategy. This is a demanding task, and therefore up till now the fitness data available were generated in controlled experiments.

Hauser et al. (1998b) intercrossed three weedy *B. rapa* populations and three *B. napus* varieties and estimated the fruit and seed set. The offspring plants were grown in the field, and monitored for survival and reproduction. Combining these fitness components into a multiplicative fitness estimate, hybrids were intermediate to *B. rapa* and *B. napus* (Hauser et al., 1998b; Figure 1).

Hybrids, *B. rapa* and *B. napus* plants, originating from crossings between two populations and two varieties in the first generation, were intercrossed to obtain F_2, backcross (on both *B. rapa* and hybrids), and pure *B. rapa* and *B. napus* seeds. The same fitness components as described above were estimated in the field for these offspring plants. On an average offspring from backcrossings and F_2 matings had a reduced fitness relative to offspring from intraspecific matings (pure species) (Hauser et al., 1998a).

A third backcross generation (BC_3) produced from backcrossing *B. rapa*-like BC_1 plants for another two generations were quantified as to their seed production and pollen fertility. With respect to these two fitness parameters there were no differences between the BC_3 plants and genuine *B. rapa* (Snow at al., 1999).

All the weedy species known to hybridize with oilseed rape have seed dormancy, a beneficial trait in an agro-ecosystem with efficient weed control. Therefore the seed germination pattern of hybrids and backcross plants is probably a key fitness parameter with a large effect on survival. Differences in seed germination between oilseed rape, wild *B. rapa* and their hybrids were investigated on seeds from controlled crosses (to get well characterised seeds of the two species and their reciprocal hybrids) as well as on seeds from a number of natural populations of weedy *B. rapa*. As expected all *B. napus* seeds germinated showing no sign of dormancy. *B. rapa* on the contrary, showed a very high degree of dormancy. As to the hybrids, both types (*napus* x *rapa*) and (*rapa* x *napus*) seeds showed almost no sign of dormancy (Landbo & Jørgensen, 1997). Lack of dormancy could be disadvantageous, as the number of hybrids reaching the flowering state could be limited due to germination under unfavorable conditions (e.g. with effective weed management) thereby limiting interspecific geneflow. The seed dormancy was restored in seeds from the first backcross with *B. rapa*.

In conclusion spontaneous hybrids and backcross plants are produced in field experiments and found in natural populations, and flow of transgenes from oilseed rape to *B. rapa* seems to be inevitable. Detailed information about transfer of oilseed rape markers is therefore valuable to identify segments of the oilseed rape genome rarely or never transferred to the wild recipient. Such segments could serve as "safe integration sites" for transgenes. For this purpose hybrids were obtained between different maternal *B. napus* cultivars/transgenic lines and *B. rapa* individuals from cultivars or a wild Danish population (Mikkelsen, 1996, Mikkelsen et al., 1996b). The hybrids were backcrossed to *B. rapa* individuals from the same population. In two different crossing combinations all 30 oilseed rape DNA markers – most of them specific to the C-genome - were transferred to the first backcross generation. The majority of the markers were transferred in a ratio that were not significantly different from 50% (Mikkelsen et al., 1996b); only one marker had a low transfer frequency (26%). Plants with 20 chromosomes and a high pollen fertility, both characters of weedy *B. rapa* were identified among backcross plants with oilseed rape markers (Mikkelsen, 1996). Metz et al. (1997) analysed the transfer of a transgene encoding phosphinothricin-tolerance in backcrosses with *B. rapa*. They found different transmission frequencies of the transgene for two *B. napus* lines, and ascribed an unexpected low transmission (26 % of the BC_1 progeny) to a position in the C-genome. In conclusion oilseed rape markers with a low transmission rate can be found, however, all markers studied were inherited to the next generation.

Brassica juncea

Spontaneous hybridisation between oilseed rape and *B. juncea* has also been reported (Frello et al., 1995, Jørgensen et al., 1998). Depending on the proportions between the parental species up till 3% of the offspring harvested on *B. juncea* were hybrids. Production of hybrids with *B. napus* as female was less successful (Jørgensen et al., 1998). Pollen fertility of the hybrids was rather low, 0-28%, and 20 *B. napus* specific RAPD markers were all transferred to the first backcross generation with *B. juncea* (Frello et al., 1995).

INTERGENERIC GENE FLOW

Hischfeldia incana

Field experiments also showed that spontaneous hybrids could be formed between oilseed rape and wild cruciferous species from other genera. When herbicide resistant oilseed rape was mixed with hoary mustard (*Hischfeldia incana* (synonym *Brassica adpressa*)) in a density of 1:625, hoary mustard:oilseed rape, 1.5% of the seeds germinated to be hybrids, on an average one intergeneric hybrid was produced per mustard plant (Lefol et al, 1996b). When male sterile oilseed rape was used as female parent in a 1:1 proportion of oilseed rape:hoary mustard, 70% of the seeds were hybrids (Lefol et al, 1996b). The hybrids plants developed better vegetatively than hoary mustard but their reproductive fitness was very low (Lefol et al, 1996b). They only produced 0.1 viable seed per plant. For the hybrids between oilseed rape and hoary mustard seed dormancy was in between that of the parents (Chadoeuf et al.,1998).

Raphanus raphanistrum

In a similar experiment the herbicide resistant oilseed rape was mixed with wild radish (*Raphanus raphanistrum*) in different densities. Hybridisation was detected at a density of 1:625, wild radish:oilseed rape, as 0.2% of the seeds derived from wild radish were intergeneric hybrids (Darmency et al., 1995). When male sterile oilseed rape functioned as the mother, 45 intergeneric hybrids were obtained from each *B. napus* plant. Grown in mixtures with wild radish each hybrid produced less than one backcross plant, but fertility improved in subsequent backcross generations with the wild parent. However, no stable integration of oilseed rape genetic material into the genome of wild radish was observed. In the third backcross generation none of the transgenic plants had the same chromosome number as the weedy parent possibly indicating that the transgene was not recombined into the weed-genome (Chevre et al., 1997, 1998). The hybrids between oilseed rape and wild radish had a seed dormancy in between that of the parents (Chadoeuf et al.,1998).

Sinapis arvensis

Successful hybridisations between oilseed rape and wild mustard (*Sinapis arvensis*) were only detected when male sterile oilseed rape was used as the female. Six hybrids were obtained from 50.000 flowers (Lefol et al., 1996a). With wild mustard as the female and a transgenic herbicide resistant *B. napus* as pollen donor no hybrids were found among 2.9 million seeds harvested.

CONCLUSION

At present we know that five relatives of oilseed rape may receive transgenes by spontaneous gene flow. However, the cross compatibility of many other wild relatives has never been thoroughly analysed. The fitness of offspring between oilseed rape and a wild relative will of course depend on the genes introgressed to the offspring and the recipient ecosystem. Except for herbicide resistance the fitness advantage provided by transgenic traits introgressed from oilseed rape is poorly known. In a non-selective environment Snow et al. (1998) analysed the fitness of herbicide tolerant BC_3 plants from backcrossing to *B. rapa*. The BC_3 plants also expressed a NPTII gene and a gene for restoration of fertility. As expected there was no cost of expression compared to non-transgenic segregants. Therefore, a transgene providing the wild plants with a selective advantage is likely to increase fitness and speed up the introgression process. With commercialisation of new transgenic oilseed rape lines, spontaneous gene flow may pyramide several advantageous transgenes in the wild recipients and create weedier weeds.

REFERENCES

Chadoeuf R; Darmency H; Maillet J; Renard M (1998). Survival of buried seeds of interspecific hybrids between oilseed rape, hoary mustard and wild radish. *Field Crops Research* **58**, 197-204.

Chèvre A-M; Eber F; Baranger A; Renard M (1997). Gene flow from transgenic crops. *Nature* **389**, 924.

Chèvre A-M; Eber F; Baranger A; Hureau G; Barret P; Picault H; Renard M (1998). Characterization of backcross generations obtained under field conditions from oilseed raps-wild radish F_1 interspecific hybrids: an assessment of transgene dispersal. *Theoretical and Applied Genetics* **97**, 90-98.

Darmency H; Fleury A; Lefol E (1995). Effect of transgenic release on weed biodiversity: Oilseed rape and wild radish. *Brighton Crop Protection Conference- Weeds-*, 433-438.

Ellstrand N C; Prentice H C; Hancock J F (1999). Gene flow and introgression from domesticated plants into their wild relatives. *ARES* **30** (in press).

Frello S; Hansen K R; Jensen J; Jørgensen R B (1995). Inheritance of rapeseed (*Brassica napus*)-specific RAPD markers and a transgene in the cross *B. juncea* x (*B. juncea* x *B. napus*). *Theoretical and Applied Genetics* **91**, 236-241.

Harberd D J (1975). *Brassica* L. In: *Hybridization and the flora of the British Isles*, ed. C A Stace. Academic Press: London.

Hauser T P; Jørgensen R B; Østergård H (1997). Preferential exclusion of hybrids in mixed pollinations between oilseed rape (*Brassica napus*) and weedy *B. campestris* (*Brassicaceae*). *American Journal of Botany* **84**, 756-762.

Hauser T P; Jørgensen R B; Østergård H (1998a). Fitness of backcross and F_2 hybrids between weedy Brassica rapa and oilseed rape (B. napus). *Heredity* **81**, 436-443.

Hauser T P; Shaw R G; Østergård H (1998b). Fitness of F_1 hybrids between *Brassica rapa* and oilseed rape (*B. napus*). *Heredity* **81**, 29-435.

Jørgensen R B; Andersen B (1994). Spontaneous hybridization between oilseed rape (*Brassica napus*) and weedy *B. campestris (Brassicaceae)*. *American Journal of Botany* **81**, 1620-1626.

Jørgensen R B; Andersen B; Landbo L; Mikkelsen T R (1996). Spontaneous hybridization between oilseed rape (*Brassica napus)* and weedy relatives. *Acta Horticulturae* **407**, 193-200.

Jørgensen R B; Andersen B; Hauser T P; Landbo L; Mikkelsen T R; Østergård H (1998). Introgression of crop genes from oilseed rape (*Brassica napus*) to related wild species – an avenue for the escape of engineered genes. *Acta Horticulturae* **459**, 211-217.

Landbo L; Andersen B; Jørgensen R B (1996). Natural hybridisation between oilseed rape and a wild relative: hybrids among seeds from weedy *B. campestris. Hereditas* **125**, 89-91.

Landbo L; Jørgensen R B (1997). Seed germination in weedy *Brassica campestris* and its hybrids with *B. napus*: implications for risk assessment of transgenic oilseed rape. *Euphytica* **97**, 209-216.

Lefol E; Danielou V; Darmency H (1996a). Predicting hybridization between transgenic oilseed rape and wild mustard. *Field Crops Research* **45**, 153-161.

Lefol E; Fleury A; Darmency H (1996b). Gene dispersal from transgenic crops. II Hybridization between oilseed rape and the wild hoary mustard. *Sexual Plant Reproduction* **9**, 189-196.

Metz P L J; Jacobsen E; Nap J P; Pereira A; Stiekema W J (1997). The impact on biosafety of the phosphinothricin-tolerance transgene in inter-specific *B. rapa* x *B. napus* hybrids and their successive backcrosses. *Theoretical and Applied Genetics* **95**, 442-450

Mikkelsen, T R (1996). Flow of engineered genes from crops to wild and weedy relatives – is there an environmental risk? A case study on oilseed rape (*Brassica napus*) and weedy *Brassica campestris. Ph.D. thesis by Thomas R. Mikkelsen, Risø national Laboratory, 1996.*

Mikkelsen T R; Andersen B; Jørgensen R B (1996a). The risk of crop transgene spread. *Nature* **380**, 31.

Mikkelsen T R; Jensen J; Jørgensen R B (1996b). Inheritance of oilseed rape (*Brassica napus*) RAPD markers in a backcross progeny with *Brassica campestris*. *Theoretical and Appied. Genetics* **92**, 492-497.

Scheffler J A; Dale P J (1994). Opportunities for gene transfer from transgenic oilseed rape (*Brassica napus*) to related species. *Transgenic Research* **3**, 263-278.

Scott S E; Wilkinson M J (1998). Transgene risk id low. *Nature* **393**, 320.

Snow A A; Andersen B; Jørgensen, R B (1999). Costs of transgenic herbicide resistance introgressed from *Brassica napus* into *Brassica rapa*. *Molecular Ecology, in press.*

Figure 1. Multiplicative fitness estimate of F_1 hybrids and pure parental species (*B. rapa* and *B. napus*). Modified from Hauser et al. (1998b).

Figure 1. Multiplicative fitness estimate of F_1 hybrids and pure parental species (*B. rapa* and *B. napus*). Modified from Hauser et al. (1998b).

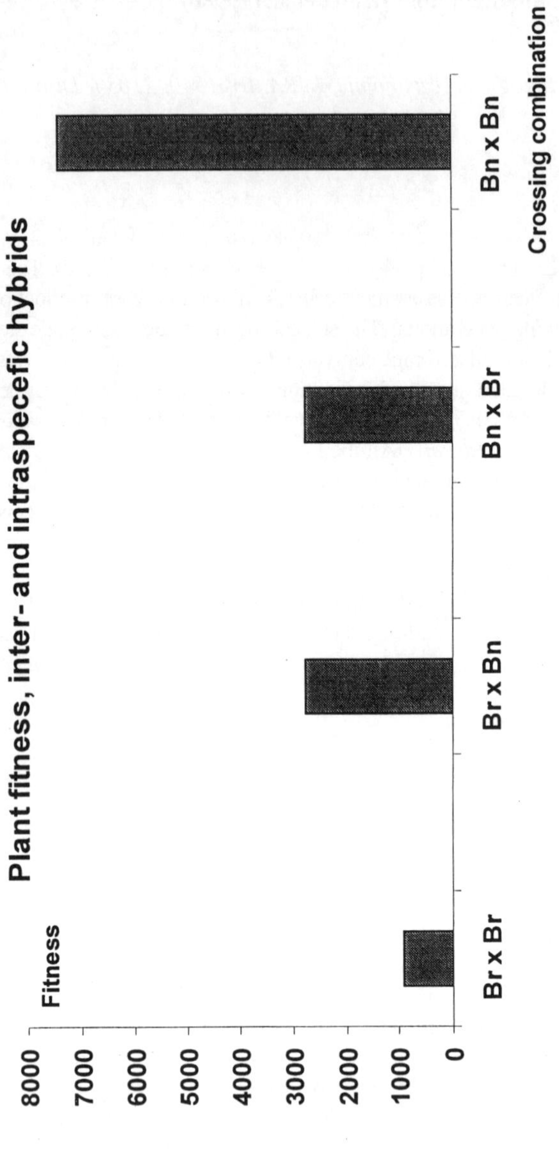
Plant fitness, inter- and intraspecefic hybrids
Fitness
8000
7000
6000
5000
4000
3000
2000
1000
0
Br x Br
Br x Bn
Bn x Br
Bn x Bn
Crossing combination

Gene flow from oilseed rape to weeds

A M Chèvre, F Eber, M Renard
Station d'Amélioration des Plantes, INRA, BP 29, 35650, Le Rheu, France

H Darmency
Unité de Malherbologie et Agronomie, INRA, BV 1540, 21034, Dijon, France

ABSTRACT

Gene flow between transgenic crops and their wild relatives is a key issue for the scientific and public acceptance of genetically modified varieties. The results reported in this paper concern the assessment of F1 interspecific hybrid production and the analysis of the following generations. The subjects of the studies were male sterile or male fertile herbicide tolerant oilseed rape cultivars (*Brassica napus*) and three weedy species largely spread in cultivated areas and presenting an overlapping flowering period. These were hoary mustard (*Hischfeldia incana*), wild mustard (*Sinapsis arvensis*) and wild radish (*Raphanus raphanistrum*). All the F1 interspecific hybrids were obtained except using wild mustard as the female but the rates of production depends on the weed and on the conditions of production. Experiments under normal agronomic conditions revealed a very low frequency of hybrids. The F1 hybrids are vigorous but their low female fertility reduces their fitness. From oilseed rape-wild radish hybrids, we studied the four following generations, under optimal conditions. We observed a reduction of the transgene transmission and of the chromosome number but an increase of the male and female fertility. However, none of the herbicide tolerant plants had the same chromosome number as the weed. Complementary experiments and strategies for reducing the gene flow are presented and discussed.

INTRODUCTION

One of the main concerns linked to the commercial release of herbicide resistant crops (HRC) is the likelihood of the spread of resistance genes to wild relatives of crops. This could create 'super-weeds' and make weed control more difficult. Since agronomists have seldom identified this kind of event in fields, the question is to estimate to what extent this phenomenon occurs naturally and how herbicide resistance can lead to troublesome consequences. It is important also to determine how to reduce the probability of occurrence of agronomic problems. In order to know if interspecific hybridisation and introgression are easy, different aspects have to be studied; (1) the existence of close relatives of the crop growing in sympatry, and the similarity between their phenology to provide overlap of flowering; (2) the nature of the mating system that allows pollen flow and allogamy; (3) the production of F1 interspecific hybrids and their survival; (4) the occurrence of fertile plants in the successive generations; (5) the possibility of chromosome recombination and spread of the gene in populations of the wild species.

There is little data available about gene flow from crops to related species. Evidence of past introgression is limited as it was not studied in detail. The question of gene flow is particularly relevant to oilseed rape (AACC, 2n=38) because this species is 20-40 % allogamous, produces a huge quantity of pollen, 5×10^{12} pollen grain per ha, and outcrossing occurs through wind and insects. Simultaneously, numerous weedy species more or less related to oilseed rape are present within the cultivated areas with an overlapping flowering period, depending on the region and of the cultivars. For instance, early flowering cultivars had flowering time ending at the onset of flowering of wild mustard in Burgundy (France), that means there is no risk of cross-pollination, while late flowering cultivars overlaps for at least 25 % of their flowers with flowering of the weed (Lefol *et al.*, 1996b). An experimental approach is essential to provide the necessary insight into the topic. We show here an example of experimental research in which we try to bring new data and advice for good management at every step of the introgression between oilseed rape and some common *Brassicaceae* weeds.

METHODS AND MATERIALS

Different oilseed rape cultivars were used: either male sterile (Ogu-INRA system) or male fertile with or without a transgene (*bar* or *pat*) which confers the tolerance to the herbicide Basta® (glufosinate). The populations of weedy species, hoary mustard (AdAd, 2n=14), wild mustard (SarSar, 2n=18) and wild radish (RrRr, 2n=18), were locally collected in Rennes and Dijon regions (France).

Field experiments were performed during several years with different ratios of plants according to the optimal or agronomic conditions analysed in Rennes and Dijon. At each generation, the presence of the transgene was detected by herbicide spraying. The chromosome number was assessed by chromosome counting and/or flow cytometry (Eber *et al.*, 1997). The morphology of the plants was observed. The male fertility was assessed as the percentage of pollen stained by an aceto-carmine solution (at least 600 pollen grains/plant) and the female fertility (number of seeds per 100 flowers and per plant) was deduced from harvesting data.

RESULTS

F1 interspecific hybrid production

Between oilseed rape and different weeds

Under optimal conditions, using male sterile oilseed rape as female and the same ratio of crop and weed, the frequency of interspecific hybrids depends on the weed used as pollinator; it was assessed as 0.18, 1.9 and 23.8 interspecific hybrid seeds per 100 flowers using wild mustard, hoary mustard and wild radish as pollinators, respectively (Eber *et al.*, 1994; Chèvre *et al.*, 1996). Where the weed is the female and the oilseed rape the pollinator, a very low frequency of hybrid production was seen with hoary mustard (Lefol *et al.*, 1996a) and wild radish (Darmency *et al.*, 1998). Hybridisation on wild mustard was not detected (Lefol *et al.*, 1996b).

Between different oilseed rape cultivars and wild radish under optimal conditions

Using different male sterile oilseed rape cultivars containing the *bar* gene at the heterozygous stage and wild radish as pollinator, we observed that the F1 interspecific hybrid rates depends on the female parent; they ranged from 1.4 to 100.4 hybrid seeds/100 flowers and from 5.3 to 1213.3 hybrid seeds/plant (Baranger *et al.*, 1995).

Between an oilseed rape cultivar and wild radish under normal agronomic conditions

One hectare of Synergy cultivar glufosinate tolerant (80% of male sterile hybrid with two copies at the hemizygous stage + 20% a pure line with two copies at the homozygous stage) was sown. Wild radish plants were transplanted at different densities:

- within the field, either as isolated plants (1 plant/20m²) with 3 replicates or in groups (100 plants/50m²) with 3 replicates,
- in the border of the field, either as isolated plants (1 plant each 10 metres) or in groups of 40 plants on lines with three replicates,
- in the margin of the field, either as isolated plants (1 plant each 10 metres) or in groups 40 plants on lines with three replicates.

At least 20 metres between each treatment were maintained. Three fields were grown, two in Dijon during two years and one in Rennes. Whatever the location of the wild radish plants, a good overlapping of the flowering period was observed.

Wild radish was harvested plant by plant and the seedlings obtained from the seeds were herbicide treated either under greenhouse or field conditions. Among the total of 189,420 seedlings treated, only one hybrid at 2n=37 (ACRrRr) was detected that was glufosinate resistant. It was harvested on an isolated wild radish plant growing in the margin of the field. With 95% confidence limit, the expected frequency of hybrid plants among the total of observed plants, whatever the location of the wild radish mother plants, ranged from 10^{-7} to 3×10^{-5}.

In the Rennes trial, oilseed rape seed was harvested around the wild radish plants developed within the field or as clusters in the border or the margin of the field. According to our previous results, only the smallest seeds (diameter <1.6mm) were analysed. Two criteria of identification were used. In the greenhouse, the chromosome number assessed by flow cytometry and in the field, the duration of the flowering period, as interspecific hybrids flower longer than oilseed rape because of their female sterility, then the chromosome number of such plants was also checked by flow cytometry. Among seedlings characterised, a total of 23 hybrids were detected; 78% of them had 28 chromosomes (ACRr) but four amphidiploids (2n=56, AACCRrRr) and one 'BC1' like plant (2n=37, ACRrRr) were also obtained. Among the 73,847 plants obtained only from the smallest oilseed rape seeds harvested in the different locations around wild radish plants, the expected frequency of hybrid plants ranged from 2×10^{-5} to 5×10^{-4}, with 95% confidence limit. However, the smallest seeds represented only 7.3 to 12.3% of the total oilseed rape seeds harvested.

Analysis of the F1 interspecific hybrids and of the following generations

From the interspecific hybrids obtained between oilseed rape and hoary mustard

From oilseed rape-hoary mustard interspecific hybrid seeds, we observed that dormancy is intermediate between that of the crop and the weed (Chadoeuf *et al.,* 1998). The morphology of the F1 interspecific hybrid is very close to oilseed rape. This could explain why hybrids were seldom identified in the past. Hybrids are as good as, or even more, competitive than the weed (Lefol *et al.*, 1995), apart that seed output is very low, 0.2 seed per plant on average

(Lefol *et al.*, 1996a). This makes the fitness of hybrids very close to 10^{-6}. This should not be a troublesome problem if good weed management practices were effectively used.

From oilseed rape-wild radish hybrids

From the high number of glufosinate tolerant F1 interspecific hybrids (generation G1) obtained under optimal conditions, the four following generations were produced always with the same field design i.e. the same ratio of hybrids and of wild radish and the harvest was performed on the hybrids. From F1 interspecific hybrids (2n=28), we observed at the following G2 generation an increase of the chromosome number (Figure 1) and of the *bar* gene transmission (Figure 2) because of the efficiency of unreduced gametes (Chèvre *et al.*, 1997; 1998). However, the female fertility remained very low (Figure 2). During the following G3, G4 (Chèvre *et al.*, 1997) and G5 generations, the chromosome number decreased (Figure 1) and at the G5 generation, 91% of the herbicide tolerant hybrids had less than 23 chromosomes. Similarly, the percentage of glufosinate tolerant plants (Figure 2) decreased whereas the male and female (Figure 2) fertility increased. Some herbicide tolerant plants had a female fertility equivalent to wild radish but none of them had 18 chromosomes as wild radish. Their fitness was very low because they showed a chlorophyll deficiency, likely due to an incompatibility between the radish nucleus and the oilseed rape cytoplasm. The reciprocal phenomenon was already described by Pelletier *et al.* (1983).

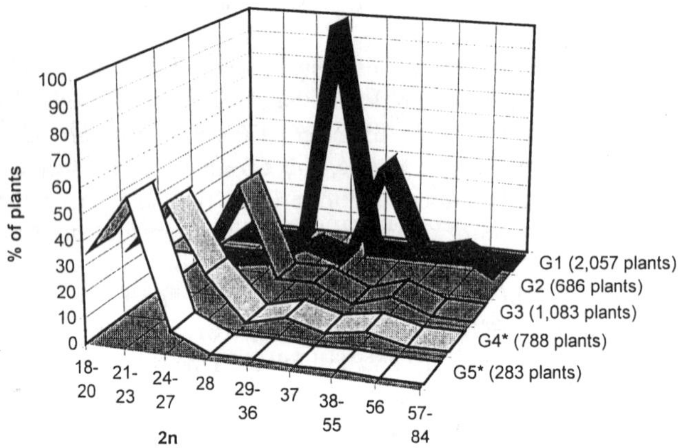

Figure 1. Evolution of the chromosome number along the different generations
* only the herbicide tolerant plants were analysed

DISCUSSION

From our experimental data, we showed that the production of interspecific hybrids between oilseed rape and weeds is possible but depends on; (1) the weed concerned and on the genotype of oilseed rape (Baranger *et al.*, 1995; Chèvre *et al.*, 1996); (2) the female parent since usually more seeds are produced using oilseed rape as female (Kerlan *et al*, 1992), (3) the variability in the weeds (Lefol *et al.*,1996a; Darmency *et al.*,1998); (4) the spatial relationship between the crop and the weed species (Lefol *et al.*,1996a; Darmency *et al.*,1998). These observations are in agreement with the results reported by other authors (Jorgensen & Andersen 1994; Bing *et al.*, 1996; Jorgensen *et al.*, 1998). However, usually, the hybrid production on the weeds and under normal agronomic conditions remains low.

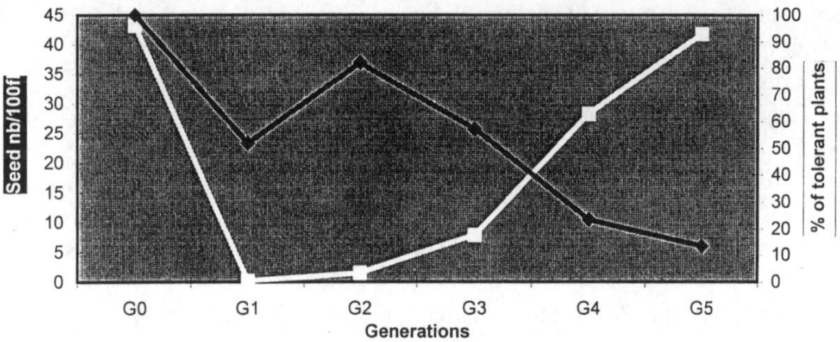

Figure 2. Evolution of the female fertility (white line) assessed by the number of seeds/100 flowers and of the percentage of herbicide tolerant plants (black line) along the different generations.

Triploid F1 interspecific hybrids are vigorous but have a low fertility. However, the female fertility of amphidiploid plants (AACCRrRr, 2n=56) which were obtained under normal agronomic conditions and which had an oilseed rape cytoplasm and a male fertility equivalent to oilseed rape (data not shown) remains to be checked.

The analysis of the following generations from oilseed rape-wild radish F1 interspecific hybrids revealed that, at the fifth generation, none of the herbicide tolerant plants had the same chromosome number as the weed. So, the transgene was not established in the genome of the weed (Chèvre *et al.*, 1997). However, only one insertion event was used to produce this material and our analysis of different specific oilseed rape loci indicated that their transmission rates are different according to the locus (Chèvre *et al.*, 1998). Further analysis is needed to investigate the effect of the initial location of the transgene on gene introgression into the genome of a close related species.

Different models are under study to integrate the different genetic parameters, the spatial and temporal effects using data from small-scale field trials. However, the main difficulty remains the validation of the data from large-scale experiments to get accurate models that could be used for prediction.

In addition, different strategies are under study for reducing the risks of gene flow which include (1) the identification of a potential 'safe insertion site' for the transferred gene, (2) modifications which could reduce the dispersal of pollen (e.g. self-fertilising cultivars) and seeds (e.g. reducing seed loss at harvesting and dormancy), (3) the adaptation of agronomic practices (e.g. management strategies for herbicide tolerant volunteers).

REFERENCES

Baranger A; Chèvre A M; Eber F; Renard M (1995). Effect of oilseed rape genotype on the spontaneous hybridization rate with a weedy species: an assessment of transgene dispersal. *Theoretical and Applied Genetics* **91**, 956-963.

Bing D J; Downey R K; Rakow G F W (1996). Hybridizations among *Brassica napus*, *B. rapa* and *B. juncea* and their two weedy relatives *B. nigra* and *Sinapis arvensis* under open pollination conditions in the field. *Plant Breeding* **115**, 470-473.

Chadoeuf R; Darmency H; Maillet J; Renard M (1998). Survival of buried seeds of interspecific hybrids between oilseed rape, hoary mustard and wild radish. *Field Crops Research* **58**, 197-204.

Chèvre A M; Eber F; Baranger A; Kerlan M C; Barret P; Vallée P; Renard M (1996). Interspecific gene flow as a component of risk assessment for transgenic Brassicas. *Ninth Crucifer genetic workshop*. ISHS. Ed. J S Dias, I Crute, A A Monteiro, *Acta Horticulturae* **407**, 169-179.

Chèvre A M; Eber F; Baranger A; Renard M (1997). Gene flow from transgenic crops. *Nature* **389**, 924.

Chèvre A M; Eber F; Baranger A; Hureau G; Barret P; Picault H; Renard M (1998). Characterization of backcross generations obtained under field conditions from oilseed rape-wild radish F1 interspecific hybrids: an assessment of transgene dispersal. *Theoretical and Applied Genetics* **97**, 90-98.

Darmency H; Lefol E; Fleury A (1998). Spontaneous hybridizations between oilseed rape and wild radish. *Molecular Ecology* **7**, 1467-1473.

Eber F; Chèvre A M; Baranger A; Vallée P; Tanguy X; Renard M (1994). Spontaneous hybridization between a male sterile oilseed rape and two weeds. *Theoretical and Applied Genetics* **88**, 362-368.

Eber F; Letanneur J C; Chèvre A M (1997). Chromosome number of oilseed rape (*Brassica napus*)-wild radish (*Raphanus raphanistrum*) spontaneous hybrids and of their progeny estimated by flow cytometry. *Cruciferae Newsletter* **19**, 17-18.

Jorgensen R B; Andersen B (1994). Spontaneous hybridization between oilseed rape (*Brassica napus*) and a weedy *B. campestris* (*Brassicaceae*): a risk of growing genetically modified oilseed rape. *American Journal of Botany* **81** (12), 1620-1626.

Jorgensen R B; Andersen B; Hauser T P; Landbo L; Mikkelsen T R; Ostergard H (1998). Introgression of crop genes from oilseed rape (*Brassica napus*) to related wild species - an avenue for the escape of engineered genes. *Brassica 97*. International symposium on Brassicas. ISHS. Ed. G Thomas, A Monteiro, *Acta Horticulturae* **459**, 211-217.

Kerlan M C; Chèvre A M; Eber F; Baranger A; Renard M (1992). Risk assesssment of outcrossing of transgenic rapeseed to related species: I. Interspecific hybrid production under optimal conditions with emphasis on pollination and fertilization. *Euphytica* **62**, 145-153.

Lefol E; Danielou V; Darmency H; Boucher F; Maillet J; Renard M (1995) Gene dispersal from transgenic crops: I. Growth of interspecific hybrids between oilseed rape and the wild hoary mustard. *Journal of Applied Ecology* **32,** 803-808.

Lefol E; Fleury A; Darmency H (1996a). Gene dispersal from transgenic crops: II. Hybridization between oilseed rape and the wild hoary mustard. *Sexual Plant Reproduction* **9**, 189-196.

Lefol E; Danielou V; Darmency H (1996b). Predicting hybridization between transgenic oilseed rape and wild mustard. *Field Crops Research* **45**, 153-161.

Pelletier G; Primard C; Vedel F; Chétrit P; Remy R; Rousselle P; Renard M (1983). Intergeneric cytoplasmic hybridization in *Cruciferae* by protoplast fusion. *Molecular Genetics* **191**, 244-250.

Gene flow from transgenic canola to wild radish - a model system to determine the risks

M A Rieger, C Preston
The Co-operative Research Centre for Weed Management Systems, University of Adelaide PMB 1 Glen Osmond, 5064 Australia

T Potter
South Australian Research and Development Institute, Struan Research Station, Naracoorte 5271, Australia

S B Powles
WA Herbicide Resistance Initiative, University of Western Australia, Nedlands, Perth 6907, Australia

ABSTRACT

Under Australian field conditions, gene flow between canola (*Brassica napus*) and a related weed species, wild radish (*Raphanus raphanistrum*) was examined. A model system without transgenic material but utilising herbicide resistance was employed. In the first year of this study a total of five putative hybrids have been discovered. The morphology, pollen fertility and seed set of these individuals is discussed.

INTRODUCTION

Crops such as wheat, barley and grain legumes dominate the Southern Australian broad-acre cropping system. Canola (*Brassica napus*) is a more recent introduction into this system and plantings have steadily increased in area over the last five years from 107,000 ha in 1993 to over 1 million ha in 1998. In some areas, canola is replacing grain legumes as a break crop in cereal rotations. Canola has been introduced into a farming system with a background of extensive herbicide resistant weed populations. The most prominent among these weeds is annual rye-grass (*Lolium rigidium*). This highly adaptable weed now has populations resistant to most herbicides, including glyphosate, used in Australian cropping systems. Another major weed is wild radish (*Raphanus raphanistrum*). Populations of wild radish resistant to acetolactate synthase (ALS) herbicides have recently been documented. These two weed threats have led to the widespread adoption of triazine tolerant canola varieties, which account for nearly 50% of all canola plantings. These lines contain a triazine-resistance gene from a weedy relative (*Brassica rapa*) that was crossed into canola.

The high adoption rate of triazine-tolerant canola varieties over the past three years demonstrates that growers are receptive to this technology despite the penalties of lower yields and oil contents. It is into this environment that other herbicide-resistant canola varieties will be introduced into Australia over the next few years. Imidazolinone tolerant canola should be introduced in 2000 or 2001, closely followed by the transgenic glufosinate resistant and glyphosate resistant varieties.

A major concern with canola is its ability to out-cross to weedy relatives that infest canola crops. For this reason, there has been intense research interest investigating the possibility of transgenic herbicide-resistant genes flowing from canola to weedy relatives (Jorgensen & Andersen, 1994; Brown & Brown, 1996; Mikkelsen *et al.*, 1996a & 1996b). By and large, much of this research has failed to mimic normal cropping situations and has often used male-sterile canola varieties (Eber *et al.*, 1994; Baranger *et al.*, 1995; Darmency *et al.*, 1995; Lefol *et al.*, 1997). The use of male-sterile varieties increases that probability of out-crossing by the elimination of competition from canola pollen. With male-fertile canola it is more difficult to determine hybrid production; however, this mimics the field situation more accurately. Further there is a need to incorporate weeds at realistic field densities to ensure the rates of hybridisation are not overestimated. This paper will focus on the likelihood of hybridisation from canola to wild radish with experimental "in-crop" field trials and using a male-fertile variety of canola.

METHODS AND MATERIALS

Field trials located at Naracoorte (South Australia) commenced in autumn 1997 and will continue over two consecutive seasons. No transgenic material was used in the field trials due to regulatory constraints. In the absence of transgenic material, an alternative marker system, resistance to sulfonlyurea and imidazolinone herbicides, was utilised. Gene flow was also examined in both directions, from canola to wild radish and from wild radish to canola.

Direction a: gene flow from canola to wild radish

A field was planted to traditionally bred imidazoline-resistant canola (cultivar 45A71) and subsequently inter-planted with susceptible wild radish at one and four plants/m^2. There were a total of eight 10m x 10m plots and wild radish seeds were hand collected from plants prior to crop harvest.

Direction b: gene flow from wild radish to canola

A field was planted to ALS-herbicide susceptible canola (cultivar Karoo) and inter-planted with ALS-resistant wild radish plants at one and four plants/m^2. There were fifteen 10m x 10m plots. After flowering was completed the resistant wild radish plants were removed. The canola seed was harvested mechanically at the regular harvest time.

Screening of the wild radish seed - direction a

The wild radish seeds collected from all plots were kept separate for each individual plant. Wild radish seeds are known to be highly dormant and were pre-treated to break dormancy. This pre-treatment involved a 10 minute soak in 6% sodium hypochlorite with an additional period soaking in water. Pre-treated seeds were then germinated on 1.6% water agar with 5×10^{-5} M ALS-inhibiting herbicide (imazethapyr) incorporated. Seeds were allowed to germinate at 25°C/15°C in an incubator for a period of three weeks. Assessment of root growth occurred after this period. The roots of susceptible wild radish are severely stunted if seed is germinated in the presence of this concentration of herbicide. Plant with roots >10 mm long were classified as resistant and transplanted to soil (screen one). Once the survivors of screen one had reached the two/three leaf stage they were treated with 48 g/ha imazethapyr *via* a belt driven laboratory spray unit. Survival was assessed at approximately one week after spray application (screen two).

Screening of the canola seed - direction b

Canola seed was machine harvested and separated for each of the 15 plots. In the following season the canola seed was planted out in plots and allowed to germinate. The number of plants per plot was determined based on the area of each plot and counting the number of plants present in twenty 10 x 10 cm quadrats. Once these plants had reached the two/three leaf stage the area was sprayed with two applications of 7.5 g/ha chlorsulfuron in alternate directions to ensure even coverage. A week later an additional application of 15 g/ha of chlorsulfuron was made. These plots were then left for a period of three weeks to allow the herbicide-susceptible seedlings to die (screen one). Surviving plants in these plots were transplanted to pots and two weeks after transplantation were again treated with 15 g/ha of chlorsulfuron (screen two).

Hybrid confirmation

A suite of verification methods will be used for hybrid confirmation including observations of morphology, flowering and cross fertility along with the use of molecular techniques. Molecular techniques will use the presence of a mutation endowing herbicide resistance in the ALS gene as the main identification tool. If the mutations within ALS are identical to those of the putative resistant parent species, other parts of the ALS gene will be examined to determine whether it came from the resistant parent.

RESULTS

Direction a

A total of 38,635 wild radish seeds were processed through screen one. Of these 19,036 (49.3%) seeds germinated and 5,631 (14.6%) were transplanted to soil. A total of 6 plants survived the transplantation step and after screen two, 2 individuals (A1, A2) remained. These two putative hybrids did not differ significantly in morphology from canola. Pollen viability was examined using the fluoresceine diacetate method. Three separate flowers were examined per plant, and at least 500 pollen grains counted per flower (Table 1). Both hybrids produced pods when crossed with wild radish and canola (Table 2).

Direction b

An estimated number of 51 million seeds were processed through screen one. Two thousand five hundred and seventy plants survived this first screen and were subsequently transplanted into pots. Most of the plants surviving screen one were visually identical with canola, but three plants (B1, B2, B3) were morphologically different. These plants had many characteristics in common with wild radish including differences in leaf shape, hairiness of leaves, flower structure, and pod constriction. These putative hybrids clearly demonstrated morphology that was intermediate between wild radish and canola. Only these three plants survived screen two. The field where screen one had been performed was subsequently sown to wheat. After two months seed from further 86 plants were collected from this field and will be screened for potential herbicide-resistance. The fluoresceine diacetate method was used to examine pollen viability of these potential hybrid plants (Table 1). All putative hybrids produced pods when crossed with both canola and wild radish (Table 2).

Table 1. The percent pollen viability of putative hybrids based on three replicates.

	% viability			% overall
A1	49	81	78	70
A2	51.5	27	52	43.5
B1	39	60	63	63
B2	53	66	69	63
B3	30	80	83	64
Canola	59	56	59	58
Wild radish	73.5	76	63	71

Table 2. The number of pods produced by putative hybrids after crossing with canola and wild radish pollen. Number of selfed pods produced also shown.

	Self	x canola	x wild radish
A1	27	44	16
A2	69	12	62
B1	88	89	55
B2	41	12	42
B3	24	9	5

DISCUSSION

The potential hybrids produced by the first year of this field trial under Australian conditions indeed suggest that hybrid production between canola and wild radish is possible. The implications of these results are two-fold. Firstly based on research conducted into the evolution of herbicide resistance, the mutation rate one would expect for ALS-resistance is 1 x 10^{-6} (Jasieniuk *et. al.*, 1996). Numbers from these initial trials suggest that the rates of hybridisation *via* gene flow are lower than selection for herbicide resistance through a mutation. This would indicate that weed species would be more likely to develop herbicide resistance *via* a mutation before hybridisation. Secondly if the plants produced by these field trials are in fact hybrids, two facets of their biology needs to be considered. All the potential hybrids produced by this study are reasonably fertile and all produced selfed seeds. The implication of this could be that once hybrids have been established through gene flow, spread of these individuals would be rapid. Instead of only one mechanism of weeds developing herbicide resistance, mutations through selection pressure, there may now be a slower but no less important mechanism, hybridisation through gene flow.

The methodology used during this trial has proved quite successful. Two years of these trials have now been conducted and only minor modifications were necessary in the second year. These modifications included increasing the number of replicates per treatment, modifying the method of collection of the wild radish seed and the addition of fungicide to avoid unnecessary seedling loss. Due to the success of this method additional trials have been planned. These future trials are to examine hybridisation between canola and several other important weed species in South Australia, namely *Brassica tournefortii* and *Diplotaxis murialis*.

ACKNOWLEDGEMENTS

We would like to thank Mr. Hugh Pederson, Miss Lynley Stone and Mr. Peter Crisp for technical and field assistance.

REFERENCES

Baranger A; Chevre A M; Eber F; Renard M (1995). Effect of oilseed rape genotype on the spontaneous hybridisation rate with a weedy species - an assessment of transgenic dispersal. *Theoretical and Applied Genetics* **91**, 956-963.

Brown J; Brown A P (1996). Gene transfer between canola (*Brassica napus* L. and *B. campestris* L.) and related weed species. *Annals of Applied Biology* **129**, 513-522.

Darmency H; Fleury A; Lefol E (1995). Effect of transgenic release on weed biodiversity: oilseed rape and wild radish. Brighton Crop Protection Conference - Weeds **2**, 433-8.

Eber F; Tanguy X; Chevre A M; Baranger A; Vallee P; Renard M (1994). Spontaneous hybridisation between a male sterile oilseed rape and two weeds. *Theoretical and Applied Genetics* **88**, 362-368.

Jasieniuk M; Brule-Babel A L; Morrison I N (1996). The evolution of genetics of herbicide resistance in weeds. *Weed Science* **44**, 176-193

Jorgensen R B; Andersen B (1994). Spontaneous hybridisation between oilseed rape (*Brassica napus*) and weedy *Brassica rapa* (*Brassiceae*) - a risk of growing genetically modified oilseed rape. *American Journal of Botany* **81**, 1620-1626.

Lefol E; Seguin-Swartz G; Downey K (1997). Sexual hybridisation in crosses of cultivated *Brassica* species with the crucifers *Erucastrum gallicum* and *Raphanus raphanistrum*: potential for gene introgression. *Euphytica* **95**, 127-139.

Mikkelsen T R; Andersen B; Jorgensen R B (1996*a*). The risk of transgenic crop spread. *Nature* **330**, 31.
Mikkelsen T R; Jensen J; Jorgensen R B (1996*b*). Inheritance of oilseed rape (*Brassica napus*) RAPD markers in a backcross progeny with *Brassica rapa*. *Theoretical and Applied Genetics* **92**, 492-497.

Fitness costs associated with transgenic glufosinate tolerance introgressed from *Brassica napus* ssp *oleifera* (oilseed rape) into weedy *Brassica rapa*

A A Snow
Department of Evolution, Ecology, and Organismal Biology, The Ohio State University, Columbus, OH, 43210 - 1293, USA

R Bagger Jørgensen
Plant Biology and Biogeochemistry Department, PBK-301, Risø National Laboratory, DK-4000 Roskilde, Denmark

ABSTRACT

Wild relatives of genetically engineered crops can acquire transgenic traits such as herbicide tolerance *via* spontaneous crop-wild hybridisation. The fitness costs associated with a transgene conferring tolerance of glufosinate were tested when introgressed into weedy *Brassica rapa*. Crosses were made between transgenic *Brassica napus* ssp *oleifera* (oilseed rape) and wild *B. rapa* from Denmark. The reproductive success of 457 BC_3 progeny representing six full-sib families raised in growth rooms was quantified (plants were pollinated by captive bumblebees). Segregation for herbicide tolerance was 1:1 overall, as expected for a dominant Mendelian trait. There were no significant differences between transgenic and non-transgenic plants in survival or the number of seeds per plant, indicating that fitness costs associated with the transgene are likely to be negligible. Pollen fertility and seed production of BC_3 plants were as great as those of *B. rapa* raised in the same growth rooms. These results suggest that transgenic herbicide tolerance is capable of introgressing into populations of *B. rapa* and persisting, even in the absence of selection due to applications of this herbicide.

INTRODUCTION

A widely acknowledged drawback of genetically engineered crops is the potential for transgenes to spread to related weeds *via* crop-weed hybridisation (e.g. Ellstrand & Hoffman, 1992; Raybould & Gray, 1993). Crops that are known to interbreed with weedy relatives under field conditions include rice, sorghum, squash, oilseed rape (canola), sunflower, sugar beet, carrot, strawberry, radish, wheat, oats, and others (Snow & Morán Palma, 1997; Zemetra *et al.*, 1997). Most of these economically important crops have been targeted for improvement by the addition of transgenic traits and several have already been released commercially, including herbicide tolerant oilseed rape. In the short term, the most obvious problem cases will likely be weeds that acquire transgenic tolerance of widely used, broad-spectrum herbicides such as glyphosate and glufosinate (Dyer, 1994). With the exception of *Lolium rigidum* in Australia (Powles *et al.*, 1998), repeated use of these particular herbicides has not yet led to the evolution of tolerant weed genotypes (Bradshaw *et al.*, 1997; Heap, 1997), as often occurs with other types of herbicides. Now, however, weedy taxa that hybridise with transgenic crops will be able to acquire tolerance of glyphosate and glufosinate following crop-to-wild gene flow.

Once transgenes conferring herbicide tolerance move into weeds, their frequency within local populations will be influenced by; 1) continued immigration due to gene flow from cultivated plants; 2) strong selection favouring herbicide-tolerant genotypes during periods of exposure to the herbicide; 3) in the absence of the herbicide, the possibility of selection *against* these genotypes if the transgene is associated with reduced fitness. Such costs could be caused by pleiotropy, physiological costs of the tolerance trait or effects of particular insertion sites within the genome (such as linkage to deleterious alleles or disruption of coding regions; Bergelson *et al.*, 1996). In commercialised transgenic crops, the authors expect that any costs of herbicide tolerance will be minimal due to economic incentives for developing high-yielding, herbicide-tolerant varieties. These costs could be different in weeds than in crops, however, due to factors such as different genetic backgrounds, genotype-by-environment interactions or crop genes that are tightly linked to the transgene insertion site and are deleterious to the weed but not the crop.

The purpose of this study was to test for fitness costs associated with transgenic tolerance of glufosinate in a widespread weed, *Brassica rapa* L. (= *B. campestris*), which hybridises with oilseed rape. The study focused on glufosinate tolerant oilseed rape as one of the first crops that could affect the abundance of an agricultural weed (as a result of crop-wild hybridisation followed by the use of this type of herbicide). A secondary goal of the study was to characterise the seed production of selected backcrossed plants (the BC_3 generation with *B. rapa* as the recurrent parent) as compared to the seed production of pure *B. rapa* in order to detect major abnormalities, if any, resulting from introgression. A more detailed description of the study can be found in Snow *et al.* (1999).

Wild *B. rapa*, commonly known as field mustard or bird rape, is an economically important weed in temperate regions of Eurasia, North America, South Africa, Australia, and New Zealand (Holm *et al.*, 1997) and is often seen in fields of oilseed rape (e.g., Jørgensen & Andersen, 1994). *Brassica napus* ssp *oleifera* (oilseed rape) is an ancient allotetraploid possessing the AA genome from *B. rapa* (2n=20) as well as the CC genome from *B. oleracea* (2n=18; Song *et al.*, 1995). Thus, F_1 hybrids between *B. napus* and *B. rapa* have a triploid AAC genome constitution of 2n=29. Because they have a full complement of the *B. rapa* genome, fertility of some F_1 progeny can be almost as high as that of pure *B. rapa* (Jørgensen & Andersen, 1994; Jørgensen *et al.*, 1996; Hauser *et al.*, 1997).

Crop-to-wild hybridisation is facilitated by the fact that *B. rapa* is an obligate outcrosser that relies on bees, other insects, and possibly wind to effect pollen transfer (Jørgensen & Andersen, 1994). Pollen from fields of *B. napus* has been detected at distances of up to 1.5 km away from the crop (e.g., Timmons *et al.*; also see Scott & Wilkinson, 1998) and further gene movement *via* human-mediated seed dispersal may be common. Spontaneous backcrossing to the weed has been demonstrated in field experiments involving both DNA markers and transgenic tolerance of glufosinate (Jørgensen *et al.*, 1996; Mikklesen *et al.*, 1996).

METHODS AND MATERIALS

Transgenic BC_2 plants were used, described in Mikkelsen *et al.* (1996), to create a BC_3 generation segregating for the presence or absence of three tightly linked transgenes: the *bar* gene for glufosinate tolerance, the *neo* gene coding for NPTII (conferring resistance to kanamycin), and the *barstar* gene, which restores male fertility. This multiple-gene construct has no detectable effects on the yield or competitive ability of Drakkar oilseed rape (Fredshavn *et al.,* 1995). The crosses made to obtain the BC_3 generation involved varying numbers of indoor-grown plants, depending on their availability at times when crosses were conducted (see Snow *et al.*, 1999). First, *B. rapa* plants from a wild population in Denmark were crossed with transgenic *B. napus*. Transgenic F_1 progeny were then crossed with *B. rapa*, and transgenic BC_1 plants were crossed with *B. rapa*. Six BC_2 progeny were each crossed with a different *B. rapa* plant. Thus, the BC_3 progeny used in this study were divided into six full-sib families: 1-3, 1-4, 1-10, 1-12, 1-13, and 9-14, with the first number denoting the seed's "maternal grandmother" and the second number referring to the *B. rapa* paternal parent from the wild population (each full-sib family had a unique maternal and paternal parent).

To test for costs associated with glufosinate tolerance, BC_3 plants were grown that segregated for the presence or absence of glufosinate tolerance and recorded the total number of fruits and seeds produced by each plant. Progeny from each family were grown under the same environmental conditions, but due to lack of space and temporary technical problems with two growth rooms, it was not possible to put all of the families in the same growth room or to distribute them equally among rooms. Therefore, seed production differences among families cannot be distinguished from differences due to environmental conditions of the growth rooms because these two factors were confounded. All plants within each family were treated identically, however, so that effects of being transgenic were tested within each family and the entire data set was analysed by nesting families within growth rooms for analyses of variance. *B. rapa* plants from a wild population in Denmark were also grown to serve as a reference for general comparisons with the BC_3 generation.

Although light levels in the growth rooms were lower than full sunlight, the plants showed no symptoms of light deprivation and attained sizes similar to those of field-grown plants. When flowering began, a hive of bumblebees was placed in each room to allow semi-natural pollination of these obligately outcrossing plants (Biobest greenhouse pollination kits, Westerlo, Belgium). During "daylight" hours, c. 10-15 bees foraged for pollen and nectar at any given time in each room, resulting in fruit set of c. 100%. The pollen fertility was characterised by using methods described in Snow *et al.* (1999). When the plants stopped flowering, the number of fruits per plant was counted and 10 mature, undehisced, randomly selected fruits were collected from each plant for seed counts. The average number of seeds per plant was then calculated (number of fruits x average number of seeds per fruit). Transgenic tolerance of glufosinate was scored by a PCR method or by leaf dot tests (Snow *et al.*, 1999). When these two methods were compared in a sample of 74 plants, the same result was obtained in 99% of the samples (all but one), leading us to conclude that both methods give the same results.

RESULTS

Overall, 51% of the BC_3 seedlings that were herbicide-tolerant, which is very close to the expected Mendelian 50% (N= 457). The proportion of seedlings that survived to maturity and produced at least one fruit was greater than 90% in all families and was not affected by whether the plant was transgenic (Chi-square tests). Survivorship of pure *B. rapa* was >95%. Analysis of variance showed that the total number of seeds per plant was affected by family and growth room, but was not influenced by whether the plants were transgenic (Figure 1; Snow *et al.*, 1999). Therefore, based on this measure of overall performance, no cost of transgenic tolerance of glufosinate was detected. Transgenic BC_3 plants produced fewer fruits per plant than non-transgenic plants in two of the six families (Figure 1), but the authors regard total seed output to be a more relevant component of overall fitness than fruit number. A nested ANOVA showed that neither trangenes (+ or -) nor the transgene-by-family interaction had a significant effect on the number of seeds produced per plant.

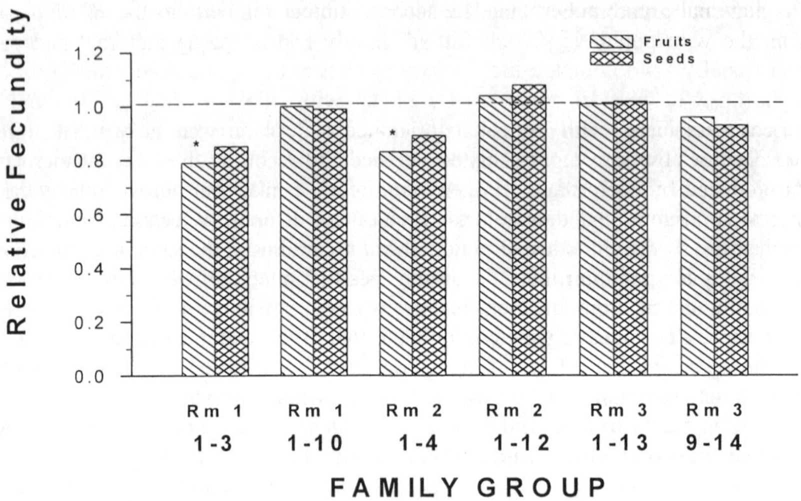

Figure 1. Relative lifetime fecundity of transgenic BC_3 progeny, calculated as the average number of seeds or fruits of transgenic plants divided by the corresponding mean for non-transgenic plants. Dotted line illustrates the expectation for equal fecundity. T-tests were used to compare fruit or seed production of transgenic *versus* non-transgenic plants within families; sample sizes were 29-51 per treatment within each family; * indicates $P<0.05$.

In comparing the fecundity of the BC_3 generation with that of pure *B. rapa*, it is important to note that the latter represent a more diverse mixture of genotypes and may have differed from the relatively small group of *B. rapa* plants involved in these crosses. Pollen fertility of plants in the BC_3 families averaged 88 - 95% and was not significantly different from the pollen fertility of *B. rapa* (92%, ANOVA, details not shown). The mean number of seeds per plant varied among BC_3 families and was higher than or similar to that of *B. rapa* plants from the same growth room (Snow *et al.*, 1999). Based on these comparisons, it is concluded that the lifetime fecundity of BC_3 plants was similar to that a mixture of pure *B. rapa* genotypes.

DISCUSSION

This study confirmed that transgenic herbicide tolerance was transmitted to the BC_3 generation at an average frequency of c. 50%, as expected for a dominant Mendelian trait. In general, costs associated with transgenic glufosinate tolerance were small or undetectable. It is concluded that under the conditions of this experiment, the costs associated with transgenic herbicide tolerance were negligible. It would be useful to carry out further studies in the field, with more plant families and several levels of competition, to determine whether any costs of transgenic glufosinate tolerance can be detected under natural conditions.

A secondary goal was to compare the survival, pollen fertility, and seed production of BC_3 plants with those of pure *B. rapa*. BC_1 plants with >95% pollen fertility and 20-21 chromosomes were intentionally selected for the subsequent crosses in order to utilise the types of individuals that are most likely to succeed in passing crop genes on to successive generations. Thus, as expected, the BC_3 generation also had high pollen fertility and high survival. One surprising result was that the seed output of these BC_3 families was similar to or greater than the seed production of pure *B. rapa* grown under the same conditions. If this pattern also holds for plants in the field, it suggests that after crop alleles have introgressed into *B. rapa* for a few generations, those that are not strongly deleterious and are not linked to other deleterious traits would be able to persist for many subsequent generations.

Although the conclusions are limited to a particular transgenic line of oilseed rape, the results support the widespread assumption that once a transgenic trait such as glufosinate tolerance has been selected for marketing, it is not likely to entail a strong fitness disadvantage when transferred to weedy *B. rapa* populations. Clearly, there is some urgency for knowing more about this particular weed-crop complex because *B. rapa* is already a serious weed of more than 20 crops in more than 50 countries (Holm *et al.*, 1997) and glufosinate-tolerant oilseed rape is now being grown commercially. If further field experiments also demonstrate little or no costs associated with transgenic tolerance of glufosinate, it is possible that the spread of this transgene to natural populations of *B. rapa* will lead to infestations of this species that are more difficult to control. This will be especially significant in fields of transgenic oilseed rape where glufosinate applications are sometimes used as a primary control method.

In the near future, it is expected that crops such as oilseed rape will possess additional fitness-related transgenes such as those conferring tolerance of lepidopteran pests, fungal diseases, and other types of herbicides such as glyphosate. Further research on ecological implications for weedy relatives should focus on the benefits as well as possible costs associated with *multiple* fitness-related traits. Within the next decade, it may be possible to limit crop-to-wild gene flow using crops with a suicide trait known as a "technology protection system" (sometimes referred to as the Terminator concept), but at present this controversial technology is still under development.

ACKNOWLEDGEMENTS

Plant Genetic Systems for the transgenic oilseed rape seeds, and B Andersen and E Ibsen for technical assistance and the support by the Danish Nature and Forestry Agency, the Novo Nordisk Foundation, and Risø's Department of Plant Biology and Biogeochemistry.

REFERENCES

Bergelson J; Purrington C B; Palm C J; López-Gutiérrez J C (1996). Costs of resistance: a test using transgenic *Arabidopsis thaliana*. *Proceedings of the Royal Society of London* **263**,1659-1663.

Bradshaw L D; Padgette S R; Kimball S L; Wells B H (1997). Perspectives on glyphosate resistance. *Weed Technology* **11**, 89-198.

Dyer W E (1994). Herbicide-resistant crops: a weed scientist's perspective. *Phytoprotection* **75** (Suppl.), 71-77.

Ellstrand N C; Hoffman C A (1992). Hybridization as an avenue of escape for engineered genes. *BioScience* **40**, 438-442.

Fredshavn J R; Poulsen G S; Huybrechts I; Ruselsheim P (1995). Competitiveness of transgenic oilseed rape. *Transgenic Research* **4**, 142-148.

Hauser T P; Jørgensen R B; Østergård H (1997). Preferential exclusion of hybrids in mixed pollinations between oilseed rape (*Brassica napus*) and weedy *B. campestris*. *American Journal of Botany* **84**, 756-762.

Heap I (1997). The occurrence of herbicide-resistant weeds worldwide. *Pesticide Science* **51**, 235-243.

Holm L G; Doll J; Holm E; Pancho J; Herberger J (1997). *World weeds: natural histories and distributions.* John Wiley and Sons, Inc., New York.

Jørgensen R B; Andersen B (1994). Spontaneous hybridization between oilseed rape (*Brassica napus*) and weedy *B. campestris* (*Brassicaceae*): a risk of growing genetically modified oilseed rape. *American Journal of Botany* **81**, 1620-1626.

Jørgensen R B; Andersen B; Landbo L; Mikkelsen T R (1996). Spontaneous hybridization between oilseed rape (*Brassica napus*) and weedy relatives. *Acta Horticulturae* **407**, 193-200.

Mikkelsen T R; Andersen B; Jørgensen R B (1996). The risk of crop transgene spread. *Nature* **380**, 31.

Powles S B; Lorraine-Colwill D F; Dellow J J; Preston C (1998). Evolved resistance to glyphosate in rigid ryegrass (*Lolium rigidum*) in Australia. *Weed Science* **46**, 604-607.

Raybould A F; Gray A J (1993). Genetically modified crops and hybridization with wild relatives: a UK perspective. *Journal of Applied Ecology* **30**, 199-219.

Scott S E; Wilkinson M J (1998). Transgene risk is low. *Nature* **393**, 320.

Snow A A; Andersen B; Jørgensen R B (1999). Costs of herbicide resistance introgressed from *Brassica napus* into weedy *Brassica rapa*. *Molecular Ecology, in press*.

Snow A A; Morán Palma P (1997). Commercialization of transgenic plants: potential ecological risks. *BioScience* **47**, 86-96.

Song K; Lu P; Tang K; Osborn T C (1995). Rapid genome change in synthetic polyploids of *Brassica* and its implications for polyploid evolution. *Proceedings of the National Academy of Sciences, USA* **92**, 7719-7723.

Timmons A M; Obrien E T; Charters Y M; Dubbels S J; Wilkinson M J (1995). Assessing the risks of wind pollination from fields of genetically modified *Brassica napus* ssp. *oleifera*. *Euphytica* **85**, 417-423.

Zemetra R S; Hansen J; Mallory-Smith C A (1997). Potential for gene transfer between wheat (*Triticum aestivum*) and jointed goatgrass (*Aegilops cylindrica*). *Weed Science* **46**, 313-317.

Gene flow from oilseed rape to *Sinapis arvensis*: Variation at the population level

C L Moyes, J Lilley, C Casais, P J Dale
Brassica and Oilseeds Research Department, John Innes Centre, Colney Lane, Norwich, NR4 7UH, UK

ABSTRACT

Sinapis arvensis is a widespread weed species which is controlled by the use of herbicides. If oilseed rape crops containing transgenes for herbicide resistence are to be released, it is important to consider the consequences for weed control. Previous work on sexual compatibility between these two species has given variable results. The aim of this work is to investigate the likelihood of transgenes escaping from rape into *S. arvensis* populations. This paper reports the preliminary findings from crosses carried out between a transgenic rape line and *S. arvensis* populations collected from around the UK and France.

INTRODUCTION

As part of the risk assessment for genetically modified crops it is important to consider the species with which the crop will hybridise. Sexual compatibility between oilseed rape (*Brassica napus*) and related species has been reviewed (Scheffler & Dale, 1994). *Sinapis arvensis* was not considered to be sexually compatible because the majority of studies found that hybridisation between *S. arvensis* and *Brassica napus* is not possible without embryo rescue or ovule culture (Bing *et al.*, 1995; Bing *et al.*, 1996a; Bing *et al.*, 1996b; Brown & Brown, 1996; Kerlan *et al.*, 1992). However, Chevre *et al.* (1996) and Lefol *et al.* (1996) have found hybrids can be produced under open pollination conditions.

S. arvensis is a very widespread weed found in disturbed sites (Fogg, 1950). This means that it is frequently found growing in cultivated areas and was a serious weed before control by herbicides was achieved (Rich, 1991; Edwards, 1980). Herbicide tolerant oilseed rape may be one of the first GM crops to be released on a commercial scale.

For these reasons it is important to clarify the ability of *S. arvensis* to hybridise with oilseed rape. The differences found between the previous studies may reflect geographical variation in hybridisation ability so seed was collected from populations across the UK and from France where hybrids were previously found (Chevre *et al.*, 1996). This paper describes the preliminary work on the sexual compatibility of these populations with oilseed rape.

METHODS AND MATERIALS

Seed were collected from all populations of *S. arvensis* encountered when travelling around the UK in September/October 1997. Seed from French populations were supplied by A. Chevre (INRA, Rennes).

The oilseed rape plants used as pollen donors were Westar 10 transformed with a construct containing the BAR and NPTII genes conferring resistance to glufosinate ammonia and kanamycin, respectively. The plants were hemizygous for the transgene. Oilseed rape plants used as female parents were from the same line backcrossed to Westar 10, therefore, half of these plants contained the transgene.

Seed from 102 *S. arvensis* populations were grown and two plants from each population were used for the pollinations. For each plant, 30 buds were emasculated just before opening and pollinated with rape pollen. Each plant was also used to pollinate 20 emasculated rape buds. In addition, 16 plants were pollinated with other members of the same species to give a figure for the maximum number of seed obtainable under the experimental conditions. The pods were allowed to develop naturally in the glasshouse and seed harvested after 8 weeks for *S. arvensis* and 6 weeks for rape.

To date half of the seed have been sown. The resulting plants have been tested for presence of the transgenes by PCR using primers for BAR and NPTII. Leaf discs were cultured in the presence of the herbicide to test for glufosinate tolerance and to back up the PCR results.

RESULTS

Table 1 gives the location of the populations sampled. Dumfries and Galloway, Lancashire, Cheshire, Gwent, Somerset, Surrey, W. Glamorgan, Mid Glamorgan, S. Glamorgan, Devon and Cornwall were not visited. No *S. arvensis* populations were found in Gwynedd, Clwyd, Powys, Staffordshire, Derbyshire, Leicestershire, Hertfordshire, Essex, Avon or Hampshire

Table 1. The location of each population sampled and the number of seed produced in the interspecific crosses. The mean seed number is given for crosses involving two *S. arvensis* plants with 30 *S. arvensis* buds and 20 *B. napus* buds pollinated per plant. The number of the resulting seed which successfully germinated is given, out of the number of seed sown. Figures in brackets give the number of seeds germinating and immediately dying.

Population	Location	Mean no. seed/bud pollinated Sinapis female	 Sinapis male	No. of plants obtained /cross Sinapis female	 Sinapis male
1	Cambridgeshire	0	0	-	-
2	Cambridgeshire	0	0	-	-
3	Suffolk	0.033	0	2/2	-
4*	Norfolk	-	-	-	-
5	Norfolk	0	0	-	-
6	Cambridgeshire	0.033	0.100	-	-
7	Cambridgeshire	0	0	-	-
8	Cambridgeshire	0	0	-	-
9	Lincolnshire	0.017	0	1/1	-
10*	Lincolnshire	-	-	-	-
11	Lincolnshire	0	0	-	-
12	Lincolnshire	0	0	-	-

Table 1. continued

Population	Location	Mean no. seed/bud pollinated Sinapis female	Sinapis male	No. of plants obtained /cross Sinapis female	Sinapis male
13	Lincolnshire	0	0	-	-
14*	Lincolnshire	-	-	-	-
15	Lincolnshire	0	0	-	-
16	Lincolnshire	0	0.175	-	2(1)/7
17	Lincolnshire	0	0	-	-
18	Cambridgeshire	0	0	-	-
19	Kent	0	0	-	-
20	Kent	0	0	-	-
21	Kent	0.017	0	0/1	-
22	Kent	0	0	-	-
23	Kent	0	0	-	-
24	West Sussex	0	0	-	-
25	East Sussex	0	0	-	-
26	East Sussex	0	0	-	-
27	East Sussex	0.183	0	11/11	-
28	Kent	0.017	0.050	-	-
29	Norfolk	0	0	-	-
30	Norfolk	0	0	-	-
31	Norfolk	0.017	0	1/1	-
32	West Sussex	0	0	-	-
33	Norfolk	0.033	0	-	-
34	Suffolk	0	0.025	-	0/1
35	Suffolk	0	0	-	-
36	Norfolk	0	0.025	-	1/1
37	Norfolk	0	0	-	-
38	Norfolk	0	0	-	-
39	Norfolk	0	0	-	-
40	Cambridgeshire	0	0	-	-
41	Norfolk	0	0.025	-	0/1
42	Cambridgeshire	0	0	-	-
43	Northamptonshire	0.017	0.250	0/1	10/10
44	Warwickshire	0	0	-	-
45	Hereford and Worcester	0	0	-	-
46	Hereford and Worcester	0	0	-	-
47	Gloucestershire	0	0.100	-	4/4
48	Wiltshire	0	0	-	-
49	Dorset	0	0	-	-
50	Dorset	0.117	0	5(2)/7	-
51	Wiltshire	0	0.050	-	2/2
52	Oxfordshire	0	0	-	-
53	Oxfordshire	0	0	-	-
54	Oxfordshire	0	0	-	-
55	Buckinghamshire	0	0.050	-	-
56*	Berkshire	-	-	-	-
57	Wiltshire	0	0	-	-
58	Oxfordshire	0.167	0	-	-
59	Buckinghamshire	0	0	-	-
60	Buckinghamshire	0	0	-	-
61	Bedfordshire	0	0.025	-	1/1

Table 1. continued

Population	Location	Mean no. seed/bud pollinated Sinapis female	Mean no. seed/bud pollinated Sinapis male	No. of plants obtained /cross Sinapis female	No. of plants obtained /cross Sinapis male
62	Bedfordshire	0	0	-	-
63	Lincolnshire	0	0.050	-	0/2
64	Nottinghamshire	0	0	-	-
65	South Yorkshire	0	0	-	-
66	Humberside	0.033	0.050	0/2	-
67	Humberside	0.183	0.025	-	-
68	Humberside	0	0.025	-	0/1
69	Humberside	0	0.025	-	0/1
70	Humberside	0	0	-	-
71	North Yorkshire	0	1.525	-	-
72	North Yorkshire	0	0.025	-	0/1
73	North Yorkshire	0	0	-	-
74	West Yorkshire	0	0	-	-
75	North Yorkshire	0	0	-	-
76	Durham	0	0	-	-
77	Northumberland	0.017	0.050	(1)/1	1/2
78	Northumberland	0.017	0	-	-
79	Northumberland	0	0	-	-
80	Borders	0	0	-	-
81	Borders	0	0.025	-	0/1
82	Lothian	0	0.175	-	-
83	Lothian	0	0	-	-
84	Fife	0	0	-	-
85	Tayside	0	0	-	-
86	Tayside	0	0.025	-	1/1
87	Grampian	0	0	-	-
88	Grampian	0	0	-	-
89	Grampian	0	0.050	-	2/2
90	Grampian	0.033	0	-	-
91	Grampian	0	0	-	-
92	Grampian	0	0	-	-
93	Grampian	0.067	0	2/5	-
94	Grampian	0.033	0	1(1)/2	-
95	Grampian	0	0	-	-
96	Highland	0	0.025	-	1/1
97	Highland	0	0	-	-
98	Highland	0	0	-	-
99	Highland	0	0.025	-	-
100	Highland	0	0	-	-
101	Highland	0	0	-	-
102	Strathclyde	0	0	-	-
103	Cumbria	0	0.475	-	1/3
104	Cumbria	0	0	-	-
105	Shropshire	0.100	0	-	-
106	Dyfed	0	0	-	-
Fr	France	0.083	0	-	-

* = populations producing no viable seed in the field in 1997.

Overall the number of seed produced per 100 buds was 1.1 for crosses where *S. arvensis* was the female and 3.4 for the reciprocal cross. The direction of the cross had no significant effect on the number of seed produced (Mann-Whitney, W=40654, p=0.157). There was no significant link between the ability of an individual *S. arvensis* plant to produce seed as the female parent and its success as the pollen donor (df=1, X^2=0.110, p>0.10). Control crosses within species produced 15,380 seed/100 buds for *S. arvensis* x *S. arvensis* and 33,360 seed/100 buds for oilseed rape x oilseed rape.

The results of the PCR and the BAR colour assay found that none of the potential hybrids produced from crosses with *S. arvensis* as the female parent contained the transgene. The rape parent was hemizygous for the transgene and so this result does not eliminate the possibility that these progeny are true hybrids. However, all of the potential hybrids had an identical morphology to that of the female parent.

DISCUSSION

S. arvensis plants were found across the country in areas where oilseed rape is grown and in areas where it is not. A very low number of seed were produced in the reciprocal crosses between *S. arvensis* and oilseed rape. These crosses involved *S. arvensis* plants from across the sampling area in regions with and without rape production. The hybrid status of these progeny has not yet been confirmed. These seed may have resulted from a number of processes other than hybridisation. They may be the result of contamination during the hand pollinations or matromorphy induced by intergeneric pollen as has been reported to occur in the Cruciferae (Banga, 1986). A range of molecular markers and root-tip chromosome counts are being used to investigate the hybrid status of these piants.

In previous studies where hybrids were obtained between these species, hybrids only resulted from crosses where oilseed rape was the female parent (Lefol *et al.*, 1996; Chevre *et al.*, 1996) and plants with a higher ploidy level are usually more successful as the female parent (Kerlan *et al.*, 1992). In this study more seed were produced when oilseed rape was the female parent but this was not significant.

The crosses discussed here form the preliminary survey of the *S. arvensis* populations and will not necessarily predict hybridisation in the field. If any of the progeny are confirmed as hybrids then more work will be carried out to determine whether hybridisation can occur without hand pollinations.

ACKNOWLEDGEMENTS

We acknowledge the contribution of BBSRC and direct financial support for this project from MAFF, project code RG0218.

REFERENCES

Banga S S (1986). Hybrid pollen-aided induction of matromrphy in *Brassica. Zeitschrift Pflanzenzuchtg* **96**, 86-89.

Bing D J; Downey R K; Rakow G F W (1995). An evaluation of the potential of intergeneric gene transfer between *Brassica napus* and *Sinapis arevensis. Plant Breeding* **114**, 481-484.

Bing D J; Downey R K; Rakow G F W (1996a). Assessment of transgene escape from *Brassica rapa* (*B. campestris*) into *B. nigra* or *Sinapis arvensis. Plant Breeding*, **115**, 1-4.

Bing D J; Downey R K; Rakow G F W (1996b). Hybridizations among *Brassica napus*, *B. rapa* and *B. juncea* and their two weedy relatives *B. nigra* and *Sinapis arvensis* under open pollination conditions in the field. *Plant Breeding* **115**, 470-473.

Brown J; Brown A P (1996). Gene transfer between canola (*Brassica napus* L. and *B. campestris* L.) and related weed species. *Annals of Applied Biology* **129**, 513-522.

Chevre A M; Eber F; Baranger A; Kerlan M C; Barret P; Festoc G; Vallee P; Renard M (1996). Interspecific gene flow as a component of risk assessment for transgenic *Brassicas. Acta Horticulturae* **407**, 169-179.

Edwards M (1980). Aspects of the population ecology of charlock. *Journal of Applied Ecology* **17**, 151-171.

Fogg G E (1950). Biological flora of the British Isles: *Sinapis arvensis* L. *Journal of Ecology* **146**, 415-429.

Kerlan M C; Chevre A M; Eber F; Baranger A; Renard M (1992). Risk assessment of outcrossing of transgenic rapeseed to related species: 1. Interspecific hybrid production under optimal conditions with emphasisi on pollination and fertilization. *Euphytica* **62**, 145-153.

Lefol E; Danielou V; Darmency H (1996). Predicting hybridisation between transgenic oilseed rape and wild mustard. *Field Crops Research* **45**, 153-161.

Rich T C G (1991). *Crucifers of Great Britain and Ireland.* Botanical Society of the British Isles: London.

Scheffler J A; Dale P J (1994). Opportunities for gene transfer from transgenic oilseed rape (*Brassica napus*) to related species. *Transgenic Research* **3**, 263-278.

SESSION 5

CONSEQUENCES OF GENE FLOW BETWEEN SPECIES – 2

Chairman: Mr J Orson
Morley Research Centre, Wymondham

Session Organiser: Dr H Smith
IACR-Broom's Barn, Bury St Edmunds

Predicting the environmental impact of transgene outcrossing to wild and weedy rices in Asia

M B Cohen, M T Jackson, B R Lu, S R Morin, A M Mortimer, J L Pham, L J Wade
International Rice Research Institute, P. O. B. 933, Manila 1099, Philippines

ABSTRACT

Asia is the center of origin of Asian cultivated rice, *Oryza sativa*, and is also home to numerous wild species of *Oryza*. Two wild species, *O. rufipogon* and *O. nivara*, are abundant in many parts of Asia and are known to hybridize with *O. sativa* under natural conditions. Numerous types of weedy rice also occur in Asia, derived from *O. sativa*, wild species, and hybrids between wild rices and *O. sativa*. Genetic engineering of rice for resistance to insects, diseases, and abiotic stresses has been rapidly advancing, although no transgenic varieties have as yet been widely field tested. We propose an interdisciplinary set of research activities to help predict the consequences of transgene outcrossing to wild and weedy rices, before transgenic varieties with enhanced stress resistance are extensively released into the environment.

INTRODUCTION

Rice is the staple food of half the world's population (Lampe, 1995). It provides 20% of global human per capita energy and 15% of per capita protein. Of 25 major rice-producing nations, 17 are located within South, Southeast and East Asia. Agricultural population densities on Asia's rice-producing lands are among the highest in the world and continue to increase at a remarkable rate. Rapid population growth puts increasing pressure on the already strained food-producing resources. By the year 2025, it is estimated that the world will need an additional 13 million tons of rough rice each year. With accelerating loss of productive riceland due to rising sea levels, salinization, erosion, and human settlements, the problem becomes one of increasing yields under ever-more trying circumstances (Dowling *et al.*, 1998).

Genetic engineering offers new opportunities to substantially increase rice yields per hectare. Extensive research on genetic engineering of rice is underway in dozens of laboratories around the world, in both the private and public sectors. Rice has been transformed with gene constructs to confer improved resistance to insects, pathogens, herbicides, and submergence, and to alter plant morphology and grain quality (Khush, 1996). Genetic engineering for other traits, including nutritional quality, and resistance to drought, salinity, and aluminum toxicity can be expected in the near future. Asia is the center of origin of Asian cultivated rice, *Oryza sativa*. Asia is also home to numerous wild and weedy species of *Oryza*, some of which are known to hybridize with *O. sativa* under natural conditions (Oka, 1991). Given the occurrence and environmental persistence of hybrids between cultivated and wild and weedy rices, the extensive amount of land that may eventually be planted to transgenic rice, and the large populations of wild and weedy rices in many rice growing areas, it must be assumed that transgenes will escape to wild and weedy relatives. In this paper, we analyze the scope of the potential problem and outline an integrated set of research activities that will help to predict the consequences of outcrossing of transgenes to wild and weedy rices in Asia.

TAXONOMY AND ECOLOGY OF CULTIVATED, WILD, AND WEEDY RICES

Rice and its wild relatives are classified in the genus *Oryza.* The 22 species of wild rice are distributed throughout the tropics of Asia, Africa, Australia, Oceanea, and Latin America (Vaughan, 1994; Bellon *et al.*, 1998). Several species complexes, based primarily on genomic identities, are recognized within the genus. The *O. sativa* complex consists of eight species which share the AA genome, including two wild species widespread in Asia, *O. rufipogon* and *O. nivara.* These two taxa are considered by some to be perennial and annual forms of the same species, and indeed the relationships among *O. sativa*, *O. nivara*, and *O. rufipogon* are complex (Oka, 1991). *O. sativa* is able to hybridize under natural conditions with other *Oryza* species that have the AA genome (Oka, 1991). The incidence of pollination of *O. rufipogon* and *O. nivara* by cultivated rice is increased by the fact that these wild species have a much higher incidence of outcrossing than *O. sativa*, which is primarily self-pollinating. Oka & Chang (1961) described "hybrid swarms" of plants consisting of F_1 hybrids and backcross progeny between *O. sativa* and *O. rufipogon*, clearly indicating that such hybrids can arise and persist under natural conditions. The ability to cross wild AA genome species with *O. sativa* has also facilitated their use in rice breeding (Bellon *et al.*, 1998). *O. nivara* is the source of a widely used gene for resistance to rice grassy stunt virus, and *O. rufipogon* has been a source of cytoplasmic male sterility for use in hybrid rice seed production.

In addition to wild rices, a diversity of weedy rices occurs in Asia. Weedy rices are becoming an increasingly severe problem as farmers switch their method of crop establishment from transplanting to direct seeding. Weedy rices are superior competitors to modern cultivars due to early vigor, greater tillering and height of plants. Close mimics of rice especially in the early stages of the crop, they are most easily recognized in the field as "off-types" at harvest by their height and by examination of panicles (easy shattering) and grain (having awns and red pericarps). Three main mechanisms may be hypothesized for the genesis of weedy rice in farmers fields: invasion and persistence of annual wild rices (aided by contaminated farmer seed, machinery movement); segregation and selection of "weedy" characters in existing cultivars; and introgression of genes from wild, annual or perennial, rice into sown cultivars. Recent genetic work supports the last two mechanisms (Suh *et al.*, 1997).

In Asia, extensive populations of *O. nivara* and/or *O. rufipogon* occur in India, Sri Lanka, Laos, Indonesia, Thailand, Cambodia, and Vietnam (Vaughan, 1994). Smaller populations can be found in China, Nepal, and Bangladesh, among other countries. *O. nivara* and *O. rufipogon* are still common vegetation in irrigation ditches, swamps, and riverbanks interspersed within rice-growing areas, although the size and number of these populations has been dramatically reduced by urbanization and agricultural intensification. In some areas, *in situ* conservation programs will be necessary to prevent local extinction of wild rices. Such programs have already begun in China and Thailand. Management of weedy rices and outcrossing of transgenes to wild and weedy rices are intertwined with conservation issues. For example, *in situ* conservation of *O. nivara* and *O. rufipogon* populations may increase local weedy rice problems, or be perceived to do so by farmers. Outcrossing of transgenes will change the genetic background of conserved wild rice populations and in some cases may change their ecology.

GENETIC ENGINEERING OF RICE FOR STRESS RESISTANCE

Much of the initial work in genetic engineering of rice, as with other crops, has concerned genes for resistance to *biotic stresses*, e.g. plant pathogens and insect pests. This is in part because, in many cases, high levels of resistance can be achieved with single genes that are already available and well studied. For example, rice has been transformed with *Bacillus thuringiensis* (Bt) toxin genes for resistance to stem borers (Lepidoptera: Pyralidae) (e.g., Cheng *et al.*, 1998) and modified plant chitinase genes for resistance to sheath blight, *Rhizoctonia solani* (Lin *et al.*, 1995). Genetic engineering for resistance to stem borers and sheath blight has been a priority not only because of their widespread importance in Asia, but also because high levels of resistance to these organisms has not been found in the gene pool of AA genome rice species. One further consequence of the lack of genes for high levels of resistance to stem borers and sheath blight in AA genome rice species is that transgenes conferring such resistance will be truly novel additions to the ecosystem. This is in contrast with genes for resistance to other insects and diseases, such as brown planthopper (*Nilaparvata lugens*) and rice blast (*Pyricularia oryzae*), for which resistance genes are common in traditional rice varieties and wild rices, and have been long deployed in improved rice varieties. Anecdotal observations of wild and weedy rice populations suggest that insects and diseases do not appreciably constrain the distribution or abundance of these plants. However, to the best of our knowledge, no systematic study of the population dynamics of insects or pathogens on wild or weedy rices has been published.

Genetic engineering for resistance to *abiotic stresses*, such as drought, submergence, and salinity has proceeded more slowly than that for biotic stresses because genes and promoters that can confer such resistances have not yet, or only recently, been identified and cloned. Progress in understanding proteins and pathways conferring resistance to submergence (Setter *et al.*, 1997) and salinity (Schachtman & Schroeder, 1994) indicates that plant transformation for resistance to these stresses is not far off. Resistance to acid sulfate/aluminum toxicity (de la Fuente *et al.*, 1997), and drought (Iuchi *et al.*, 1996) will take longer, but eventual success is highly probable. The adaptive range of rice species and cultivars is largely determined by water regime, and thus ecosystems for rice are defined on the basis of agrohydrology, e.g. irrigated, rainfed lowland, floodprone/deepwater, and upland (Garrity *et al.*, 1986). In rainfed conditions, the greatest yield losses are due to drought and submergence. *O. rufipogon* and *O. nivara* occur widely in rainfed lowland rice growing areas and *O. rufipogon* is abundant in some areas with acid sulfate soils, although these two species do not generally occur in saline soils. Because of the importance of water availability and soil chemistry on the distribution and abundance of rices, it is probable that the outcrossing of transgenes for abiotic stresses will have significant effects on the ecology of some populations of *O. rufipogon* and *O. nivara*.

A PROPOSED RESEARCH AGENDA

To predict the consequences of transgene outcrossing to wild and weedy rices will require an interdisciplinary research program involving germplasm specialists, plant ecologists and physiologists, and pest management and social scientists. Some of this research will also contribute to assessing the origins of weedy rices, the current conservation status of wild rices, and the potential for success of *in situ* conservation programs. We propose the following five linked research activities.

1. Survey the ecogeographical distribution and diversity of target species

There is a need for an updated and more comprehensive inventory of the distribution and conservation status of *O. rufipogon* and *O. nivara* populations. This will require updating information on previously-identified populations and visits to areas that have been inadequately surveyed but are likely to harbor wild rice populations. Data collection at each site should include climate and micro-climate, topography, associated vegetation, soil type, estimated conservation status, use and potential use of the plants by local people, local land use and agricultural practices. The transgenic traits that are likely to be most in demand by local farmers and the transgenic varieties that are likely to be popular in the area should also be identified. Cuttings and seeds should be collected for herbarium specimens and for analysis with genetic markers.

2. Identify key factors that determine the distribution and abundance of target species

To predict the impact of the outcrossing of a particular transgenic trait into a wild or weedy rice population, it is necessary to understand the ecological factors that constrain the distribution and abundance of that population. This will require monitoring the habitat and demographic characteristics of selected populations on at least a monthly basis, quantifying biotic and abiotic factors, e.g. identity and density of associated plant species, and of insect herbivores and plant pathogens occurring on the target species; soil nutrient status, water depth, shading, and micro-climate. This information can be used to identify candidate factors that determine population levels. However, because this approach cannot explain the mechanisms by which species abundance is actually regulated, there is also a need to manipulate habitat characteristics experimentally to confirm their role as key factors influencing population growth rate. This will involve deliberate additions and removals of plants from populations to examine density compensation mechanisms, and experimental alterations of habitat variables. These variables should be chosen on the basis of the initial one or two years of surveys, and are likely to include manipulation of insect herbivore density and abiotic variables such as water and nutrient levels.

3. Analyze gene flow among wild, weedy, and cultivated rices

Previous work with DNA and allozyme markers has indicated that patterns of gene flow among wild, weedy, and cultivated rices differs among locations (e.g. Suh *et al.*, 1997). There is a need to expand gene flow studies to additional sites. Studies using markers for the mitochondrial and chloroplast genomes, in addition to the nuclear genome, could further elucidate the genetic origins of weedy rices and how transgenes might spread among wild and weedy populations.

4. Determine fitness of hybrids between improved varieties and target species

To estimate fitness of the hybrids between improved varieties and target wild and weedy species, a comparison of selected vegetative and reproductive characters of the parents, F_1 hybrids, and F_2 and backcrossing progenies will be conducted under controlled greenhouse conditions, and under field conditions. These are particularly important to measure for hybrids between perennial wild rices and cultivated (annual) rices. If the hybrids are not perennial but have limited capability of sexual reproduction, then they will have little ability to spread or persist in wild populations.

5. Quantify tolerance of target species to selected abiotic stresses

An indirect approach can be taken to study the possible effects of transgene outcrossing on wild and weedy rices, by quantifying the existing tolerance of wild and weedy rices to selected abiotic stresses. The initial focus should be on drought and submergence because of their great agronomic importance and the potential for environmental impact of transgenes conferring resistance to these stresses. There are several kinds of drought and submergence stress. These result from the continuum of water regimes that occur in rice growing areas: from drought on continuously aerobic soils in the uplands, to the drought-prone rainfed lowlands where drought generally follows ponding of water on previously anaerobic soils; to favorable rainfed and irrigated water regimes, and then to short-term flooding in submergence-prone rainfed lowland, and finally to deepwater in flood-prone ecosystems. Greenhouse experiments can be used to test the tolerance of wild and weedy rices and F_1 hybrids to water stress regimes characteristic of the areas in which the plants are found. Dependent upon progress and results from the drought and submergence tests, treatments of salinity and/or acid sulfate/aluminum toxicity could be added to the water regimes in a factorial design.

PERSPECTIVES

Regulations for field testing of genetically modified organisms have been established in China, Thailand, and the Philippines, among other countries in Asia. These regulations generally require that proposals for field testing provide information about the potential for, and the possible harmful effects of, outcrossing. To the best of our knowledge, field tests of transgenic rice have so far taken place only in Japan and China, but it is likely that the number of field tests will soon begin to increase rapidly in several countries. While initial field testing of transgenic rice can be confined to areas where wild and weedy relatives do not occur or have been removed, or held in plots isolated by distance and bordering vegetation, it clearly will not be practical to enforce such restrictions once large scale field tests begin or transgenic varieties are released to farmers. Outcrossing was a focus of the first conference on rice biosafety, where it was concluded that "Further study of the ecology and distribution of wild and weedy rices is recommended, but this need not and should not delay field tests in areas where the ecology is well known" (Clegg *et al.*, 1993). Hopefully, when promising transgenic rice varieties are ready for extensive field testing and release to farmers, sufficient information on the consequences of transgene outcrossing will be available to permit regulatory bodies to make well-informed decisions about possible risks in all affected rice-growing areas.

REFERENCES

Bellon M R; Brar D; Lu B R; Pham J L (1998). Rice genetic resources. In: *Sustainability of Rice in the Global Food System*, eds N G Dowling, S M Greenfield & K S Fischer, pp. 251-283. Pacific Basin Study Center: Davis, Calif. (USA), and International Rice Research Institute: Manila, Philippines..

Cheng X; Sardana R; Kaplan H; Altosaar I (1998). *Agrobacterium*-transformed rice plants expressing synthetic *cryIA(b)* and *cryIA(c)* genes are highly toxic to striped stem borer and yellow stem borer. *Proceedings of the National Academy of Sciences, USA* **95**, 2767-2772.

Clegg M T; Giddings L V; Lewis C S; Barton J H (eds.) (1993). Report of the International Consultation on Rice Biosafety in Southeast Asia, September 1-2, 1992. Technical Paper, Biotechnology Series No. 1. World Bank: New York.

de la Fuente J M; Ramirez-Rodrigues V; Cabrera-Ponce J L; Herrera-Estrella L (1997). Aluminum tolerance in transgenic plants by alteration of citrate synthesis. *Science* **276**, 1566-1568.

Dowling N G; Greenfield S M; Fischer K S (1998). *Sustainability of Rice in the Global Food System*. Pacific Basin Study Center: Davis, Calif. (USA), and International Rice Research Institute: Manila, Philippines.

Garrity D P; Oldeman L R; Morris R A (1986). Rainfed lowland rice ecosystems: characterization and distribution. In: *Progress in Rainfed Lowland Rice*, pp. 3-23. International Rice Research Institute: Manila, Philippines.

Iuchi S; Yamaguchi-Shinozaki K; Urao T; Terao T; Shinozaki K (1996). Novel drought-inducible genes in the highly drought-tolerant cowpea: Cloning of cDNAs and analysis of the expression of the corresponding genes. *Plant Cell Physiology* **37**, 1073-1082.

Khush G S (ed.) (1996). *Rice Genetics III. Proceedings of the Third International Rice Genetics Symposium, 16-20 Oct 1995.* International Rice Research Institute: Manila, Philippines.

Lampe K (1995). Rice research: food for 4 billion people. *GeoJournal* **35**, 253-259.

Lin W; Anuratha C S; Datta K; Potrykus I; Muthukrishnan S; Datta S (1995). Genetic engineering of rice for resistance to sheath blight. *Bio/Technology* **13**, 686-691.

Oka H I; Chang W T (1961). Hybrid swarms between wild and cultivated rice species *Oryza perennis* and *O. sativa*. *Evolution* **15**, 326-339.

Oka H I (1991). Genetic diversity of wild and cultivated rices. In: *Rice Biotechnology* eds G S Khush & G H Toeniessen, pp. 55-81. International Rice Research Institute: Manila, Philippines.

Schachtman D P; Schroeder J I (1994). Structure and transport mechanism of a high-affinity potassium uptake transporter from higher plants. *Nature* **370**, 655-658.

Setter T L; Ellis M; Laureles E V; Ella E S; Senadhira D; Mishra S B; Sarkarung S; and Datta S (1997). Physiology and genetics of submergence tolerance in rice. *Annals of Botany* **79**, 67-77.

Suh H S; Sato Y I; Morishima H (1997). Genetic characterization of weedy rice (*Oryza sativa* L.) based on morpho-physiology, isozymes, and RAPD markers. *Theoretical and Applied Genetics* **94**, 316-321.

Vaughan D A (1994). *The Wild Relatives of Rice: a Genetic Resources Handbook.* International Rice Research Institute: Manila, Philippines.

The production of herbicide-resistant jointed goatgrass (*Aegilops cylindrica*) x wheat (*Triticum aestivum*) hybrids in the field by natural hybridization and management strategies to reduce their occurrence

S S Seefeldt, F L Young
USDA/ARS, Washington State University, Pullman, WA 99164 USA

R Zemetra
Department of Plant, Soil, and Entomological Sciences, University of Idaho, Moscow, ID 83844-2339 USA

S S Jones
Department of Crop and Soil Sciences, Washington State University, Pullman, WA 99164 USA

ABSTRACT

Two imazamox-resistant hybrids from a cross between *A. cylindrica* and an imazamox-resistant wheat ('cv' FS-4 IR wheat) were discovered in a research plot where FS-4 IR wheat seed had been replanted from the harvest of an efficacy study conducted the year before. Seven seeds (BC_1) were produced by these plants. Six of the seven seedlings from these seeds survived an application of 72 g ai ha^{-1} imazamox. Management strategies to reduce the occurrence of herbicide-resistant *A. cylindrica* x wheat hybrids can be incorporated before, during, and after the crop year when the herbicide-resistant wheat is planted. Strategies to control hybrids that do occur are discussed. These strategies take advantage of the biology of the *A. cylindrica* x wheat hybrid plant.

INTRODUCTION

A. cylindrica (2N = 4x = 28) is an important winter annual grass weed problem in winter wheat (2N = 6x = 42) in the western United States. Because *A. cylindrica* shares the D genome with winter wheat (Kimber & Sears, 1987) (Figure 1), no selective herbicides have been developed. Recently, a soft red winter wheat (cv. 'Fidel') was mutagenized and plants resistant to imazamox, an imidazolinone herbicide, were selected from the mutagenized population (cv FS-4 IR wheat) (White & Morrison, 1998). Field experiments in Washington, USA and Oregon, USA (Ball *et al.*, 1999) have documented high levels of resistance to imazamox by the FS-4 IR wheat.

A concern in the development of herbicide-resistant crops is the potential transfer of the resistance gene to other plant species (Harrison, 1992). Studies in Idaho, USA have demonstrated that glufosinate-resistant canola could out-cross to *Brassica rapa* to produce glufosinate-resistant hybrids in both greenhouse (Brown & Brown, 1996) and field studies (Brown *et al.*, 1995). It has been known for 70 yr that *A. cylindrica* crosses with wheat to produce hybrids that are phenotypically intermediate and sterile (Johnston & Parker, 1929; Mayfield, 1927). Only recently has it been shown that the *A. cylindrica* x wheat hybrids will produce viable seed with two to three percent fertility (Zemetra *et al.*, 1998).

In a 1995-96 herbicide efficacy field study, *A. cylindrica* pollen fertilized some FS-4 IR wheat plants at the Palouse Conservation Field Station near Pullman, WA (Ball *et al.*, 1999). Seed harvested from this experiment was cleaned by hand of *A. cylindrica* spikelets and replanted for a second imazamox efficacy trial (1996-91) at a second location in the Palouse Conservation Field Station. Just before harvest, two hybrid plants (SSS201 and SSS405) from a winter wheat x *A. cylindrica* cross were discovered in plots where imazamox at the two highest rates (53 and 69 g ai ha^{-1}) was applied. All imazamox applications included 0.25 % R-ll (v:v) a surfactant, and 380 ml ha^{-1} 'Prefer'. Imazamox at the two high rates killed all *A. cylindrica*, but had no observable effect on the FS-4 IR wheat. The hybrid plants were growing in the wheat row, indicating that the hybrid seed was in the seed lot, had been planted, and resembled winter wheat in appearance. The hybrid plants were 15 cm taller and more robust than the surrounding FS-4 IR wheat and had 8 and 10 tillers. About 2 % of the spikelets on these hybrid plants had viable BC_1 seed (four from one plant and three from the other). The objective of this research was to document that the gene which imparts resistance to imazamox had been transferred to hybrids between *A. cylindrica* and wheat and to outline management strategies which would reduce this occurrence.

MATERIALS AND METHODS

The seven BC_1 seeds were germinated and mitotic chromosomes were analyzed in root tip cells as described in Seefeldt *et al.* (1998). After the root-tip squashes, the seven BC_1 seedlings were transplanted to small pots (0.5 L) and placed in a greenhouse (22 C and 16:8 h day:night). At the 3 to 4 leaf stage, the BC_1 plants, six FS-4 IR wheat plants, and six *A. cylindrica* plants were sprayed with 72 g ai ha^{-1} imazamox with 0.25 % R-11[1] (v:v) and 380 ml ha^{-1} 'Prefer' in a spray chamber. The plants were then vernalized for 8 wk at 4°C in a growth chamber.

DISCUSSION

Transfer of the resistance gene

Winter wheat/*A. cylindrica* hybrids will have 35 chromosomes, (21 from wheat and 14 from *A. cylindrica*) (Figure 1) and the BC_1 generation may result from backcrosses with wheat or *A. cylindrica* (Zemetra *et al.*, 1998). Fertility in the hybrid is reduced compared to wheat because of the lack of bivalent formation with the three unpaired genomes. When the hybrid backcrosses with wheat, subsequent selfing would tend to return the chromosome number to 42, whereas backcrosses with *A. cylindrica* would tend to return the chromosome number to 28. From the two sets of root squashes, the chromosome counts were 2n = 39, 39, 42, and 45 for SSS201 and 39, 40 and 54 for SSS405 (Seefeldt *et al.*, 1998).

All FS-4 IR wheat plants and six of the seven BC_1 plants survived imazamox at 72 g ai ha^{-1}. All of the *A. cylindrica* plants were killed. The BC_1 plant that did not survive (chromosome count 2N=45) may have been a backcross to a susceptible *A. cylindrica* plant or the resistance gene may have been lost during meiosis (Figure 1). It is clear that the FS-4 IR wheat was originally pollinated by *A. cylindrica*, the gene for resistance to imazamox was in the hybrids, and the majority of the seed produced by the hybrids possessed the resistance gene.

There was a range of phenotypic variation in the six surviving BC_1 plants, with two similar to the FS-4 IR wheat, one similar to *A. cylindrica*, and three intermediate in appearance. Although F_1 hybrids are more vigorous than the parent wheat or *A. cylindrica*, the BC_1 plants, after being treated with imazamox, were less vigorous, particularly at the seedling stage, than the treated FS-4 IR wheat plants. The BC_1 plants were allowed to self-pollinate. However no seed was produced indicating that the BC_1 plants from this study were not self-fertile (Seefeldt *et al.*, 1998).

Management strategies

The gene for resistance to imazamox in FS-4 IR wheat is thought to be located in the D genome which occurs in both wheat and *A. cylindrica*. This location would increase the potential for transfer of the resistance gene from wheat to *A. cylindrica* (Figure 1). If the resistance gene was located in the wheat A or B genome there is reduced probability of establishment of the resistance gene in *A. cylindrica* backcrosses as the unpaired A and B chromosomes tend to get lost with each additional backcross (Zemetra *et al.*, 1998). There is however, a low probability that the gene could persist through translocation from the A or B genome to the C genome in *A. cylindrica*. Since the D genome has homologous pairs of chromosomes, it is unlikely that translocation would occur within this genome. It is possible that a pair of A or B chromosomes could undergo reduced meiosis which would result in a substitution line (2N=28) or an addition line (2N=30 or 2N=32). *A. cylindrica* plants that result from either scenario would probably be sterile or non-competitive. However, under herbicide selection, stabilization of the A or B genome chromosome pair is possible. While the 30 and 32 chromosome plant may be self-fertile because of normal bivalent formation, resistance transfer by outcrossing to *A. cylindrica* would be reduced due to the differences in chromosome number in the two plants (30 or 32 and 28). Backcrosses to FS-4 IR would, of course, remain resistant.

The rapid transfer of a herbicide-resistance trait from wheat to *A. cylindrica* is a problem with serious implications for growers who would use herbicide-resistant wheat in areas infested with *A. cylindrica*. However, it might be possible to mitigate the problem with management strategies that take advantage of aspects of the hybridization process. These strategies are listed below with reference to the planted resistant wheat.

Preplant

1. If the weed infestation is heavy consider a one-time burn on non-highly erosive land to kill *A. cylindrica* seed on the surface followed by a spring crop or fallow to further reduce the *A. cylindrica* seed bank. If the land is highly erodible one could chisel plow at intervals to reduce erosion.

Crop year

2. Growers must use herbicide-resistant wheat that comes from 'certified' wheat fields where there is no *A. cylindrica* within half a mile of the field.
3. Ensure that the herbicide-resistant wheat is a variety that is competitive against *A. cylindrica.*
4. Use high seeding rates and narrow row spacing to increase crop competitiveness.
5. The herbicide should be applied for maximum efficacy to the field and to the field borders to reduce cross-pollination of off-site *A. cylindrica.*

Harvest

6. As much wheat and *A. cylindrica* seed as possible should be harvested to remove the resistance gene from the field.
7. Grain trucks hauling herbicide-resistant seed from the field should be covered to prevent herbicide-resistant seed from blowing off the truck and being introduced to other fields and roadsides.
8. All farm machinery should be thoroughly cleaned before moving off the field.

Following year

9. A non winter wheat crop should be planted that allows the use of alternative methods to control any *A. cylindrica*, herbicide-resistant wheat, and herbicide-resistant hybrids.
10. Because the increased size of the F_1 hybrids makes them easy to see in the field, hand-weeding should be used where practical to remove hybrids in order to prevent backcrosses with *A. cylindrica* and wheat. Removed plants should be destroyed.
11. Adding an additional year of spring crops before may drive the *A. cylindrica* seed bank to near extinction.
12. The next winter wheat crop should be one that is not herbicide resistant which will limit backcrosses to susceptible genes.
13. Any small infestations of *A. cylindrica* should be killed.

Growers, once made aware of the ease of transfer of the resistance gene to winter wheat/*A. cylindrica* hybrids and the progeny of those hybrids, will want to adopt strategies that reduce the movement of the resistance gene to *A. cylindrica*. These strategies will ensure that herbicide-resistant wheats will be useful tools for the control of problem weeds such as *A. cylindrica* for years to come.

ACKNOWLEDGEMENTS

The authors thank Kim Hemmesch and Xiwen Cai for technical assistance.

REFERENCES

Ball D A; Young F L; Ogg, A G Jr. (1999). Selective control of *A. cylindrica* with imazamox in herbicide-resistant wheat. *Weed Technology* (In review).

Brown J; Brown A P (1996). Gene transfer between canola (*Brassica napus* L. and *B. campestris* L.) and related weed species. *Annals of Applied Biology* **129,** 513-522.

Brown J; Thill, D C; Mallory-Smith C; Brown A P; Brammer T A; Nair H S (1995). Gene transfer between canola (*Brassica napus* L.) and related weed species. In: *Proceedings of the USDA/ARS Biological Risk Conference. Biotechnology Risk Assessment: USEPA/USDA/Environment Canada/Agriculture and Ag. Food Canada,* pp 55-74. USDA/ARS: Pensecola, FL.

Harrison H F Jr. (1992). Developing herbicide-tolerant crop cultivars: Introduction. *Weed Technology* **6,** 613-614.

Johnston C O; Parker J H. (1929). *Aegilops cylindrica* Host., a wheat-field weed in Kansas. *Transcripts of Kansas Academic Science* **32,** 80-84.

Kimber, G; Sears E R (1987). Evolution in the genus *Triticum* and the origin of cultivated wheat. In *Wheat and Wheat Improvement,* ed E G Heyne, pp 154-164. Agronomy Monograph No. 13. Tri-Societies: Madison, WI.

Mayfield L (1927). Goat grass - A weed pest of central Kansas wheat fields. *Kansas Agricultural Student* **7,** 40-41.

Seefeldt S S; Zemetra R; Young F L; Jones S S (1998). Production of herbicide-resistant jointed goatgrass (*Aegilops cylindrica*) x wheat (*Triticum aestivum*) hybrids in the field by natural hybridization. *Weed Science* **46,** (in press).

White R H; Morrison R A (1998). IMI® Wheat: New Weed Management Technology. *Proceedings of the Western Society of Weed Science* **51,** 106-107.

Zemetra R S; Hansen J; Mallory-Smith C A (1998). Potential for gene transfer between wheat (*Triticum aestivum*) and jointed goatgrass (*Aegilops cylindrica*). *Weed Science* **46,** 313-317.

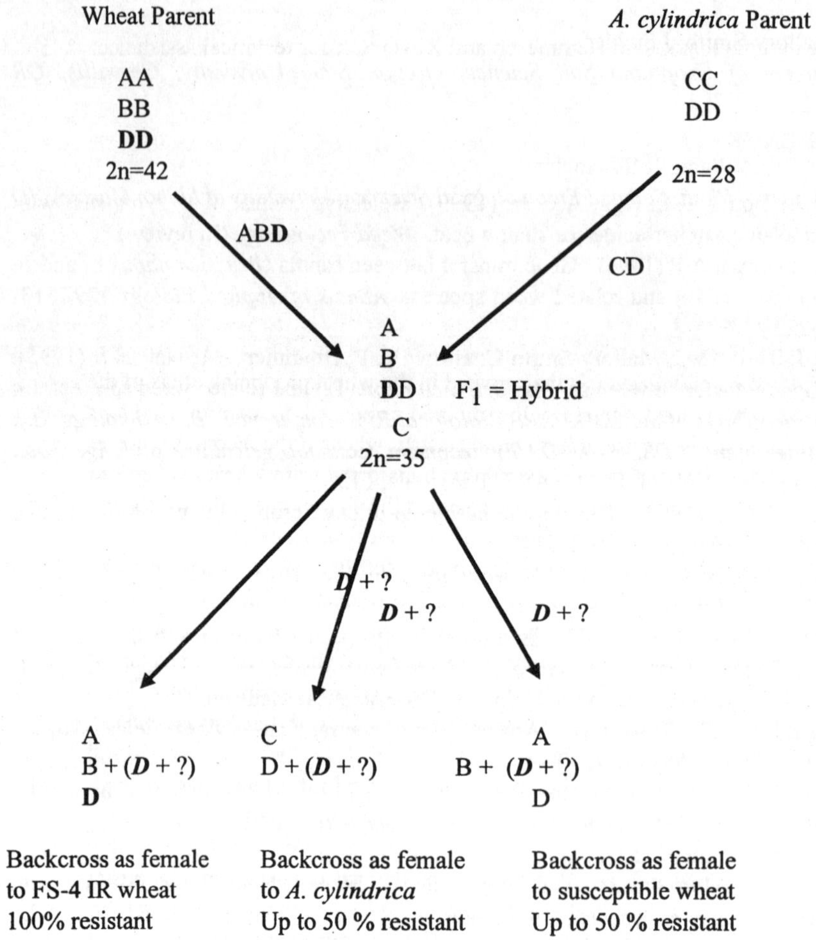

Figure 1. Movement of resistance gene from wheat to *A. cylindrica* where **D** = resistance gene in the D genome and ***D*** = 50 % probability of resistance gene.

Potential for gene flow between wheat (*Triticum aestivum*) and jointed goatgrass (*Aegilops cylindrica*) in the field

C A Mallory-Smith, J Snyder
Department of Crop and Soil Science, Oregon State University, Corvallis, OR 97331, USA

J L Hansen, Z Wang, R S Zemetra
Department of Plant, Soil and Entomological Sciences, University of Idaho, Moscow, ID 83844, USA

ABSTRACT

Aegilops cylindrica is a major weed in the wheat producing areas of the western United States. One method proposed to control *Aegilops cylindrica* is the use of herbicide resistant wheat. Of concern is whether a herbicide resistance gene could be transferred from wheat to *Aegilops cylindrica*. *Aegilops cylindrica* shares the D genome with wheat and interspecific hybrids between the two species occur in the field. In previous greenhouse research, partial self-fertility was restored in the BC_2 generation after backcrossing to *Aegilops cylindrica*. The objective of this research was to determine if backcrossing of the wheat x *Aegilops cylindrica* hybrid could occur in the field. Hybrids were transplanted with various combinations of wheat and *Aegilops cylindrica* plants. Field derived BC_1 plants were transplanted into stands of *Aegilops cylindrica*. Both BC_1 and BC_2 seeds were produced in the field by natural backcrossing. Seed harvested off the hybrid plants had a high germination rate. Four BC_1 plants were found to be partially self-fertile. The production of BC_2 seed in the field and the low level of self-fertility of BC_1 plants indicates that hybrids could serve as a bridge in the transfer of genes between wheat and *Aegilops cylindrica* in the field. Any release of a herbicide resistant wheat should be accompanied by a management plan that would minimize the potential for gene movement between these species.

INTRODUCTION

Aegilops cylindrica is a major weed of winter wheat in the western United States. Its similarity to wheat makes control very difficult once *Aegilops cylindrica* is introduced into a wheat field (Donald & Ogg, 1991). Herbicide-resistant wheat may provide a means to selectively control *Aegilops cylindrica* in wheat fields. An imidazolinone herbicide-resistant wheat has been developed and is currently being field tested in the Pacific Northwest of the United States (Mallory-Smith *et al.*, 1996). While the use of herbicides in combination with herbicide resistant wheat may provide a control method for *Aegilops cylindrica*, the sexual transfer of the herbicide resistance gene from wheat to *Aegilops cylindrica* would eliminate this benefit.

Hybrids can occur between *Aegilops cylindrica* and wheat in the field because they are related species that share a common genome (Kimber & Sears, 1987). The shared D genome between wheat and *Aegilops cylindrica* allows hybrids between the species to be produced in the field. These hybrids were believed to be sterile but viable seed was found on the hybrids (Mallory-Smith *et al.*, 1996). In the greenhouse, the interspecific hybrids showed a low level of female fertility (2%) that allowed for backcrossing to occur between the hybrid and either *Aegilops cylindrica* or wheat (Zemetra *et al.*, 1998). Partial female fertility increased with subsequent backcrosses with partial self-fertility restored after two backcrosses to *Aegilops cylindrica*. The restoration of self-fertility after two backcrosses raises the question whether the hybrid could serve as a bridge to transfer a herbicide resistance gene from wheat to *Aegilops cylindrica*. Because the current source of imidazolinone herbicide resistance in the wheat being field tested is controlled by a single, semidominant gene (Newhouse *et al.*, 1992), the risk of transfer is increased if backcrossing can occur. The objective of this research was to determine if BC_1 and BC_2 seed could be produced in the field and if the rate of seed production in the backcross generations was similar to that observed in the greenhouse.

MATERIALS AND METHODS

Hybrid and backcross seed was produced in the greenhouse by crossing a native collection of *Aegilops cylindrica* with the soft white winter wheat cultivar 'Madsen' (Allan *et al.*, 1989). The wheat was used as the female parent and the approach method was used for pollination. For the field BC_1 seed production trial, hybrid seed was germinated and transplanted into the field in Corvallis, Oregon in the fall of 1995 and 1996. There were four replications of one, five, or ten hybrids transplanted into plots seeded with one of five mixtures of *Aegilops cylindrica* and Madsen wheat (150/0, 100/50, 75/75, 50/100, and 0/150). Hybrid heads were collected in July, 1996 and July, 1997 and a maximum of 100 heads per plot were hand-threshed to determine percent seed set. Seed found in these trials was tested to determine the germination rate by germinating the seed in the dark at 24 C for one week. The BC_1 plants from the germination trial were planted in the greenhouse and heads were collected to determine level of self-fertility. For the field BC_2 production trial, a set of BC_1 seed from the 100% *Aegilops cylindrica* plots were germinated and transplanted into plots of 100% *Aegilops cylindrica* in Corvallis, Oregon in the fall 1996 and 1997. Heads were harvested off the BC_1 plants and hand-threshed to determine number of seed set. Standard ANOVA was run on BC_1 seed production trial for percent seed set and percent germination to determine if differences existed among the wheat - *Aegilops cylindrica* combinations.

RESULTS AND DISCUSSION

Seed set occurred on the hybrid plants in both 1996 and 1997 (Table 1). The mean percent seed set was 2.2% in 1996 and was 3.7% in 1997. This is equal or greater than that observed for seed set on hybrids in the greenhouse (Zemetra *et al.*, 1998). No differences for percent seed set were observed among the different plots of wheat and *Aegilops cylindrica* in 1996 or 1997. This demonstrates BC_1 seed can be produced in the field and confirms our greenhouse results that percent seed set is related to the partial female fertility of the *Aegilops cylindrica* x wheat hybrid and not the species providing the pollen.

BC_1 seed produced in the 1995-96 and 1996-97 field trials were germinated and grown out in greenhouses in Oregon and Idaho. The average germination was 85% in 1995-96 and 89% in 1996-97. No differences were found among the wheat - *Aegilops cylindrica* combinations for percent germination. This indicates that the seed set on the hybrid plants in the field have a high likelihood of germinating in the field the next growing season.

Table 1. Percent seed set on *Aegilops cylindrica* x wheat hybrids grown among various mixtures of *Aegilops cylindrica* and wheat.

		Number of *Aegilops cylindrica* x wheat hybrids					
Aegilops		1		5		10	
cylindrica	Wheat	1996	1997	1996	1997	1996	1997
150	0	2.3	4.0	2.2	3.5	2.2	2.9
100	50	2.9	2.1	2.3	3.6	2.3	3.8
75	75	3.2	1.8	2.2	4.4	2.1	2.8
50	100	1.6	3.3	1.7	4.2	2.4	4.0
0	150	1.2	5.7	2.1	4.8	2.5	5.2

Heads of the greenhouse grown BC_1 plants were checked for seed to determine if any of the BC_1 plants were self-fertile. Four BC_1 plants were found to have a high level of self-fertility (Table 2). Two of the four plants appeared to be phenotypically like wheat, one like *Aegilops cylindrica* and the fourth plant was more similar to the hybrids in appearance. The presence of self-fertility in the BC_1 generation is one generation earlier than observed in the greenhouse (Zemetra et al. 1998) but this may be related to the greater number of plants evaluated in the field trial.

Table 2. The restoration of fertility in wheat x *Aegilops cylindrica* F_1 hybrids and their backcross progenies.

	Chromosome mean	Number range	Number of plants	Number of plants setting seed	Seed set (%) Mean	Seed set (%) Range
F_1 and backcross progenies used as male parent						
F_1	35	35	10	0	0	0
BC_1	45	41-49	4	1	2.4	0-9.5
BC_2	34	28-39	6	3	13.0	0-38.5
F_1 and backcross progenies used as female parent						
F_1	35	35	10	10	0.9	/*
BC_1	45	41-49	4	4	4.4	0.8-13.2
BC_2	34	28-39	5**	5	26.8	5.6-81.1

* F_1 plants were harvested in bulk.
** One of BC_2 plant grew poorly and was only used as male parent.

In the 1996-97 BC_2 seed production trial, the mean frequency of seed set on the BC_1 plants was 2.5% and ranged from 0.2% to 5.2% In the 1997-98 BC_2 seed production trial, the mean frequency of seed set on the BC_1 plants was 7.5% and ranged from 1.6% to 20.3%. In 1996-97, the mean frequency of seed set on the field grown BC_1 plants was lower than that observed in the greenhouse but was higher in the 1997-98 field trial. The field harvested BC_2 seeds are currently being germinated and the BC_2 plants will be grown out in the greenhouse to determine the level of self-fertility for this backcross generation.

Based on the results of this study BC_2 plants, originating from a wheat x *Aegilops cylindrica* hybrid can be produced in the field. Based on our greenhouse study results where self-fertility was restored in the BC_2 generation of *Aegilops cylindrica* parents and the low number of partially self-fertile BC_1 plants found in this study, there is a good potential for gene movement from wheat to *Aegilops cylindrica* in the field. Management practices and genome placement of herbicide resistance genes will need to be taken into account to minimize the movement of herbicide resistance from wheat to *Aegilops cylindrica*.

REFERENCES

Allan R E; Peterson C J; Rubenthaler G L; Line R F; Roberts D E (1989). Registration of 'Madsen' wheat. *Crop Science,* **26**, 1575.

Donald W W; Ogg A G (1991). Biology and control of jointed goatgrass (*Aegilops cylindrica*), a review. *Weed Technology,* **53**, 3-17

Kimber G; Sears E R (1987). Evolution in the genus *Triticum* and the origin of cultivated wheat. In: *Wheat and Wheat Improvement. Agronomy Monograph No. 13*, ed. E G Heyne, pp. 154-164. ASA-CSSA-SSSA: Madison, Wisconsin.

Mallory-Smith C A; Hansen J; Zemetra R S (1996). Gene transfer between wheat and *Aegilops cylindrica.* Proceedings Second International Weed Control Congress, Copenhagen, Denmark. pp. 441-445.

Newhouse K.E; Smith W A; Starrett M A; Schaffer T J; Singh B J (1992). Tolerance to imidazolinone herbicides in wheat. *Plant Physiology,* **100**, 882-886.

Zemetra R S; Hansen J; Mallory-Smith C A (1998). Potential for gene transfer between wheat (*Triticum aestivum*) and jointed goatgrass (*Aegilops cylindrica*). *Weed Science,* **46**(3), 313-317.

Assessing potential risks of transgene escape from fiber plantations

S P DiFazio, S Leonardi, S Cheng, S H Strauss
Department of Forest Science, Oregon State University, Corvallis, OR 97331-7501, USA

ABSTRACT

Woody crops such as trees pose difficult obstacles for genetic risk analysis because of their long generation times, large size, and potential for long distance dispersal of pollen and seed. This necessitates reliance on established stands for retrospective studies of gene flow, and simulation models to assess the consequences of introducing new genes into the environment. We are using these approaches to assess potential risks of large-scale cultivation of transgenic hybrid poplar (*Populus spp.*). To predict gene flow from hybrid plantations, it is necessary to understand both the dynamics of gene movement within and between populations, and the ability of hybrid trees to produce fit progeny. We have found that up to 3.8% of progeny of wild females growing near hybrid poplar plantations were fathered by hybrid males. Also, hybrid seedlings established and grew at about the same rate as wild seedlings. Data from these field studies will be integrated in a spatial simulation model which is currently under development. The model will be used to identify the genetic and ecological parameters most important for predicting the spread of herbicide- and insect-resistance transgenes over space and time. Potential economic and agronomic significance of transgene spread will also be assessed.

INTRODUCTION

Forests cover more than one-third of the terrestrial landscape, and wood provides heat, shelter, and an array of other products for the majority of the world's inhabitants. Wood consumption has increased concomitantly with world population, and is projected to continue growing for the foreseeable future. For example, per capita consumption of fiber products increased 50% in the developed world and 300% in the developing world between 1970 and 1994 (FAO 1997). However, forests also harbor the majority of the earth's terrestrial species, and increasing pressure on native ecosystems is eroding that biodiversity (Wilson, 1992). Therefore, it is imperative to increase the productivity of managed forests and fiber plantations to meet burgeoning demand and alleviate pressure on wild forests (Dekker-Robertson & Libby, 1998). Genetic improvement of growing stocks is one strategy for enhancing productivity, and genetic engineering has the potential to make large contributions toward this goal.

However, despite its great promise, there are some risks associated with genetic engineering. These include the potential for increased invasiveness of the engineered organism, transfer of the transgene to wild relatives (leading to either increased invasiveness or reduced fitness), direct effects of the new gene product on nontarget organisms, accelerated development of resistance to herbicides or insecticides by agricultural pests (*e.g.*, Rogers & Parkes, 1995). Whether any of these possibilities pose threats significantly greater than are commonly assumed in conventional agronomic or forest management systems is, however, unclear (*e.g.*, National Research Council, 1989).

Trees present some special complications for assessing risks of transgenics compared to herbaceous agricultural crops. For example, most trees have extended juvenile periods, long life spans, large stature, high fecundity, extensive gene flow, vegetative persistence or spread, and wild relatives that are spatially and genetically close to planted stock (James *et al.,* 1997). Therefore, risk assessments for trees must necessarily encompass long time frames and extensive areas, and spread may be highly dependent on rare events that favor colonization. We propose that simulation modeling must be a key tool for exploring the risk of transgene escape from tree plantations. Similar approaches have been taken to study invasions of exotic trees (Higgins & Richardson,1998).

We are studying potential gene escape from transgenic hybrid poplar (*Populus spp.*, including aspens and cottonwoods) plantations. Poplar is a model organism for tree biotechnology, and is likely to be the first commercially released transgenic forest tree. Field trials are underway for clones transformed with herbicide resistance, insect-resistance, disease resistance, wood modification, and flowering control transgenes (Strauss *et al.,* 1998).

Transgene spread will depend in part on dispersal distances, establishment rates, and competitiveness of transgenic progeny relative to wild trees. For this reason we are gathering data on current levels of seed and pollen flow from hybrid poplar plantations, rates of seedling establishment in the vicinity of plantations, and competitiveness of hybrid progeny compared to wild seedlings. These studies will provide baseline gene dispersal data and allow simulation of the potential spread of transgenes in two areas of poplar cultivation in the Pacific Northwest, USA. A simulation model will serve to integrate data from field studies, information on growth and management of plantations and surrounding areas, and expected effects of the transgenes themselves on fitness. Simulation results will then be evaluated in terms of potential impacts and means for mitigation. Results from these studies will be help clarify whether stringent measures for transgene containment, such as genetically engineered sterility, are warranted.

METHODS AND MATERIALS

Study sites

We studied gene flow in the vicinity of two industrial plantations of hybrid poplar clones (*Populus trichocarpa* x *Populus deltoides*) in western Oregon: River Ranch and Clatskanie. The plantations contained large monoclonal blocks of reproductively mature male and female clones (Figure 1).

Figure 1. Study sites. A: Clatskanie; B: River Ranch. Circles represent female trees for which progeny were analyzed. Squares represent establishment plots. Symbol size represents the proportion of progeny with hybrid parentage (indicative of pollen or seed flow from plantations). o: 0% hybrid pollination; •: 1% hybrid pollination; ●: 3% hybrid pollination; .: 10% hybrid seedlings; ■: 70% hybrid seedlings. Black blocks were mature stands of hybrid male cottonwood.

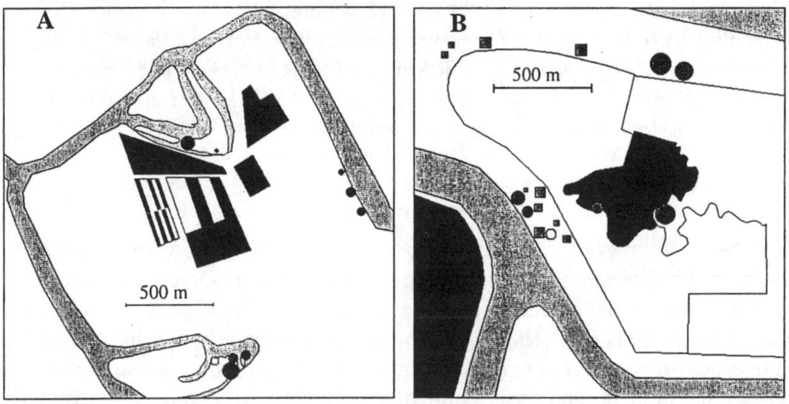

Identification of hybrid progeny

We identified progeny of hybrid parents using a three-tiered approach. First, we examined leaf morphology of seedlings to identify putative hybrids. *P. deltoides* does not occur in western Oregon, and there are several morphological characters that distinguish this species from the native *P. trichocarpa* (Eckenwalder, 1996). For example, abaxial leaf color and the shape and length of leaf petioles are quite distinct between the parental species, and hybrid progeny are intermediate. Therefore, we used leaf morphology to screen large numbers of progeny for hybrid parentage.

Second, we confirmed hybrid parentage using molecular markers. We identified five RAPD loci (Williams *et al.,* 1990) and one simple sequence repeat locus (SSR, Powell *et al.*, 1997) that were fixed in *P. deltoides* and absent in *P. trichocarpa.* We extracted DNA from foliage and analyzed RAPDs using protocols developed in our lab (http://www.fsl.orst.edu/tgerc/dnaext.htm; Aagaard *et al.*, 1995). We are now performing paternity analyses on hybrid progeny using a set of highly polymorphic SSR loci developed with Dr. Toby Bradshaw, University of Washington, USA).

Gene flow by pollen

We examined pollen flow from hybrid plantations by collecting seeds from eight wild *P. trichocarpa* females growing in close proximity to the hybrid plantations (Figure 1). In order to estimate maximum expected levels of gene flow, we selected the closest female trees (2 to 500 m from plantations) that overlapped extensively in flowering phenology with hybrid clones. We germinated 100 to 300 seeds per female in a greenhouse and transplanted seedlings to the field in the spring following collection. We assessed leaf morphology following one year of growth, and extracted DNA from all putative hybrids for molecular analyses.

Gene flow by seed

We placed 20 seed traps directly adjacent to one of the plantations (River Ranch) to gauge the magnitude of hybrid gene flow by seed (Figure 1). Seeds were germinated in a greenhouse and hybrid progeny were identified as described above. We also examined seedling establishment in the vicinity of the seed traps using artificially disturbed 1 m^2 plots. Poplars depend on disturbance to become established; we therefore removed competing vegetation and watered during dry periods to ensure some seedling survival. We then measured the growth and survival of all seedlings, and identified hybrid progeny using RAPD and SSR markers. For comparison, we also measured growth and survival of seedlings from seed traps that had been germinated in the greenhouse and grown in randomized field plots in Corvallis, OR.

Establishment in the wild

We surveyed natural poplar establishment in the vicinity of flowering plantations in three locations: the western Columbia River near Clatskanie, Oregon; the Skagit River near Sedro Wooley, Washington; and the Fraser River west of Vancouver, British Columbia. These areas were chosen to provide the best opportunity to observe hybrid seedling establishment in the region. We measured the approximate area and density of seedling recruitment, and examined leaf morphology to identify hybrids. We also collected foliage from 96 seedlings at each site for molecular analyses.

RESULTS AND DISCUSSION

Gene flow by pollen

Assessments of hybridity based on seedling leaf morphology matched RAPD results more than 90% of the time. Frequency of hybrid progeny averaged 0.7% (range 0 to 3.6%) at Clatskanie and 1.9% (range 0 to 3.8%) at River Ranch. There was no apparent relationship with direction or distance, though the sampling was not designed to explicitly test for these effects (Figure 1).

Gene flow by seeds

On average, 18% of seedlings from seed traps were of hybrid origin (range 5 to 65%), while an average of 42% of seedlings in plots were hybrids (range 9 to 75%) (Figure 1). Overall, more hybrid seedlings established in plots than expected based on seeds collected in adjacent traps (Wilcoxon Signed-Rank Test, $P = 0.001$). Hybrids also had higher overwinter survival than wild seedlings, though the difference was not significant (Wilcoxon Signed-Rank Test, $P = 0.06$. In addition, hybrid seedlings were significantly larger than wild seedlings at the end of the growing season (ANOVA, $P= 0.0001$) (Figure 2). However, this relationship did not hold for seedlings derived from seed traps and grown under controlled conditions: there was no significant difference in size of wild and hybrid seedlings after one year of growth in the field (ANOVA, $P=0.24$) (Figure 2).

Figure 2 Least square mean of seedling volume index (diameter^2 * height) of seedlings derived from seed traps and grown under controlled conditions, and seedlings naturally established in the experimental plots in the field.

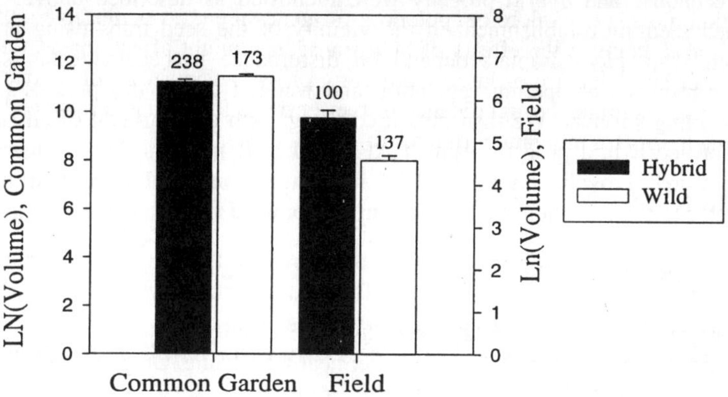

Establishment in the wild

We observed a very low rate of hybrid establishment under wild conditions in the vicinity of plantations. All plantations had been flowering for at least three years prior to the survey, and one for as long as 15 years (Fraser River). However, areas of poplar establishment were relatively small, and the frequency of hybrids ranged from 0 to 1% (Table 1), despite the fact that we sampled as close to plantations as possible. These plantations were all surrounded by extensive, heavily flowering stands of wild cottonwood which may have provided a dilution effect for hybrid propagules.

Table 1. Surveys of seedling establishment in the vicinity of hybrid plantations

Site	Area (ha)[a]	Plantation Area (ha)[b]	Seedling Area (ha)[c]	Seedlings Examined	RAPD Hybrids[d]	Total Hybrids[e]
Fraser R.	100	118	0.05	~2500	1 (1%)	2
Columbia R.	2000	250	0.02	1200	1 (1%)	3
Skagit R.	56	44	0.25	>100000	0 (0%)	0

a, Approximate area surveyed for poplar establishment
b, Area of flowering hybrid poplar plantations in the survey zone
c, Approximate area of seedling establishment
d, Number (%) hybrid seedlings determined by RAPD analysis, based on 96 samples
e, Total number of hybrids, including those originally identified by leaf morphology

CONCLUSIONS

Populus trichocarpa x deltoides hybrids are capable of producing progeny that are fit enough to grow and compete in the wild in the Pacific Northwest of the United States. There is therefore little evidence that these intersectional hybrids are effectively 'domesticated.' However, because of their stringent habitat requirements, opportunities for establishment are rare. Also, the abundance of seed and pollen from wild stands may provide a powerful diluton effect, even very close to hybrid plantations. Thus, despite the physical potential for long distance gene flow from these trees, the rate of effective gene flow is likely to be very low. We will use our simulation models to explore the potential for gene spread over multiple generations. We will also examine the effects of how transgenes might affect spread rates in the presence of herbicide use or insect herbivory, or by causing male or female sterility. We will then examine the significance of these effects for wild and managed systems, including means for mitigation, if needed.

REFERENCES

Aagaard J E; Vollmer S; Sorensen F; Strauss S H (1996). Mitochondrial DNA products among RAPD profiles are frequent and strongly differentiated between races of Douglas-fir. *Molecular Ecology,* **4**, 441-447.

Dekker-Robertson DL; Libby W J (1998). American forest policy- global ethical tradeoffs. *Bioscience,* **48**, 471-477.

Eckenwalder J E (1996). Systematics and evolution of Populus. In: *Biology of Populus and its implications for management and conservation*, eds R F Stettler, H D; Bradshaw Jr.; P E Heilman; T M Hinckley. pp. 7-32. National Research Council Canada: Ottawa, ON.

Food and Agricultural Organization of the United Nations (FAO) (1997). *State of the World's Forests, 1997.* FAO: Rome, Italy.

Higgins S I; Richardson D M (1998). Pine invasions in the southern hemisphere: modelling interactions between organism, environment, and disturbance. *Plant Ecology,* **135**, 79-93.

James R; DiFazio S P; Brunner A; Strauss S H (1998). Environmental effects of genetic engineering of woody biomass crops. *Biomass & Bioenergy* **14**, 403-414.

National Research Council (1989). *Testing genetically modified organisms: Framework for decisions.* National Academy Press: Washington, D.C.

Powell W; Machray G C; Provan J (1996). Polymorphism revealed by simple sequence repeats. *Trends Plant Science,* **1**, 215-222.

Rogers H J; Parkes H C (1995). Transgenic plants and the environment. *Journal of Experimental Botany* **46**, 467-488.

Strauss S H; Meilan R; DiFazio S; Leonardi S; Brunner A; Mohamed R; Skinner J; (1998). *tree genetic engineering research cooperative (TGERC) annual report: 1997-1998.* Forest Research Laboratory, Oregon State University: Corvallis, OR.

Williams J G; Kubelik A R; Livak K J; Rafalski J A; Tingey S V (1990). DNA polymorphisms amplified by arbitrary primers are useful as genetic markers. *Nucleic Acids Research,* **18**, 6531-6535.

Wilson E O (1992). *The diversity of life.* Belknap Press: Cambridge, MA.

ACKNOWLEDGEMENTS

This research was supported by USDA Biotechnology Risk Assessment Grant 97-39210-5022, The US EPA STAR Fellowship Program, the US DOE Biofuels and Feedstocks Program, and the Tree Genetic Engineering Research Cooperative (http://www.fsl.orst.edu/tgerc/index.htm).

SESSION 6

CONSEQUENCES OF GENE FLOW BETWEEN HIGHER PLANTS AND OTHER ORGANISMS

Chairman & Session Organiser: Dr I Williams
IACR-Rothamsted, Harpenden

Insecticidal transgenes into nature: gene flow, ecological effects, relevancy, and monitoring

C Neal Stewart, Jr.
Department of Biology, University of North Carolina, Greensboro, NC 27402-6174 USA

ABSTRACT

Highly effective genes conferring pest resistance have been and are being engineered into crop plants. There is a strong likelihood that these transgenes will be transferred from agronomic ecosystems into natural ecosystems. There will be ecological risks ranging from creating more invasive weeds to affecting beneficial insects. I argue for the need of relevancy in choosing appropriate experimental systems for assessing ecological risks of commercial transgenic insecticidal crops. Finally, I will describe a transgene monitoring system based upon green fluorescent protein (GFP), and how it may be used commercially.

THE CURRENT INSECTICIDAL CROP PIPELINE

During the past 10 years, transgenic crop development effort has increased exponentially. In the USA alone, there are now four different transgenic pest resistant crops that have been deregulated (commercialised): corn, cotton, potato, and tomato. These four, all containing *Bacillus thuringiensis* crystal toxin (Bt *cry*) transgenes, are the first of many pest-resistant crops to flow from the industrial R&D pipeline. To date, there have been eighteen plant species transformed with Bt cry genes and field tested in the USA: alfalfa, *Amelanchier*, apple, belladonna, cabbage, cranberry, corn, cotton, eggplant, grape, oilseed rape, *Populus*, potato, rice, soybean, tobacco, tomato, and walnut (USDA APHIS permits, Oct., 1998). This represents a 50% increase of plant species during the last three years. Most of these plants are slated for commercialisation with insecticidal genes. While the arsenal of insecticidal genes is growing, Bt will be the primary insect resistance transgene on the market for the next 10 years.

Bacillus thuringiensis crystal endotoxins

The first Bt transgenic plants were produced over 12 years ago (Vaeck *et al.,* 1987). These first plants contained native Bt genes that were not expressed very well. Now the genes have been codon-optimized for high expression in plants and have proven to be very effective to controlling specific insects (Perlak *et al.,* 1991; Adang *et al.,* 1993). Indeed, exclusively synthetic, codon-optimized Bt genes are now used for plant nuclear transformation. The class of compounds that are responsible for insecticidal activity are crystalline proteins also known as Cry proteins or delta-endotoxins. The mode of action of Bt endotoxins is the disruption of cellular membranes in the midgut. Endotoxins are proteolytically converted into small polypeptides in the midgut. These bind to glycoprotein receptors and disrupt osmotic processes (Adang, 1991). Bt Crys have high specificity of toxicity, a highly desirable trait.

As with any insecticide that has extensive use, insects can acquire resistance to Bt. Many have warned that Bt resistance genes will become fixed in insect populations rendering both Bt transgenics and Bt sprays ineffective (Whalon & McGaughey, 1993; Tabashnik, 1994; Whalon,

1994), although there are a minority of scientists who do not seem to be very concerned about Bt resistance management (Altman *et al.,* 1996). Both diamondback moth (*Plutella xylostella*) and tobacco budworm (*Heliothis virescens*) biotypes have been found to be resistant to several Bt toxins (Gould *et al.,* 1992; Heckel, 1994; Gould *et al.,* 1997; Tabashnik *et al.,* 1997b; Liu *et al.,* 1998). Indeed, the genetic loci responsible for resistance in both species have been mapped (Heckel *et al.,* 1997; Tabashnik *et al.,* 1997a). Thus, several strategies have been proposed, including refugia, high dose, mosaics, rotations, and transgene combinations (reviewed in Tabashnik, 1994). Currently, high dose (high Bt expression) and refugia are used to manage resistance in all Bt transgenic crops grown in the USA (Gould, 1998), but industry is moving toward transgene combinations (pyramiding) to manage Bt resistance. Pyramiding, in this case combining Bt genes with other transgenes has been proposed as a necessary strategy to prevent the development of Bt resistance (Wilson *et al.,* 1992; Boulter, 1993; McGaughey 1994). A notable advantage to pyramiding is that it would be transparent to the grower and not have the associated yield loss the refugia strategy demands.

New genes for insect resistance

There are three main reasons why several groups are racing to discover novel insecticidal genes. The first is to engineer plants insecticidal to control insects that are not susceptible to Bt Cry. The second is to discover candidate genes to pyramid with Bt. A third reason, related to the first, is to capture a unique market niche. The benchmark of any new gene candidate in transgenic plants that the gene product should have at least equal toxicity compared with Bt. However, one potential problem is that new toxins have less specificity than Bt. This means that insecticidal transgenic plants of the future will likely have an increase of non-target effects. I will briefly review the best candidate genes for commercialisation in transgenic plants. More extensive reviews have been published recently (Estruch *et al.,* 1997; Gatehouse & Gatehouse, 1998; Jouanin *et al.,* 1998; Schuler *et al.,* 1998).

Cholesterol oxidase

Cholesterol oxidase (CO) from *Streptomyces* culture filtrate has been found to be highly toxic to boll weevil (*Anthonomus grandis*) (Purcell *et al.,* 1993). Monsanto is presumably developing CO for the control of this economic insect on cotton (Purcell *et al.,* 1993; Greenplate *et al.,* 1995). However, it also has activity against southern corn rootworm (*Diabrotica undecimpunctata*), tobacco budworm (*Heliothis virescens*), and yellow mealworm (*Tenebrio molitor*) (Shen *et al.,* 1997). The mode of action is the lysis of midgut epithelial cells (Purcell *et al.,* 1993).

Vegetative Insecticidal Proteins

In contrast to Bt endotoxins, which are accumulated to high amounts when Bt sporulates, Bt also produces vegetative insecticidal proteins (Vips) when it is not sporulating (Estruch *et al.,* 1996; Yu *et al.,* 1997). While Vips come from Bt, they are unrelated to Bt endotoxins (Estruch *et al.,* 1996). Similar to cholesterol oxidase, the mode of action is midgut cell lysis (Yu *et al.,* 1997). The Vip3A insecticidal protein has been shown toxic to black cutworm, (*Agrotis ipsilon*), fall armyworm (*Spodoptera frugiperda*), beet armyworm, (*Spodoptera exigua*), tobacco budworm (*H. virescens*), and corn earworm, (*Helicoverpa zea*), a broad range of hosts (Estruch *et al.,* 1996). Novartis is apparently developing Vips for controlling corn insects.

Photorhabdus luminescens toxins

Newly discovered toxins from the bacterium *Photorhabdus luminescens*, which make their living in gut of entomophagous nematodes have been shown to be toxic to several orders of insects (Bowen *et al.*, 1998). Similar to cholesterol oxidase and Vips, midgut epithelial cells are damaged in insects that have consumed the toxins. DowAgroSciences has acquired an exclusive license for the use of *P. luminescens* toxins, and will presumably move towards the goal of commercialisation of transgenic plants expressing toxin-encoding genes.

Other-bioactive proteins for insect control

There are several insect-control proteins that have been proposed to be pyramided with Bt. None of these have the toxicity of Bt Cry proteins, cholesterol oxidase, Bt Vips, or *P. luminescens* toxins. The most studied of such proteins are proteinase inhibitors. The other insect resistance proteins that have been received attention are lectins and chitinases.

Proteinase inhibitors (PIs) from plants have been studied as candidates for insect control for over 20 years (Ryan, 1990) The mode of action is the overstimulation of the production of trypsin, chymotrypsin and other proteases in the insect gut (Broadway & Duffey,1986). Most PIs inhibit insect growth but are not antibiotic. Recent data supports the hypothesis that insect digestive physiology has memory of sorts. That is, certain insects can alter their arsenal of digestive enzymes (e.g., trypsins) to overcome specific PIs. However, if an insect does not feed on a certain plant, it may not be able to readily overcome the PI and as a result, its growth and development will be inhibited (Jongsma *et al.*, 1995; Broadway, 1995; Broadway & Villani, 1995). Therefore, certain insects may be preadapted for resistance to PIs. This fact and their relatively low toxicity to insects hamper their commercial feasibility.

Lectins, carbohydrate-binding proteins, from various plant species such as the snowdrop (*Galanthus nivalis*) also decrease insect growth (Powell *et al.*, 1993). Lectins bind brush border membrane proteins of various insects, but there is no apparent relationship between the ability to bind and toxicity to the host (Harper *et al.*, 1995). Lectins are generally more effective in transgenic plants than PIs. Lectins in transgenic crop plants might also be of special biosafety concern because of reports they can act as mitogens to human T-cells (Peumans *et al.*, 1997).

Chitinases have also received attention recently as possible insect control agents that could be used in transgenic plants (Kramer & Muthukrishnan, 1997). Chitinases from insects digest chitin, an important constituent of insect exoskeletons and gut linings. Similar to PIs and lectins, chitinases affect a broad range of insects. They have apparent toxicity between PIs and Bt Cry for insects they affect. Unlike PIs, chitinases seem to be effectively synergistic with Bt Cry toxins (Kramer & Muthukrishnan, 1997; Santos *et al.*, 1997).

TRANSGENE FLOW FROM CROPS TO WEEDS

There are several crops in the US and Europe that have the potential to hybridise with wild relatives. The crops that have sexually-compatible wild relatives growing in proximity to them are at risk for receiving fitness-enhancing transgenes such as insecticidal genes, which could alter ecological parameters. Some examples in the USA and Europe are rice (Langevin *et al.*,

1990), sorghum (Paterson *et al.,* 1995), sugar beet (Bartsch *et al.,* 1996; Bartsch & Pohlorf, 1996; Bartsch & Schmidt, 1997), and sunflower (Whitton *et al.,* 1997). There are several incidental crops, vegetables and fruits that also have neighbouring wild relatives (reviewed by Raybould & Gray,1993). Of course, the largest concern in the USA and Europe has been over oilseed rape (OR) (*Brassica napus*), a crop with numerous wild relatives and increasing world-wide cultivation. OR has been the subject of extensive research, is relatively easy to transform, and has been the model of choice for biotechnology risk research. I will briefly review its breeding and biology and some of the research that has been recently accomplished in OR transgene flow and insecticidal transgenic OR.

The case of *Brassica napus* (oilseed rape)

The mustard (Brassicaceae or Cruciferae) family, to which the genus *Brassica* belongs, contains many important crop plants and weeds. *Brassica napus*, is an amphiploid species (AACC 2n=38) which putatively arose from naturally occurring interspecific hybridization between *Brassica oleracea* (AA) and *Brassica rapa* (CC). Both winter and spring forms exist within *B. napus*. Subsequent growth patterns differ depending on the climate of the production region and the form grown. Winter forms are fall-seeded and spring-harvested, while spring forms may be either grown as a spring-seeded annual crop in temperate regions or as a fall-seeded crop in milder climates, such as the southeastern USA. *Brassica napus* is a self-pollinating species that outcrosses readily with the assistance of wind and insect pollinators. Outcrossing frequencies as high as 30% for directly adjacent plants have been reported (Robbelin & Downey, 1989). *Brassica napus* can be a volunteer in other crops and along roadsides but it is not considered to be a frequent invader of non-disturbed ecosystems (Rich ,1991; Crawley *et al.,* 1993).

OR is known to be interfertile with wild *B. napus, B. rapa, B. oleracea, Brassica nigra, Brassica kaber* (*Sinapsis arvensis*) *Brassica juncea*, *Brassica adpressa*, and *Raphanus raphanistrum* (Bing, 1991; Kerlan *et al.*, 1992, 1993; Scheffler & Dale, 1994; Eber *et al.*, 1994; Darmency *et al.*, 1995; Mikkelsen *et al.*, 1996; Chevre *et al.*, 1997, Metz *et al.*, 1997). One conclusion of these studies is that there is significant maternal effect in the efficacy of the crosses, with highest hybridisation potential being when *B. napus* is used as the pollen recipient. Agronomically-important transgene gene flow from OR will likely initially occur with OR as the pollen donor. Even though hybridisation frequencies were low, all the authors warned of significant introgression possibilities, as several of the resultant hybrids were fertile, especially under open pollinating conditions. The complicated taxonomy and tractable biotechnology of OR and its relatives have made this system a popular one for researching the risks of agricultural biotechnology.

Transgenic insecticidal OR, wild relatives and herbivory

There are at least three foreseeable effects of the gene flow of insecticidal genes from crop to wild relative. First, the persistence of transgenes may skew transgenic crop:regufia areas, which could slightly affect resistance management (Wearing & Hokkanen, 1995). Second, insecticidal transgenes could decrease beneficial and/or non-target insect populations. The published studies of Bt side-effects to beneficial insects are sparse. However, where they have been studied in the field, significant effects to beneficial insects have not been found (Flint *et al.,* 1995; Sims, 1995; Arpaia, 1996; Orr & Landis, 1997). One example contrary to this is a recent report (Hilbeck, *et al.*, 1998) that involved laboratory experiments using Bt corn, the pest/prey insect *Ostrinia nubilalis* (European corn borer, ECB), and the lepidopterous predator *Chrysoperla carnea.* While no direct effect of Bt was found on *C. carnea*, the predator grew more slowly and had greater mortality when fed Bt-exposed prey. The authors hypothesised that the Bt-stunted ECB were sick, leading to indirect toxicity to the non-target insect. While it is doubtful that direct effects of Bt on non-target pests will be obtained, the toxins entering the pipeline may have increased side-effects. Since they have a broader host range, there is a higher probability of increased risks on non-target insects; risks that will need to be explored on a case-by-case basis. The third and potentially the largest effect of gene flow of insecticidal transgenes into wild relatives will be the alteration of fitness or increased invasiveness of the plant host. As above, the OR/wild relative system is our best model.

There are very limited data on insect herbivory on the fitness of OR relatives. One study in small field plots in England showed that foliar insecticide did not increase fitness of wild radish (Rees & Brown, 1992). However, ambient insect levels were low and little damage (<5% defoliation) occurred on no-insecticide-treatment plants. However, because of higher ambient slug populations, higher fitness was observed on molluscicide-treatment plants (Rees & Brown, 1992). These results show that insecticidal (Bt) *R. raphanistrum* and *B. rapa* could be new risks that should be assessed. Certainly others have warned that insect resistance in weeds would not be a desirable modification (Kareiva *et al.,* 1994; Stewart *et al.,* 1997). My lab is in the process of testing the ecological performance of Bt-transgenic *B. rapa* in the field; the litmus test of gene flow/transgene effect in OR.

Relevancy of biotech risk research

Three field trials have been performed by companies (AgrEvo and Calgene) in 1998 using insecticidal OR in the USA (USDA APHIS permits, Oct., 1998). These companies tested lepidopteran-resistant transgenic OR with unspecified genes (confidential business information) in California. We can assume that these field trials are a precursor to eventual commercial releases. It is evident that any commercial transgenic OR will donate genes to wild relatives grown in proximity. It is also evident that insecticidal genes with broader pest toxicity will be used more often in the future. Insecticidal proteins in transgenic plants that affect a greater number of pests will lead to greater uncertainty in two risk factors. The first is the effect to non-target or beneficial insects. The second is herbivore-mediated selection effects that favour insecticidal transgenic plants over non-transgenic plants. These effects could likely interact with each other. Added to these will be the effects of scale (Dale,1997). Commercialisation area is much larger than field trial area. Simply, the ability for the transgenic plant to affect negatively more organisms will lead to greater complexity of risks.

The interaction of gene flow and pest resistance, therefore, needs to be studied using relevant insecticidal genes, promoters, and hosts. I believe that the scientific community has responded appropriately for the most part. However, the public availability of genes coding for cholesterol oxidase, Vips, and *P. luminescens* toxins will likely be limited because of intellectual property concerns and other business reasons. Therefore, I predict that increased co-operation between industry and public researchers will be necessary to thoroughly test new transgenic insecticidal plants for ecological risks. The days of using PI genes with the *CaMV* 35S promoter are over. These experiments are no longer highly relevant to tomorrow's agbiotech products. Can we envision the day when multiple transgenes of broad effects will put be into crop plants such as

OR? Will we be able to even thoroughly test for and predict complex ecological interactions that are set-off from such a GMO introduction? The answer is both yes and no. To manage for unintended effects of gene flow, it seems to be desirable to use monitoring tools to track transgenes. Transgene monitoring may not be able to mitigate for increased side-effects to non-target pests, but it would enable the management of potentially problematic genes flowing out of the host crop. One group has suggested an alternative to monitoring (Daniell *et al.,* 1998) consisting of transferring transgenes into chloroplasts, which are assumed to be maternally inherited. This approach is a controversial solution to transgene containment that is based upon faulty assumptions (Cummins, 1998; Stewart & Prakash, 1998). However, the most pragmatic pitfall to this possible solution is that tobacco is the only plant amenable to chloroplast transformation today. Since the ability to transform any crop using chloroplast transformation is at least a decade away, it should not be considered a commercially viable alternative.

TRANSGENE MONITORING

In the past, studies of gene flow and transgene persistence in the environment have been monitored largely using phenotypic (Manasse, 1992; Luby & McNicol, 1995), biochemical (Klinger *et al.,* 1992; Arias & Reiseberg, 1994), or molecular markers (Jorgensen & Andersen, 1994; Rogers *et al.*, 1996). These often require expensive substrates (in the case of GUS (Jefferson, 1989)), are not suitable for real-time and *in vivo* detection, and are not universal. However, with the recent availability of green fluorescent protein (GFP) encoding genes, a tractable monitoring system is feasible and has been recently introduced (Stewart, 1996). In this method GFP in transgenic plants can be visualised on a macroscopic scale non-destructively using ultraviolet light. GFP from jellyfish is a 27 kD monomer that has the unique characteristic of fluorescing green when excited with ultra-violet (360-400 nm) or blue (440-480 nm) light. This fluorescence depends only upon elemental oxygen, necessary for double-bond formation between the a and b bond of Tyr 66 (Ormo *et al.,* 1996; Yang *et al.,* 1996), and UV or blue light. Therefore, the protein does not need any substrate, enzyme, or co-factor for fluorescence, making it a genuine, *in vivo* marker (Niedz *et al.,* 1995).

Transgenic plants expressing GFP can be detected using only a hand-held UV light *in vivo* and in real time. Transgenic plants appear green under ultra-violet light amidst non-transgenic plants that appear red due to the red autofluorescence of chlorophyll when excited with UV light. Green fluorescent protein is stably inherited in progeny and is useful as a tag to mark transgenic plants *in vivo*. My lab (Stewart, 1996; Leffel *et al.,* 1997) and others (Pang *et al.,* 1996; Haseloff *et al.,* 1997) have shown that whole plant fluorescence with GFP is a powerful tool, and that GFP is stably inherited (Leffel *et al.,* 1997). We are now employing an improved chimeric gene with a endoplasmic reticulum retention signal (Haseloff *et al.,* 1997), *mgfp5-er,* which has proven to be much improved over *mgfp4* as a whole plant marker. (Mabon *et al.*, submitted). This procedure could be of great benefit in a larger transgenic crop management system (Marshall, 1998).

For this method to be successful a GFP gene is linked or fused to a gene of interest, e.g., an insecticidal transgene on the same plasmid, prior to introduction into plants. The gene of interest is then monitored by visual observation of GFP in the plant. My lab has shown that this procedure functions as expected (Mabon *et al.*, submitted). In addition, there is no evidence that GFP is toxic or has any costs to the host plants in the field (Harper *et al.*, submitted). These two facts greatly enhance the usability of a GFP monitoring system. One can envision releasing plants with blue, cyan, green and yellow fluorescent proteins linked to genes of different agronomic traits. One can also envision remote sensing technology and precision agricultural applications of the technology.

CONCLUSION

In conclusion, it is clear that the pipeline of pest-resistant transgenic crops is full. These crops have the potential to greatly decrease conventional pesticide inputs into the environment, and to improve agronomic efficiency. There are, however, real risks that arise from the novel gene introductions and gene flow from crop to wild relatives that will persist in the environment. It is clear these risks need to be assessed, and science-based decisions made with regards to real biosafety issues. It also seems prudent to develop monitoring methods to track transgenes in the environment. Monitoring can serve as a long-term ecological research tool, and as a means to control for unintended problems that might arise from commercial releases.

ACKNOWLEDGMENTS

Thanks to collaborators and colleagues for their intellectual and technical contributions to projects that are related to this review. Funding from the USDA, NSF, and industry is greatly appreciated.

REFERENCES

Adang MJ (1991). *Bacillus thuringiensis* insecticidal crystal proteins: Gene structure, action, and utilization. In E. Maramorosch (Ed.), *Biotechnology for Biological Control of Pests and Vectors.* (pp. 3-24). Boca Raton, Florida: CRC Press.

Adang MJ; Brody MS; Cardineau G; Eagan N; Roush RT; Shewmaker CK; Jones A; Oakes JV; McBride KE (1993). The construction and expression of a *Bacillus thuringiensis cryIIIA* gene in protoplasts and potato plants. *Plant Molecular Biology* **21**, 1131-1145.

Altman DW; Benedict JH; Sachs ES (1996). Transgenic plants for the development of durable insect resistance. *Engineering Plants for Commercial Products and Applications* **792**, 106-114.

Arias DM; Reiseberg LH (1994). Gene flow between cultivated and wild sunflowers. *Theoretical and Applied Genetics* **89**, 655-660.

Arpaia S (1996). Ecological impact of Bt-transgenic plants: 1. Assessing possible effects of CryIIIB toxin on honey bee (*Apis mellifera* L.) colonies. *Journal of Genetics and Breeding* **50**, 315-319.

Bartsch D; Pohlorf M (1996). Ecological aspects of transgenic sugar-beet - transfer and expression herbicide resistance in hybrids with wild beets. *Euphytica* **91**, 55-58.

Bartsch D; Schmidt M (1997). Influences of sugar beet breeding on populations of *Beta vulgaris* ssp. *maritima* in Italy. *Journal Of Vegetation Science* **8**, 81-84.

Bartsch D; Schmidt M; Pohlorf M; Haag C; Schuphan I (1996). Competitiveness of transgenic sugar-beet resistant to beet necrotic yellow vein virus and potential impact on wild beet populations. *Molecular Ecology* **5**, 199-205.

Bing DJ (1991). Potential of Gene Transfer Among Oilseed *Brassica* and Their Weedy Relatives. University of Saskatchewan, Saskatchewan. Unpublished MS Thesis.

Boulter D (1993). Insect pest control by copying nature using genetically engineered crops. *Phytochemistry* **34**, 1453-1466.

Bowen D; Rocheleau TA; Blackburn M; Andreev O; Golubeva E; Bhartia R; ffrench-Constant RH (1998). Insecticidal toxins from the bacterium *Photorhabdus luminescens*. *Science* **280**, 2129-2132.

Broadway RM (1995). Are insects resistant to plant proteinase inhibitors? *Journal of Insect Physiology* **41**, 107-116.

Broadway RM; Duffey SS (1986). Plant proteinase inhibitors: Mechanism of action and effect on the growth and digestive physiology of larval *Heliothis zea* and *Spodoptera exiqua*. *Journal of Insect Physiology* **32**, 827-833.

Broadway RM; Villani MG (1995). Does host range influence susceptibility of herbivorous insects to non-host proteinase inhibitors? *Entomolgia Experimentalis et Applicata* **76**, 303-312.

Chevre AM; Eber F; Baranger A; Renard M (1997). Gene flow from transgenic crops. *Nature* **389**, 924

Crawley MJ; Hails RS; Rees M; Kohn D; Buxton J (1993). Ecology of transgenic oilseed rape in natural habitats. *Nature* **363**, 620-623.

Cummins JE (1998). Chloroplast-transgenic plants are not a gene flow panacea. *Nature Biotechnology* **16**, 401

Dale PJ (1997). Potential impacts from the release of transgenic plants into the environment. *Acta Physiologiae Plantarum* **19**, 595-600.

Daniell H; Datta R; Varma S; Gray S; Lee S-B (1998). Containment of herbicide resistance through genetic engineering of the chloroplast genome. *Nature Biotechnology* **16**, 345-348.

Darmency H; Fleury A; Lefol E (1995). Effect of transgenic release on weed biodiversity: oilseed rape and wild radish. *Brighton Crop Protection Conference-Weeds* **5A-2**, 433-438.

Eber F; Chevre AM; Baranger A; Vallee P; Tanguy X; Renard M (1994). Spontaneous hybridization between a male-sterile oilseed rape and two weeds. *Theoretical and Applied Genetics* **88**, 362-368.

Estruch JJ; Carozzi NB; Duck NB; Warren GW; Koziel MG (1997). Transgenic plants: An emerging approach to pest control. *Nature Biotechnology* **15**, 137-141.

Estruch JJ; Warren GW; Mullins MA; Nye GJ; Craig JA; Koziel MG (1996). VIP3A, a novel *Bacillus thuringiensis* vegetative insecticidal protein with a wide spectrum of activities against lepidopteran insects. *Proceedings of the National Academy of Sciences of the United States of America* **93**, 5389-5394.

Flint HM; Henneberry TJ; Wilson FD; Holguin E; Parks N; Buehler RE (1995). The effects of transgenic cotton, *Gossypium hirsutum* L, containing *Bacillus thuringiensis* toxin genes for the control of the pink-bollworm, *Pectinophora gossypiella* (Saunders) Lepidoptera, Gelechiidae and other arthropods. *Southwestern Entomologist* **20**, 281-292.

Gatehouse AMR; Gatehouse JA (1998). Identifying proteins with insecticidal activity: Use of encoding genes to produce insect-resistant transgenic crops. *Pesticide Science* **52**, 165-175.

Gould F (1998). Sustainability of transgenic insecticidal cultivars: integrating pest genetics and ecology. *Annual Review of Entomology* **43**, 701-726.

Gould F; Anderson A; Jones A; Sumerford D; Heckel DG; Lopez J; Micinski S; Laster M; Leonard R (1997). Initial frequency of alleles for resistance to *Bacilllus thuringiensis* toxins in field populations of *Heliothis virescens*. *Proceedings of the National Academy of Sciences of the United States of America* **94**, 3519-3523.

Gould F; Martinez-Ramirez A; Anderson A; Ferre J; Silva FJ; Moar WJ (1992). Broad-spectrum resistance to *Bacillus thuringiensis* toxins in *Heliothis virescens*. *Proceedings of the National Academy of Science, USA* **89**, 7986-7634.

Greenplate JT; Duck NB; Pershing JC;Purcell JP (1995). Cholesterol oxidase: An oostatic and larvicidal agent active against the cotton boll weevil, *Anthonomus grandis*. *Entomologia Experimentalis et Applicata* **74**, 253-258.

Harper BK; Leffel SM; Stewart CN Jr. (submitted). The fitness of transgenic plants of expressing high levels of mutant genes encoding green fluorescent protein. *Molecular Ecology*.

Harper SM; Crenshaw RW; Mullins MA; Privalle LS (1995). Lectin-binding to insect brush-border membranes. *Journal of Economic Entomology* **88**, 1197-1202.

Haseloff J; Siemering KR; Prasher DC; Hodge S (1997). Removal of a cryptic intron and subcellular localization of green fluorescent protein are required to mark transgenic *Arabidopsis* plants brightly. *Proceedings of the National Academy of Sciences of the United States of America* **94**, 2122-2127.

Heckel DG (1994). The complex genetic-basis of resistance to *Bacillus thuringiensis* toxin in insects. *Biocontrol Science and Technology* **4**, 405-417.

Heckel DG; Gahan LC; Gould F; Anderson A (1997). Identification of a linkage group with a major effect on the resistance to *Bacillus thuringiensis* Cry1Ac endotoxin in the tobacco budworm (Lepidoptera: Noctuidae). *Journal of Economic Entomology* **90**, 75-86.

Hilbeck A; Baumgartner M; Fried PM; Bigler F (1998). Effects of transgenic *Bacillus thuringiensis* corn-fed prey on mortality and development time of immature *Chrysoperla carnea* (Neuroptera: Chrysopidae). *Environmental Entomology* **27**, 480-487.

Jefferson RA (1989). The GUS reporter gene system. *Nature* **342**, 837-838.

Jongsma MA; Bakker PL; Peters J; Bosch D; Stiekema WJ (1995). Adaption of *Spodoptera exigua* larvae to plant proteinase inhibitors by induction of gut proteinase activity insensitive to inhibition. *Proceedings of the National Academy of Science, USA* **92**, 8041-8045.

Jorgensen RB; Andersen B (1994). Spontaneous hybridization between oilseed rape (*Brassica napus*) and weedy *B. campestris* (Brassicaceae): a risk of growing gentically modified oilseed rape. *American Journal of Botany* **81**, 1620-1626.

Jouanin L; Bonade-Bottino M; Girard C; Morrot G; Giband M (1998). Transgenic plants for insect resistance. *Plant Science* **131**, 1-11.

Kareiva P; Morris W; Jacobi CM (1994). Studying and managing the risk of cross-fertilization between transgenic crops and wild relatives. *Molecular Ecology* **3**, 15-21.

Kerlan MC; Chevre AM; Eber F (1993). Interspecific hybrids between a transgenic rapeseed (*Brassica napus*) and related species: cytogenetical characterization and detection of the transgene. *Genome* **36**, 1099-1105.

Kerlan MC; Chevre AM; Eber F; Baranger A; Renard M (1992). Risk assessment of outcrossing of transgenic rapeseed to related species: I. interspecific hybrid production under optimal conditions with emphasis on pollination and fertilization. *Euphytica* **62**, 145-153.

Klinger T; Arriola PE; Ellstrand NC (1992). Crop-weed hybridization in radish (*Raphanus sativus*): Effects of distance and population size. *American Journal of Botany* **79**, 1431-1435.

Kramer KJ; Muthukrishnan S (1997). Insect chitinases: Molecular Biology and potential use as biopesticides. *Insect Biochemistry and Molecular Biology* **27**, 887-900.

Langevin SA; Clay SA; Grace JB (1990). The incidence and effects of hybridization between cultivated rice and its related weed, red rice, *Oryza sativa* L. *Evolution* **44**, 1000-1008.

Leffel SM; Mabon SA; Stewart CN, Jr. (1997). Applications of green fluorescent protein in plants. *BioTechniques* **23**, 912-918.

Liu Y-B; Tabashnik BE; Pusztai-Carey M (1998). Field-evolved resistance to *Bacillus thuringiensis* toxin Cry1C in diamondback moth (Lepidoptera: Plutellidae). *Journal of Economic Entomology*, **89**, 798-804.

Luby JL; McNicol RJ (1995). Gene flow from cultivated to wild raspberries in Scotland: developing a basis for risk assessment for testing and deployment of transgenic cultivars. *Theoretical and Applied Genetics* **90**, 1133-1137.

Manasse RS (1992). Ecological risks of transgenic plants: Effects of spatial dispersion on gene flow. *Ecological Applications* **2**, 431-438.

Mabon SA; Harper BK; Moyer KA; Halfhill MD; Richards HA; Stewart CN Jr. (submitted). Green fluorescent protein as an indicator of gene movement and expression in transgenic plants. Molecular Ecology.

Marshall G (1998). Herbicide-tolerant crops-- real farmer opportunity or potential environmental problem. *Pesticide Science* **52**, 394-402.

McGaughey WH (1994). Problems of insect resistance to *Bacillus thuringiensis*. *Agriculture, Ecosystems, & Environment* **49**, 95-102.

Metz PLJ; Jacobsen E; Nap J-P; Pereira A; Stiekema WJ (1997). The impact of biosafety of the phosphinothricin-tolerance transgene in inter-specific *Brassica rapa* x *Brassica napus* hybrids and their successive backcrosses. *Theoretical and Applied Genetics* **95**, 442-450.

Mikkelsen TR; Andersen B; Jorgensen RB (1996). The risk of crop transgene spread. *Nature* **380**, 31

Niedz RP; Sussman MR; Satterlee JS (1995). Green fluorescent protein: an *in vivo* reporter of plant gene expression. *Plant Cell Reports* **14**, 403-406.

Ormo M; Cubitt AB; Kallio K; Gross LA; Tsien RY; Remington J (1996). Crystal structure of the *Aequorea victoria* green fluorescent protein. *Science* **273**, 1392-1395.

Orr DB; Landis DA (1997). Oviposition of european corn borer (Lepidoptera: Pyralidae) and impact of natural enemy populations in transgenic verses isogenic corn. *Journal Economic Entomology* **90**, 905-909.

Pang S-Z; DeBoer DL; Wan Y; Ye G; Layton JG; Neher MK; Armstrong CL; Fry JE; Hinchee MA; Fromm ME (1996). An improved green fluorescent protein gene as a vital marker in plants. *Plant Physiology* **112**, 893-900.

Paterson AH; Schertz KF; Lin YF; Liu S-C; Chang Y-L (1995). The weediness of wild plants: Molecular analysis of genes influencing dispersal and persistence of johnsongrass, *Sorghum halepense* (L.) Pers. *Proceedings of the National Academy of Science USA* **92**, 6127-6131.

Perlak FJ; Fuchs RL; Dean DA; McPherson SL; Fischhoff DA (1991). Modification of the coding sequence enhances plant expression of insect control protein genes. *Proceedings of the National Academy of Science, USA* **88**, 3324-3328.

Peumans WJ; Winter HC; Bemer V; Van Leuven F; Goldstein IJ; Truffa-Bachi P; Van Damme EJ (1997). Isolation of a novel plant lectin with an unusual specificity from *Calystegia sepium*. *Glycoconjagates Journal* **14**, 259-265.

Powell KS; Gatehouse AMR; Hilder VA; Gatehouse JA (1993). Antimetabolic effects of plant-lectins and plant and fungal enzymes on the nymphal stages of 2 important rice pests, *Nilaparvata lugens* and *Nephotettix cinciteps*. *Entomologia Experimnetalis et Applicata*, **66**, 119-126.

Purcell JP; Greenplate JT; Jennings MG; Ryerse JS; Pershing JC; Sims SR; Prinsen MJ (1993). Cholesterol oxidase: A potent insecticidal protein active against boll weevil larvae. *Biochemical and Biophysical Research Communications* **196**, 1406-1413.

Raybould AF; Gray AJ (1993). Genetically modified crops and hybridization with wild relatives: a UK perspective. *Journal of Applied Ecology* **30**, 199-219.

Rees M; Brown VK (1992). Interaction between invertibrate herbivores and plant competition. *Journal of Ecology* **80**, 353-360.

Rich TC (1991). *Crucifers of Great Britain and Ireland.* London: Botanical Society of the British Isles.

Robbelin G; Downey RK (1989). *Brassica* species. In G. Robbelen, R.K. Downey, & A. Ashri (Eds.), *Oil Crops of the World.* (pp. 339-362). New York: McGraw-Hill.

Rogers HJ; Matharu B, Parkes HC (1996). Monitoring releases of transgenic plants: theoretical and practical considerations. BCPC Symposium Proceedings No 65: Diagnostics in Crop Production :39-46.

Ryan CA (1990). Protease inhibitors in plants: Genes for improving defenses against insects and pathogens. *Annual Review of Phytopathology* **28**, 425-429.

Santos MO; Adang MJ; All JN; Boerma HR; Parrott WA (1997). Testing transgenes for insect resistance using *Aribidopsis*. *Molecular Breeding* **3**, 183-194.

Scheffler JA; Dale PJ (1994). Opportunities for gene transfer from transgenic oilseed rape (*Brassica napus*) to related species. *Transgenic Research* **3**, 263-278.

Schuler TH; Poppy GM; Kerry BR; Denholm I (1998). Insect-resistant transgenic plants. *Trends in Biotechnology* **16**, 168-175.

Shen Z; Corbin DR; Greenplate JT; Grebenok RJ; Galbraith DW; Purcell JP (1997). Studies on the mode of action of cholesterol oxidase on insect midgut membranes. *Archives of Insect Biochemistry and Physiology* **34**, 429-442.

Sims SR (1995). *Bacillus thuringiensis* var *kurstaki* CryIA(c) protein expressed in transgenic cotton-effects on beneficial and other nontarget insects. *Southwestern Entomologist* **20**, 493-500.

Stewart CN, Jr. (1996). Monitoring transgenic plants using in vivo markers. *Nature Biotechnology* **14**, 682

Stewart CN, Jr.; All JN; Raymer PL; Ramachandran S (1997). Increased fitness of transgenic insecticidal rapeseed under insect selection pressure. *Molecular Ecology* **6**, 773-779.

Stewart CN, Jr.; Prakash CS (1998). Chloroplast-transgenic plants are not a gene flow panacea. *Nature Biotechnology* **16**, 401

Tabashnik BE (1994). Evolution of resistance to *Bacillus thuringiensis*. *Annual Review of Entomology* **39**, 47-79.

Tabashnik BE; Liu Y-B; Finson N; Masson L (1997a). One gene in diamondback moth confers resistance to four *Bacillus thuringiensis* toxins. *Proceedings of the National Academy of Sciences of the United States of America, USA* **94**, 1640-1644.

Tabashnik BE; Liu B-L; Malvar T; Heckel DG; Masson L; Ballester V; Granero F; Ferre J; Mensua JL (1997b). Global variation in the genetic and biochemical basis of diamondback moth resistance to *Bacillus thuringiensis*. *Proceedings of the National Academy of Sciences of the United States of America* **94**, 12780-12785.

Vaeck M; Reynaerts A; Höfte H; Jansens S; De Beuckeler M; Dean C; Zabeau M; Van Montagu M; Leemans J (1987). Transgenic plants protected from insect attack. *Nature* **238**, 33-37.

Wearing CH; Hokkanen HMT (1995). Pest resistance to Bacillus thuringiensis: ecological crop assessment for Bt gene incorporation and strategies of management. In H.M.T. Hokkanen & J.M. Lynch (Eds.), *Biological Control: Benefits and Risks.* (pp. 236-252). New York: Cambridge University Press.

Whalon ME (1994). Problems of insect resistance to *Bacillus thuringiensis*. *Agriculture, Ecosystems and Environment* **49**, 95-102.

Whalon ME; McGaughey WH (1993). Insect resistance to *Bacillus thuringiensis*. In L. Kim (Ed.), *Advanced Engineered Pesticides.* (pp. 215-231). New York: Marcel Dekker.

Whitton J; Wolf DE; Arias DM; Snow AA; Rieseberg LH (1997). The persistence of cultivar alleles in wild populations of sunflowers five generations after hybridization. *Theor Appl Genet*, **95**, 33-40.

Wilson FD; Flint HM; Deaton RW; Fischoff DA; Perlak FJ; Armstrong TA; Fuchs RL; Berberich SA; Parks NJ; Stapp BR (1992). Resistance of cotton lines containing a *Bacillus thuringiensis* toxin to pink bollworm (Lepidoptera: Gelechiidae) and other insects. *Journal of Economic Entomology*, **85**, 1516-1521.

Yang F; Moss LG; Phillips GN, Jr. (1996). The molecular structure of green fluorescent protein. *Nature Biotechnology*, **14**, 1246-1251.

Yu C; Mullins MA; Warren GW; Koziel MG; Estruch JJ (1997). The *Bacillus thuringiensis* vegetative insecticidal protien Vip3A lyses midgut epithelium cells of susceptible insects. *Applied and Environmental Microbiology*, **63**, 532-536.

Gene flow from virus-resistant transgenic crops to wild relatives or to infecting viruses

P-Y Teycheney and M Tepfer
Laboratoire de Biologie Cellulaire, INRA-Versailles, F-78026 Versailles cedex, France

ABSTRACT

The most important potential ecological risks associated with virus-resistant transgenic plants center on questions of gene flow. Gene flow from the transgenic plant to wild relatives could potentially confer a selective advantage on the wild plants, and perhaps increase their weediness. Gene flow from the transgene to an infecting virus by recombination could lead to new viral genomes. These questions need to be examined carefully, so that the potential risks can be correctly situated in a global risk/benefit analysis, which will take into account numerous other elements.

INTRODUCTION

Gene flow from transgenic crops to wild relatives has been widely discussed in recent years. Attention has been primarily focused on certain economically important crops that have been transformed with herbicide resistance transgenes. For example, Mikkelsen *et al.*, (1996) have shown that a glufosinate tolerance transgene could be transmitted under field conditions from genetically engineered oilseed rape (*Brassica napus*) to its weedy relative, *Brassica rapa* (*B. campestris*). This raises the concern that dispersal of oilseed rape pollen over long distances (Timmons *et al.*, 1996) could lead to efficient dispersal of transgenes to wild species and by conferring herbicide resistance on the wild plants, limit the use of the herbicides involved. It has been proposed that biological containment of transgenes within transgenic crops can be achieved for some species through genetic engineering of the chloroplast genome (Daniell *et al.*, 1998). Unfortunately, chloroplasts, which are usually transmitted maternally, can also be inherited through pollen at a low frequency in many species, and therefore the use of chloroplast-transgenic plants cannot entirely prevent transgene flow (Stewart & Prakash, 1998). It is also true that gene flow will occur naturally in both directions, and that the result of pollination of rapeseed by *B. rapa* pollen will be exactly the same as that of the inverse cross. In any case, until a truly effective and universal system for genetic isolation of transgenic plants is developed, it remains pertinent to consider potential risks associated with gene flow between transgenic plants and sexually compatible wild relatives, but only for the few crop species where this is possible.

The concept of pathogen-derived resistance (Sanford & Johnson, 1985) applied to plant viruses paved the way to the development of virus-resistant transgenic (VRT) plants expressing various viral sequences, of which the first were reported in 1986 (for a review, see Beachy, 1997). The most widely used and best characterized viral transgenes are ones encoding a viral coat protein (CP). Nevertheless, many other viral genes, including ones encoding defective movement proteins, polymerases, proteases or helper components, as well as untranslatable viral sequences, have also provided high levels of resistance to viral infections when expressed in transgenic plants (Lomonosoff, 1995). In a relatively rapid move from the laboratory to the field, several VRT crop varieties have already been field tested and authorized for unrestricted commercial use in China and in the USA, and several more are to be released in the USA in the near future (White, 1999).

In preparation for deliberate small-scale and then unrestricted commercial release of VRT plants, several questions of potential risk have been raised (for a complete current overview, see Tepfer & Balázs, 1997). Ecological risks associated with VRT plants include synergism with or heteroencapsidation of an infecting virus, which could lead to increased symptom severity or to spread of non-transmissible viruses, respectively. These risks can be considered to have potential effects that are simply phenotypic in nature, since they would remain restricted to the vicinity of the VRT crops and would disappear with no persistent effects on the environment if the VRT plants involved ceased being used. They will therefore not be further discussed here. VRT plants could also be a source of permanent genotypic modifications: (i) transmission of a resistance gene from transgenic crops to wild relatives could modify their natural competitiveness, or (ii) gene flow from transgenic plants to an infecting virus could lead to the emergence of recombinant viruses with modified biological properties, such as host specificity or symptomatology. Because of their potential for long-term, irreversible effects, these two potential genotypic risks will be discussed in more detail.

GENE FLOW FROM TRANSGENIC VIRUS-RESISTANT CROPS TO WILD RELATIVES

Considering the potential impacts of gene flow is of course pertinent for the rare VRT plant species that can be expected to be cultivated in proximity to compatible wild relatives. The situation with VRT plants is quite different from that of herbicide-resistant ones, where selection pressure, i.e. herbicide use, is entirely determined by human activity. The selection pressure exerted by viruses, and by other biological or environmental stresses, is entirely beyond human control and has rarely been studied. The critical question is, if gene flow from VRT plants to wild relatives does occur, to what extent will this confer a selective advantage on the VRT wild plants?

Questions relative to gene flow from VRT crops to wild plants are currently being studied with two experimental systems, involving either sugar beet or squash. There is clear evidence for natural gene flow between cultivated sugar beets (*Beta vulgaris* ssp *vulgaris* provar. *altissima* Döll) and wild beets (*B. vulgaris* ssp *maritima* Arcang.), which are perfectly compatible (Boudry *et al.,* 1993), and also between the cultivated squash, *Cucurbita pepo* var. *ovifera*, and the wild squash, *C. pepo* var. *texana* (Fuchs & Gonsalves, 1997). As would be expected, Bartsch *et al.* (1996) have shown that transgenic sugar beet resistant to beet necrotic yellow vein virus (BNYVV) has an ecological advantage over untransformed controls when grown under field conditions, particularly when there was heavy virus infestation. The effects of gene flow in the squash system have been studied in more detail. Fuchs & Gonsalves (1997) have shown that transgene flow does occur between VRT cultivated squash expressing the coat protein genes of cucumber mosaic virus (CMV), zucchini yellow mosaic virus (ZYMV) and watermelon mosaic virus 2 (WMV2) and its wild relative under conditions of low disease pressure. Both F1 hybrid squash plants and plants of the BC1 generation exhibited greater fitness than wild-type plants when tested under conditions of high disease pressure, but not under ones of low disease pressure. They also found that certain BC1 individuals were phenotypically indistinguishable from wild squash, and thus could be predicted to survive under non-cultivated conditions.

The crucial (and still unresolved) question is whether a virus-resistant wild squash or beet can become a significantly greater threat as an invasive weed. Unfortunately, very few studies on the impact of viral infections on wild plants have been undertaken. Among the few examples, one can cite the experiments showing that under controlled glasshouse conditions CMV reduces the competitive capacity of both purslane (*Portulaca oleracea*) and chickweed (*Stellaria media*) through an inhibition of growth and/or a reduction of reproductive potential (Friess & Maillet, 1996; 1997). Even less is known about the effects on fitness of wild plants when grown under natural conditions. The study of Kelley (1993) on *Anthoxanthum odoratum* infected with barley yellow dwarf virus (BYDV) is perhaps the unique example. He showed that even though BYDV infection did not induce visible symptoms on *A. odoratum*, it did have a clear negative effect on fitness, as reflected in fecundity.

These results show that viral infection can decrease plant fitness, and tend to support the hypothesis that a virus-resistant phenotype, whether natural or acquired through genetic engineering, could confer a selective advantage. However, the situation under non-cultivated conditions may be more complex. For instance, even though *C. pepo* var. *texana* exhibits pronounced symptoms when infected in the glasshouse or experimental field, the prevalence of the target viruses is apparently extremely low in native populations (Fuchs & Gonsalves, 1997). Similarly, in a modest sampling of wild beet populations in Italy, no BNYVV was detected (Bartsch, 1997). If more detailed studies show that virus is generally absent from these species, for reasons that remain to be determined (other forms of resistance, absence of vectors, lack of source plants, etc.), one would predict that in fact introgression of a virus resistance gene would be expected to confer no selective advantage, and would have no effect on the weediness of the wild plants.

GENE FLOW FROM TRANSGENIC PLANTS TO AN INFECTING VIRUS AS A SOURCE OF VARIABILITY IN VIRAL GENOMES

Analysis of viral genomic sequence data often shows genetic features that can be best attributed to RNA recombination events between viruses (Revers *et al.,* 1996) or between viruses and plant RNAs (Mayo and Jolly, 1991; Masuta *et al.,* 1992), indicating that RNA recombination is a normal feature of virus evolution. Thus, the use of viral sequences in transgenic plants immediately raised the question of whether recombination could occur between viral transgene sequences and the genome of an infecting virus, and whether this could have an impact on virus evolution.

In 1994, Greene & Allison showed that recombination between transgene-derived mRNA encoding the 3' two-thirds of the viral coat protein (CP) of cowpea chlorotic mosaic virus (CCMV) and an isolate of CCMV that had been crippled by the deletion of the 3' one-third of its CP gene, could give rise to viable viral progeny *via* RNA recombination. The viral genomes recovered were apparently created by imprecise homologous recombination, and also had minor mutations surrounding the site of crossover. When tested on a range of host plants, several of the recombinant strains caused novel symptoms (Allison *et al.*, 1997), but none was more fit than the parental strain when co-inoculated with wild-type virus (Allison *et al.*, 1999). These results demonstrate that recombination can occur between transgene-derived viral sequences and the genome of an infecting virus. However, these recombinant strains were obtained under conditions radically different from those that would occur in the field, since conditions of high selection pressure was exerted in favor of recombination events that restore viability. Working under conditions of moderate selection pressure, i.e. with an intact cauliflower mosaic virus (CaMV) genome as the infecting strain, Wintermantel *et al.* (1996) showed that inoculation of transgenic *N. bigelovii* plants expressing D4-CaMV ORF IV with W260-CaMV, which is systemic in solanaceous hosts, generated recombinant viruses that had a distinct competitive advantage in *N. bigelovii* when compared to the parental W260 strain. These results demonstrate that gene flow from transgenic plants to infecting viruses can occur, although at very low levels and under particular conditions.

In order to estimate the potential risks of VRT plants regarding virus recombination, it has often been proposed to apply a classic formula of risk assessment, in which risk = hazard X frequency (Hull, 1994). In the context of recombination in VRT plants, application of this formula would require evaluation of the pathogenicity of recombinant viruses (hazard), and also determination of the frequency of their occurrence.

Creating recombinant viruses *in vitro* by exchanging RNA segments of different strains or of related viruses and evaluating their biological properties is currently the best way to study the potential hazard associated with recombinant viruses. Certain artificial recombinant cucumoviruses show striking changes in symptomatology (Ding *et al.*, 1996;

Salánki *et al.*, 1997; Carrère *et al.*, 1999), but their fitness has not been compared to that of the parental viruses. There is however one report of a recombinant cucumovirus that shows an increased fitness when compared to its parental strains (Fernández-Cuartero *et al.*, 1994). This particular recombinant appeared spontaneously after several years of maintenance in the greenhouse of a pseudo-recombinant strain composed of CMV RNAs 1 and 2 and tomato aspermy virus (TAV) RNA3, which is rather far from the conditions encountered during field cultivation of VRT plants. None of these reports can give a clear picture of the impact of recombinants between the transcripts of a viral transgene expressed in a VRT plant and a viral genome, in particular because artificial recombinants are created by exchanging homologous restriction fragments between cloned viruses. Creating random recombinants between the transcripts of a viral transgene and the genome of a potential infecting virus, then comparing the properties (aggressivity, host range, fitness) of recombinants to those of the parental viruses would be the better way to estimate the hazards of potential recombinants between VRT plants and viruses. To this end, we are currently exploring techniques for creating a random library of recombinant cucumoviral genomes.

The frequency of recombination is also difficult to establish, particularly since strictly speaking it can only be determined under conditions of the lowest possible selection pressure. Otherwise, counter-selection of non-viable recombinants leads to a systematic underestimation of frequency. In addition, since recombination between viral genomes is a normal phenomenon, the frequency of its occurrence in virus-infected VRT plants should be compared to that in doubly-infected non-transgenic plants. To this end, molecular strategies have been developed to detect recombinant cucumoviruses under little or no selection pressure (Aaziz & Tepfer, 1999). Using these techniques, recombinants between genomic RNAs of two cucumoviruses, CMV and TAV, have recently been observed in coinfected plants in the absence of selection pressure (Aaziz & Tepfer, submitted). The objective of current work is to study RNA recombination in the corresponding CP-CMV plants.

CONCLUSIONS

Upon re-examination of the initially proposed risk assessment formula (risk = hazard X frequency), it becomes apparent that in fact this may not be the appropriate model for the types of risk under consideration here. Compared for instance to most industrial risks, or to the risk of damage to an automobile in an accident, the potential genotypic risks associated with VRT plants differ by giving rise to changes that could be irreversible. Thus, we would propose that the only important questions are: can the suspected event occur, and if so, what are the consequences? The frequency of the event will in fact only affect the rapidity of appearance of its effects. In the types of gene flow under consideration here, this would mean that once it has been determined that the frequency is greater than zero, the important points would be to determine if the transgene can confer enhanced fitness, but also whether there is indeed selection pressure in nature that would give an advantage to the plant or virus with the novel genotype.

The potential risks associated with VRT plants are all due to natural phenomena. Gene flow between certain crop species and wild relatives has always occurred, as has recombination between viral genomes. It is thus of great importance to select the most appropriate baseline against which to compare the potential risks. For instance, not knowing the baseline impact of virus infection on a wild species, we cannot predict the possible impact of introgression of a resistance gene from a related VRT crop plant. Here, as in the area of recombination between viral genomes, more thorough understanding of the basics is necessary.

In closing, we would like to consider how to make the appropriate decision if future studies lead to the conclusion that VRT plants do present some level of ecological risks due to gene flow. Since zero risk does not exist in this domain, as in any other, we need to have the means to determine if the risk in question is acceptable or not. This broader evaluation has been made more difficult by the widely held misconception that the principle of precaution implies acceptance of only zero risk. The decision on deployment of a given VRT plant clearly involves a risk/benefit analysis, which will not only include evaluation of risk as determined in as scientifically rigourous a fashion as possible, but will also include consideration of the alternatives to not deploying the VRT plants. In a region of the world that suffers more from agricultural overproduction than from shortage, the need for more and better virus resistance is not always obvious. Nonetheless, even in Europe there are serious viral diseases against which natural resistance genes are not available. In these cases, deployment of VRT plants would make it unnecessary to use the pesticides that are currently used to eliminate the organisms that are vectors of viral disease (insects, fungi and nematodes), and would also lead to enhanced quality of the agricultural products. Viral diseases also have an important impact on the yield of certain major crops, such as rice. It should also be remembered that in a perhaps not so distant future, the global demographic pressure may make it necessary to use all available means for increasing agricultural productivity, and this may well include VRT plants.

ACKNOWLEDGEMENTS

P-Y Teycheney is supported by a fellowship from the Association pour la Recherche sur les Nicotianées.

REFERENCES

Aaziz R; Tepfer M (1999). Strategies for detection of recombination in virus-infected plants expressing a viral transgene. In: *Proceedings of the 5th International Symposium on Biosafety Results of Field Tests of Gentically Modified Plants and Microorganisms.* eds J Schiemann & R Casper. Braunschweig (in press).

Allison RF; Greene AE; Schneider WL (1997). Significance of RNA recombination in capsid protein-mediated virus-resistant transgenic plants. In: *Virus-resistant transgenic plants: potential ecological impact.* eds M Tepfer & E Balázs, pp. 40-44. INRA and Springer Verlag: Versailles and Heidelberg.

Allison RF; Schneider WL; Deng M (1999). Risk assessment of virus-resistant transgenic plants. In: *Proceedings of the 5th International Symposium on Biosafety Results of Field Tests of Gentically Modified Plants and Microorganisms.* eds J Schiemann & R Casper. Braunschweig (in press).

Bartsch D; Schmidt M; Pohl-Orf M; Haag C; Schuphan I (1996). Competitiveness of transgenic sugar beet resistant to beet necrotic yellow vein virus and potential impact on wild beet populations. *Molecular Ecology* **5**, 199-206.

Bartsch D (1997). Ecological impact of transgenic virus-resistance in crop, weed, and wild plant populations (due to potential alterations of plant invasiveness). In: *Virus-resistant transgenic plants: potential ecological impact.* eds M Tepfer & E Balázs, pp. 107-113. INRA and Springer Verlag: Versailles and Heidelberg.

Beachy RN (1997). Mechanisms and applications of pathogen-derived resistance in transgenic plants. *Current Opinion in Biotechnology* **8**, 215-220.

Boudry P; Moerchen M; Saumitou-Laprade P; Vernet P; Van Dijk H (1993). The origin and evolution of weed beets: consequences for the breeding and release of herbicide-resistant sugar beets. *Theoretical and Applied Genetics* **87**, 471-478.

Carrère I; Tepfer M; Jacquemond M (1999). Recombinants of cucumber mosaic virus (CMV): determinants of host range and symptomatology. *Archives of Virol*ogy, in press.

Daniell H; Datta R; Varma S; Gray S; Lee SB (1998). Containment of herbicide resistance through genetic engineering of the chloroplast genome. *Nature Biotechnology* **16**, 345-348.

Ding SW; Shi BJ; Li WX; Symons RH (1996). An interspecies hybrid RNA virus is significantly more virulent than either parental virus. *Proceedings of the National Academy of Sciences USA* **93**, 7470-7474.
Fernández-Cuartero B; Burgyán J; Aranda MA; Salánki K; Moriones E; Garcia-Arenal F (1994). Increase in the relative fitness of a plant virus RNA associated with its recombinant nature. *Virology* **203**, 373-377.
Friess N; Maillet J (1996). Influence of cucumber mosaic virus infection on the intraspecific competitive ability and fitness of purslane (*Portulaca oleracea*). *New Phytologist* **132**, 103-111.
Friess N; Maillet J (1997). Influence of cucumber mosaic virus infection on the intraspecific competitive ability and reproduction of chickweed (*Stellaria media*). *New Phytologist* **135**, 667-674.
Fuchs M; Gonsalves D (1997). Risk assessment of gene flow associated with the released of virus-resistant transgenic crop plants. In: *Virus-resistant transgenic plants: potential ecological impact.* eds M Tepfer & E Balázs, pp. 114-120. INRA and Springer Verlag: Versailles and Heidelberg.
Greene AE; Allison F (1994). Recombination between viral RNA and transgenic plant transcripts. *Science* **263**, 1423-1425.
Hull R (1994). Risks in using transgenic plants? *Science* **264**, 1649-1650.
Kelley SE (1993). Viruses and the advantage of sex in *Anthoxanthum odoratum*: a review. *Plant Species Biology* **8**, 217-223.
Lomonossoff GP (1995). Pathogen-derived resistance to plants. *Annual Review of Phytopathology* **33**, 323-343.
Masuta C; Uyeda S; Suzuki M; Uyeda I (1998). Evolution of a quadripartite hybrid virus by interspecific exchange and recombination between replicase components of two related tripartite RNA viruses. *Proceedings of the National Academy of Sciences USA* **95**, 10487-10492.
Mayo MA; Jolly CA (1991). The 5'-terminal sequence of potato leafroll virus RNA: evidence of recombination between virus and host RNA. *Journal of General Virology* **72**, 2591-5.
Mikkelsen TR; Andersen B; Jørgensen RB (1996). The risk of crop transgene spread. *Nature* **380**, 31.
Revers F; Le Gall O; Candresse T; Le Romancer M; Dunez J (1996). Frequent occurrence of recombinant potyvirus isolates. *Journal of General Virology* **77**, 1953-1965.
Stewart CN; Prakash CS (1998). Chloroplast-transgenic plants are not a gene flow panacae. *Nature* **16**, 401.
Salánki K; Carrère I; Jacquemond M; Balázs E; Tepfer M (1997). Biological properties of pseudorecombinant and recombinant strains created with cucumber mosaic virus and tomato aspermy virus. *Journal of Virology* **71**, 3597-3602.
Sanford JC; Johnson SA (1985). The concept of parasite-derived resistance: deriving resistance genes from the parasites own genome. *Journal of Theoretical Biology* **113**, 395-405.
Tepfer M; Balázs E, eds (1997). *Virus-resistant transgenic plants: potential ecological impact.* 126 pp. INRA and Springer Verlag: Versailles and Heidelberg.
Timmons, AM; Charters YM; Crawford JW; Burn D; Scott SE; Dubbels SJ; Wilson NJ; Robertson A; O'Brien ET; Squire GR; Wilkinson MJ (1996). Risks from transgenic crops. *Nature* **380**, 487.
White JL (1999). An overview on cultivation of virus-resistant crops in the United States. In: *Proceedings of the 5th International Symposium on Biosafety Results of Field Tests of Genetically Modified Plants and Microorganisms.* eds J Schiemann & R Casper. Braunschweig (in press).
Wintermantel WM; Schoelz JE (1996). Isolation of recombinant viruses between cauliflower mosaic virus and a viral gene in transgenic plants under conditions of moderate selection pressure. *Virology* **223**, 156-164.

Interactions between insect tolerant genetically modified plants and natural enemies

T H Schuler, G M Poppy, R P J Potting, I Denholm and B R Kerry

IACR-Rothamsted, Harpenden, Hertfordshire, AL5 2JQ, UK

ABSTRACT

Insect natural enemies play an important role in the regulation of many pest species. The effects of genetically modified (GM) crops on natural enemies and other non-target species need to be evaluated to avoid disruption of natural biological control and ecosystem dynamics. This paper reviews progress with studying effects of GM plants on tritrophic systems involving parasitoids and predators. Experiments at Rothamsted using mixtures of GM and wildtype oilseed rape plants in large cages have shown no differences in parasitism rates of the aphid *Myzus persicae* by the parasitoid *Diaeretiella rapae* on GM and untransformed plants. Ongoing experiments are investigating the effect of GM oilseed rape on the parasitisation and host searching behaviour of *Cotesia plutellae*, a larval parasitoid of the diamondback moth (*Plutella xylostella*).

INTRODUCTION

All currently commercialized insect tolerant genetically modified (GM) crops (cotton, maize and potato) express δ-endotoxin genes derived from the bacterium *Bacillus thuringiensis* (*Bt*). In addition to *Bt* genes, an increasing number of other potential insect resistance genes are being transferred into crop cultivars, such as genes coding for cholesterol oxidase, lectins (e.g. the snowdrop lectin, GNA) and inhibitors of digestive enzymes such as proteinase inhibitors (e.g. the cowpea trypsin inhibitor, CpTI) and amylase inhibitors. Constitutive promoters have been used to drive gene expression in the majority of currently existing insect tolerant GM plants leading to gene expression in most tissues of the plant (Schuler *et al.*, 1998). Most herbivorous insects feeding on GM plants are therefore likely to ingest these heterologous proteins.

Agricultural crops support not only pest insects but also beneficial insects which feed on these herbivores and which play an important role in the regulation of pest populations (van Driesche & Bellows Jr, 1996). Current pest control measures usually involve the use of broad spectrum synthetic insecticides which act on contact and are often highly toxic to these beneficial insects. Insect tolerant GM plants are expected to be more benign for beneficial insects, but they represent a new technology and interactions between these plants and the third trophic level have to be tested to avoid any unnecessary disruption of natural biological control or ecosystem dynamics.

The impact of insect tolerant plants on the population dynamics of natural enemies will depend on many factors such as the level of tolerance of the plant to the pest species, the scale on which the GM crop is grown and the susceptibility, mobility and host range of the natural enemy. The most obvious way highly tolerant crops can affect natural enemies is by severely depleting their supply of prey or hosts. Such effects will be most profound for specialist natural enemies that feed exclusively on insects the GM plant is designed to kill, a fact that applies to all pest control measures and which is by no means unique to GM plants. When highly tolerant plants are grown on a larger scale the abundance of some natural enemies may therefore decline due to prey depletion, but their persistence is not necessarily threatened if other nearby crops or weeds support acceptable prey species, or if their host range includes species other than the target pests. However, not all herbivorous insects feeding on GM plants will receive lethal doses, either because (a) a heterologous protein only provides a low level of tolerance, (b) the expression level of the heterologous protein is low or variable, or (c) a herbivore is not very susceptible to the protein (either inherently as a species or because it has developed resistance). Thus, in addition to the effects of prey depletion, the potentially more subtle effects of sublethal doses on tritrophic interactions involving predators and parasitoids need to be investigated especially if herbivores can accumulate the toxin without deleterious effects.

PARASITOIDS

Parasitoids are mainly restricted to the insect orders Hymenoptera and Diptera. They complete their larval development on or in a single host insect which is killed in the process (Godfray, 1994). Compared with predators, parasitoids are relatively host-specific; some species even depend solely on a particular stage of a single host species.

Since parasitoid larvae depend on one host individual, any pest control measure, including GM plants, that kills the host before the parasitoid has completed its development will result in the death of the parasitoid. *Bt* oilseed rape, for example, can cause 100% mortality of *Bt*-susceptible diamondback moth (*Plutella xylostella*) larvae (Stewart Jr, 1996) and, as a direct consequence, larval parasitoids of this pest, such as the braconid *Cotesia plutellae*, are unable to complete their development. In contrast, preliminary results suggest that survival of *C. plutellae* in *Bt*-resistant *P. xylostella* was not affected by the host having fed on *Bt* oilseed rape (Figure 1).

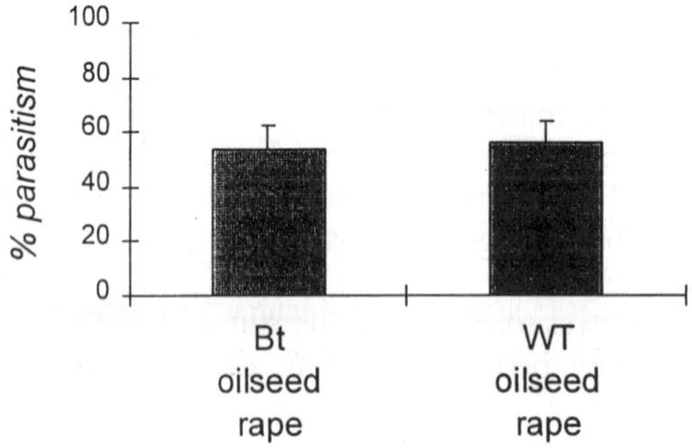

Figure 1. Parasitism of *Bt*-resistant diamondback moth larvae by *Cotesia plutellae* on *Bt* oilseed rape compared to wildtype (WT) oilseed rape of the same cultivar

It is known from studies using conventionally-bred tolerant plants and also from work with microbial *Bt* insecticides, that sublethal doses have the potential for synergistic, antagonistic or neutral interactions with parasitoids. For example, less vigorous host defence caused by sublethal poisoning potentially increases oviposition opportunities for parasitoids. There is also the possibility that sublethal toxin levels improve levels of parasitism by impairing the host's immune response to parasitism. In contrast, sublethal doses could lead to a reduced host quality, and thereby to a reduced fitness of parasitoids developing on such hosts. Parasitoid larvae might also be exposed directly to the heterologous protein when consuming host tissue. Further detailed studies are required to unravel these complex interactions.

Parasitoids of non-target pest species have received little attention so far but GM plants have the potential to affect these as well. Aphid populations, for example, are not controlled by *Bt* plants. Parasitoids are important natural antagonists of aphids and any negative side effects on these parasitoids could lead to increased problems with aphid pests. We are, therefore, investigating the effect of mixtures of *Bt* oilseed rape and wildtype oilseed rape on the aphid parasitoid *Diaeretiella rapae* on a population scale in large laboratory cages. Our results so far indicate that aphid parasitoids are as efficient in controlling aphid populations on *Bt* oilseed rape as on wildtype oilseed rape (Figure 2).

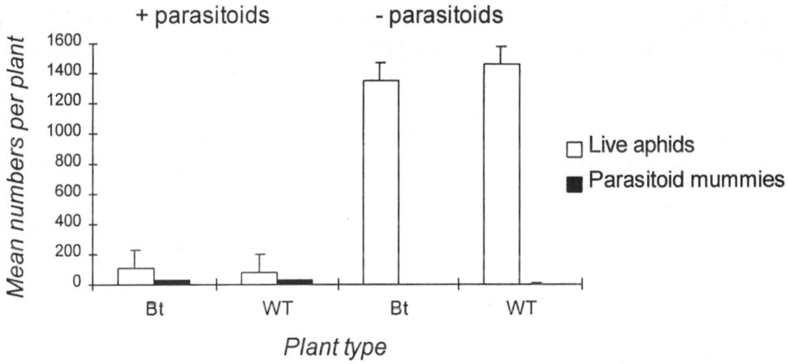

Figure 2. Population size of *Myzus persicae* and parasitism by *Diaeretiella rapae* reared for one parasitoid generation on mixtures of *Bt* oilseed rape and wildtype (WT) oilseed rape in large cages (1.7m x 1.2m x 1.0m) in the laboratory. Populations were initiated by placing fifteen apterous adult aphids on each plant (four plants of each type per cage) and, two days later, releasing seventy-two *D. rapae* females into each cage ('+ parasitoids'). No parasitoids were added to the control cages ('- parasitoids').

When locating hosts, female parasitoids use volatile stimuli released from plants, especially when the plant is damaged by herbivores (Vet & Dicke, 1992). The quantitity and/or composition of the emitted plant volatiles can vary between plant species and between genotypes of the same plant species (Dicke & Vet, 1998). It has been demonstrated that several parasitoid species use these volatile cues to discriminate in-flight between different plant species or even individual cultivars (Vet & Dicke, 1992). The genetic modification of plants could conceivably affect the volatile profile released by the plant, and subsequently influence the attractiveness for beneficial insects such as parasitoids. To investigate whether parasitoids differ in their response to GM plants, we have begun to test the response of the parasitoid *C. plutellae* to uninfested oilseed rape leaves in a windtunnel. *C. plutellae* adults were attracted to both *Bt* oilseed rape and oilseed rape transformed with the proteinase inhibitor OC-I. The parasitoids did not distinguish between control and these GM plants when offered in a choice situation (Potting, unpublished data).

A number of field experiments have investigated the effect of *Bt* plants on parasitoids on a larger scale. Orr and Landis (1997) planted microplots of untransformed maize within large plots of *Bt* maize and infested them artificially with eggs of the European cornborer (*Ostrinia nubilalis*). This method was used since *O. nubilalis* larvae do not survive on *Bt* maize. Levels of parasitism of *O. nubilalis* larvae collected from these microplots did not differ between *Bt* and untransformed plots. Mascarenhas and Luttrell (1997) studied the combined effect of sublethal exposure to *Bt* cotton and natural enemies on the survival of bollworm larvae (*Helicoverpa zea*). Sublethal exposure was achieved by rearing larvae for one to four days on *Bt* cotton in the laboratory before transferring the surviving larvae to untransformed cotton in the field. There was no difference in survival of *H. zea* larvae from GM and untransformed plants when natural enemies were excluded. However, when natural enemies were present, larvae exposed to sublethal doses of *Bt* cotton survived at lower rates than larvae reared entirely on untransformed cotton. Field trials with low-expressing *Bt* tobacco lines indicated a synergistic interaction between the transgenic plants and the ichneumonid wasp *Campoletis sonorensis* in controlling tobacco budworm larvae (*Heliothis virescens*), probably partly due to a slower larval development of the host (Johnson, 1997; Johnson *et al.*, 1997).

PREDATORY INSECTS

Predatory insects can be found in most insect orders including the Coleoptera (e.g. Carabidae, Coccinellidae), Heteroptera (e.g. Anthocoridae, Nabidae), Neuroptera (e.g. Chrysopidae), Hymenoptera (e.g. Vespidae, Formicidae) and Diptera (e.g. Syrphidae) (van Driesche & Bellows Jr, 1996). Predatory species of importance in a crop environment need to consume large numbers of pest insects to complete their life cycle. Both the larval and the adult stages are mobile and in some species both are predatory while in other species only the larvae are carnivorous while the adults feed on flowers. Most predators feed on a range of different insect species.

A number of laboratory and field studies have been conducted to investigate interactions between GM plants and predatory insects. Dogan *et al.* (1996) studied the effect of *Bt* potatoes on the ladybird *Hippodamia convergens* in the laboratory. No significant effects on survival, consumption of aphids (*Myzus persicae)*, development or reproduction in ladybirds were observed. Pilcher *et al.* (1997) fed pollen of *Bt* maize to the ladybird *Coleomegilla maculata*, the anthocorid bug *Orius insidiosus* and the lacewing *Crysopa carnea*, and found no acute detrimental effects on preimaginal development and survival. In a laboratory study by Sims (1995) a concentration of 20 $\mu g\ ml^{-1}$ activated *Bt* toxin applied to prey or in honey solution, did not affect *C. carnea* or *H. convergens*, respectively. Hilbeck *et al.* (1998), however, reported that feeding prey that had been reared on *Bt* maize to *C. carnea* in the laboratory increased larval mortality of the predator from 37% (control) to 62%. In this study, two species of prey were tested, the European cornborer (the susceptible target species) and the Egyptian leafworm, *Spodoptera littoralis* (a less susceptible species).

Microbial *Bt* insecticides are generally perceived as safe for beneficial insects and are an essential component of many integrated pest management programmes. Some studies have, however, demonstrated that microbial *Bt* insecticides can be detrimental to certain natural enemies (reviewed by Croft and Flexner, 1990).

Birch *et al*. (in press) investigated potential side effects of potatoes expressing the lectin GNA on the two spotted ladybird (*Adalia bipunctata*). GNA potatoes provide low levels of resistance to a range of pest insects including aphids. Peach potato aphids (*Myzus persicae*) were reared on GNA potatoes and then fed to ladybirds. Compared to control plants the GNA potatoes resulted in a 38% reduced fecundity of ladybird females; these females only lived half as long as females on control plants and the viability of their eggs was reduced three to four-fold. It remains to be clarified whether these negative effects were caused by a direct effect of the GNA on the ladybirds or if they were due to a reduction in prey quality.

In the field no negative side effects on predatory insects have been reported and in some cases significantly greater numbers of predators were found on *Bt* plants compared to untransformed plants. Hoffmann *et al.* (1992) monitored predatory bugs of the family Nabidae on *Bt* and CpTI tobacco in the field. Numbers of Nabidae were greater on both types of GM plants compared to the control plots but the differences were not significant. Similarly, no detrimental effects were observed on the abundance of coccinellid, anthocorid and chrysopid predators of *O. nubilalis* on transgenic *Bt* maize in a two-year field experiment (Pilcher *et al.*, 1997). In another large scale field study, Orr and Landis (1997) found no negative effects of *Bt* maize on egg predation or predator densities, and on some sampling occasions numbers of lacewings and ladybirds were significantly higher on *Bt* maize compared to untransformed maize. The negative side effects on lacewings described in the laboratory (Hilbeck *et al.*, 1998) have so far not been confirmed in the field. This might be due to the fact that in maize crops *C. carnea* tends to feed mainly on aphids while the stem-boring *O. nubilalis* larva is hardly exposed to predation by lacewings. Monitoring of *Bt* cotton has so far also failed to show any significant effects on predators, including *C. carnea* (Flint *et al.*, 1995; Wilson *et al.*, 1992).

CONCLUSIONS

So far, reports of negative side effects of GM plants on natural enemies have been confined to small scale tests in the laboratory but field studies have not shown any negative effects on beneficial insects. Future research needs to include work on more tritrophic systems, clarification of the mechanisms behind observed detrimental effects, further studies to relate data from laboratory bioassays to the population and field level and, finally, comparison of effects of GM plants with those of conventional pest control measures. Vigorous and standardised risk assessment methodologies have to be applied. Post-release monitoring of natural enemies in GM crops will also be essential.

REFERENCES

Birch A N E; Geoghegan I E; Majerus M; McNicol J W; Hackett C; Gatehouse A M R; Gatehouse J A (in press). Ecological impact on predatory 2-spot ladybirds of transgenic potatoes expressing snowdrop lectin for aphid resistance. *Journal of Molecular Breeding.*

Croft B A; Flexner J L (1990). Microbial Pesticides. In: *Arthropod Biological Control Agents and Pesticides,* ed. B A Croft, pp. 269-303. John Wiley and Sons: New York.

Dicke M; Vet L E M (1998). Plant-carnivore interactions: evolutionary and ecological consequences for plant, herbivore and carnivore. In: *Herbivores, Plants and Predators,* eds H Olff; V K Brown; R H Drent. Proceedings of the 38th Symposium of the British Ecological Society.

Dogan E B; Berry R E; Reed G L; Rossignol P A (1996). Biological parameters of convergent lady beetle (Coleoptera: Coccinellidae) feeding on aphids (Homoptera: Aphididae) on transgenic potato. *Journal of Economic Entomology* **89**, 1105-1108.

Flint H M; Henneberry T J; Wilson F D; Holguin E; Parks N; Buehler R E (1995). The effects of transgenic cotton, *Gossypium hirsutum* L., containing *Bacillus thuringiensis* toxin genes for the control of the pink bollworm, *Pectinophora gossypiella* (Saunders) Lepidoptera, Gelechiidae and other arthropods. *Southwestern Entomologist* **20**, 281-292.

Godfray H C J (1994). *Parasitoids: Behavioral and Evolutionary Ecology.* Princeton: Princeton University Press, pp. 473.

Hilbeck A; Baumgartner M; Fried P M; Bigler F (1998). Effects of transgenic *Bacillus thuringiensis* corn-fed prey on mortality and development time of immature *Chrysoperla carnea* (Neuroptera: Chrysopidae). *Environmental Entomology* **27**, 480-487.

Hoffmann M P; Zalom F G; Wilson L T; Smilanick J M; Malyj L D; Kiser J; Hilder V A; Barnes W M (1992). Field evaluation of transgenic tobacco containing genes encoding *Bacillus thuringiensis* delta-endotoxin or cowpea trypsin inhibitor: efficacy against *Helicoverpa zea* (Lepidoptera: Noctuidae). *Journal of Economic Entomology* **85**, 2516-2522.

Johnson M T (1997). Interaction of resistant plants and wasp parasitoids of tobacco budworm (Lepidoptera: Noctuidae). *Environmental Entomology* **26**, 207-214.

Johnson M T; Gould F; Kennedy G G (1997). Effects of natural enemies on relative fitness of *Heliothis virescens* genotypes adapted and not adapted to resistant host plants. *Entomologia Experimentalis et Applicata* **82**, 219-230.

Mascarenhas V J; Luttrell R G (1997). Combined effect of sublethal exposure to cotton expressing the endotoxin protein of *Bacillus thuringiensis* and natural enemies on survival of bollworm (Lepidoptera: Noctuidae) larvae. *Environmental Entomology* **26**, 939-945.

Orr D B; Landis D A (1997). Oviposition of European corn borer (Lepidoptera: Pyralidae) and impact of natural enemy populations in transgenic versus isogenic corn. *Journal of Economic Entomology* **90**, 905-909.

Pilcher C D; Obrycki J J; Rice M E; Lewis L C (1997). Preimaginal development, survival, and field abundance of insect predators on transgenic *Bacillus thuringiensis* corn. *Environmental Entomology* **26**, 446-454.

Schuler T H; Poppy G M; Kerry B R; Denholm I (1998). Insect-resistant transgenic plants. *Trends in Biotechnology* **16**, 168-175.

Sims S R (1995). *Bacillus thuringiensis* var. *kurstaki* (CryIA (C)) protein expressed in transgenic cotton: effects on beneficial and other non-target insects. *Southwestern Entomologist* **20**, 493-500.

Stewart Jr C N; Adang, M J; All J N; Raymer P L; Ramachandran S; Parrott W A (1996). Insect control and dosage effects in trangenic canola containing a synthetic *Bacillus thuringiensis cryIAc* gene. *Plant Physiology* **112**, 115-120.

van Driesche R G; Bellows Jr T S (1996). *Biological Control.* New York: Chapman & Hall, pp. 539.

Vet L E M; Dicke M (1992). Ecology of infochemical use by natural enemies in a tritrophic context. *Annual Review of Entomology* **37**, 141-172.

Wilson F D; Flint H M; Deaton R W; Fischhoff D A; Perlak F J; Armstrong T A; Fuchs R L; Berberich S A; Parks N J; Stapp B R (1992). Resistance of cotton lines containing a *Bacillus thuringiensis* toxin to pink bollworm (Lepidoptera: Gelechiidae) and other insects. *Journal of Economic Entomology* **85**, 1516-1521.

Non-target effects of proteinase inhibitors expressed in potato as an anti-nematode defence

Susan E. Cowgill, David Coates* & Howard J. Atkinson.
*Centre for Plant Sciences and *School of Biology, University of Leeds, Leeds LS2 9JT, UK*

ABSTRACT

The survival and rate of growth of nymphs of *Myzus persicae* and *Macrosiphum euphorbiae* were adversely affected by ingestion of cystatins in artificial diet assays. Prolonged exposure to cystatins in diets produced a reduction in the rate of development, survival time and fecundity of *M. persicae* adults. However, when aphids were caged on transgenic potato expressing chicken egg white cystatin (CEWc) there was no adverse effect on aphid fitness. Similarly, there were no significant differences in field populations of aphids between control and CEWc-expressing potato grown under field conditions, although the levels of expression were sufficient to provide control of potato cyst nematodes. The results highlight that tissue-specific levels of expression are important in determining the effects of cystatins on different invertebrates and it is suggested that aphids feeding on transgenic plants may provide a route via which various natural enemies are exposed to cystatins.

INTRODUCTION

Nematodes cause global crop losses that have been valued at over $100 billion per year. In Britain control of nematodes on potato relies heavily on the use of nematicides, some of which are known to pose a risk to groundwater and drinking water supplies under certain soil types (Gustafson, 1993). Consequently there is a need to develop improved methods of nematode control. Transgenic plants expressing proteinase inhibitors (PIs) have the potential to fulfil this role.

PIs have been shown to affect members of the three principal groups of economic nematodes and may represent a broad-range resistance strategy for control of nematode pests. A modified version of a cysteine proteinase inhibitor from rice, Oc-IΔD86, has been shown to limit the growth of *Heterodera schactii*, beet cyst-nematode and *Meloidogyne incognita* (a root knot nematode) when expressed in *Arabidopsis thaliana* (Urwin *et al*., 1997).

Transgenic plants expressing serine or cysteine PIs have negative effects on the growth and survival of several insect pests (Hilder *et al*., 1987; Leple *et al*., 1995; Xu *et al*., 1996). As part of our study of the environmental consequences of a PI-based anti-nematode defence we have examined the invertebrate associates of potato crops to establish the range of common beneficial and non-economic insects commonly associated with this crop in UK. Histochemical studies and information from published literature were used to select a subset of insects for further study. Work has concentrated on those species which were identified to contain cysteine proteinases or those which had been shown to be affected by exposure to PIs in previous studies. This paper describes the results of a range of laboratory and field studies to quantify the effects on exposure to cysteine PIs on two aphid species *Myzus persicae* and *Macrosiphum euphorbiae*.

MATERIAL & METHODS

The *M. persicae* clone was initially introduced in our laboratory as a single individual collected from potato in 1996. A culture of *M. euphorbiae* was obtained from a laboratory clone maintained at IACR-Rothamsted.

Cloned genes for cystatins (Oc-IΔD86, chicken egg white cystatin, maize cystatin, sunflower cystatin) were expressed in *E. coli* using a pQE vector and the QIAexpress^{TM} expression system. Trans-epoxysuccinyl-L-leucylamido-(4-guadino) butane (E64) was purchased from Sigma Chemical Company.

Growth and survival of aphids on artificial diets containing cysteine PIs (cystatins)

Assays were carried out on 0-24h aphids over a 9 day period on artificial diets containing purified cystatin. The diet was a modified version of that described by Dadd & Mittler (1966). Feeding chambers containing 200μl artificial diet were made using UV-sterilised Parafilm stretched over a PVC ring. The diet sachets were replaced every 48 hours. The length and breadth of the nymphs was measured on alternate days using a video camera linked to a Leica Quantimet image analysis package.

The survival and fecundity of *M. persicae* feeding on artificial diets containing cystatins

0-24h old *M. persicae* nymphs were transferred from Chinese cabbage (*Brassica rapa* L. ssp. *pekinensis*(Lour.)) plants to individual diet cylinders with artificial diets containing either OcIΔ86 (200ppm) or CEWc (200ppm). The survival of nymphs was recorded daily. From day 8 onwards the number of newly deposited nymphs was also recorded. The nymphs were removed daily and the mortality of the adult aphids monitored. Values for the intrinsic rate of increase (r_m) were estimated by the method of Wyatt and White (1977) using the following equation:

$$r_m = 0.738.\ \log_e (M_d)/\ d$$

where M_d is the mean number of progeny produced in a time equal to the prereproductive period (d). The parameter provides a theoretical way of comparing the aphids reproductive potential on different diet types. The data were also used to calculate the mean generation time ($T =\ d/0.738$) and the doubling time ($DT = \log_e (2)/\ r_m$).

The survival and fecundity of *M. persicae* feeding on transgenic plants

Transgenic potatoes were generated expressing CEWc constitutively. The level in leaf tissue was 60ppm which is sufficient level to achieve nematode control in roots. *M. persicae* females of known age were placed individually in clip cages on either transformed or control plants. The aphids were caged on the abaxial leaf surface throughout the canopy, avoiding immature or senescing leaves. For each plant line, 3 plants were used and 7 clip cages were set up per plant. After 24 hours, the adults were removed together with all but 2 first instar nymphs. This day was labelled day 0. The remaining 2 nymphs per clip cage were left undisturbed until adulthood. At the teneral adult stage, one aphid per cage was left so that the daily reproduction of all individually-caged second generation adult aphids could be recorded. The assays were carried out under controlled environment conditions of 20°C, 16h light, 8h dark regime.

Aphid populations on transgenic plants in the field

Four CEWc-expressing lines and an ex-tissue culture control line of the same cultivar, cv Desiree were planted in a randomised block design with 4 replicates. Each plot contained nine plants. The number of invertebrates on the above ground parts of five plants per plot were censused at weekly intervals from 1 July to 3 October 1998.

RESULTS

Growth and survival of aphids on artificial diets containing cystatins

Figures 1 and 2 show that both the survival and rate of growth of *M. persicae* nymphs was adversely affected by ingestion of cystatins from artificial diet. The cystatins varied in their toxicity and growth inhibiting properties. Mortality was most pronounced in nymphs reared on maize cystatin. Surviving nymphs reared on E64, a low molecular weight, irreversible inhibitor of papain-like cysteine proteinases, were significantly smaller than those reared on the cystatins from plants or chicken egg white.

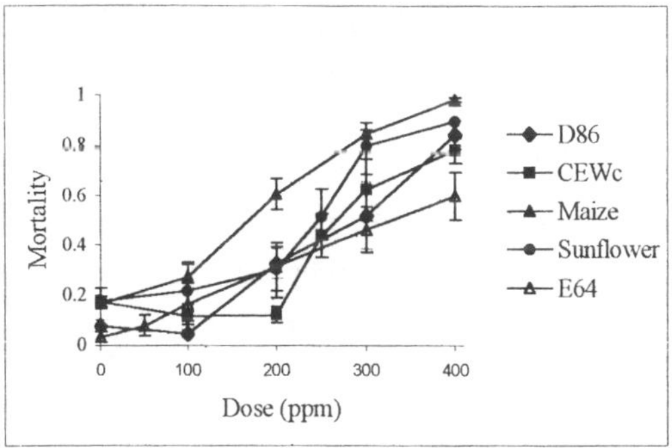

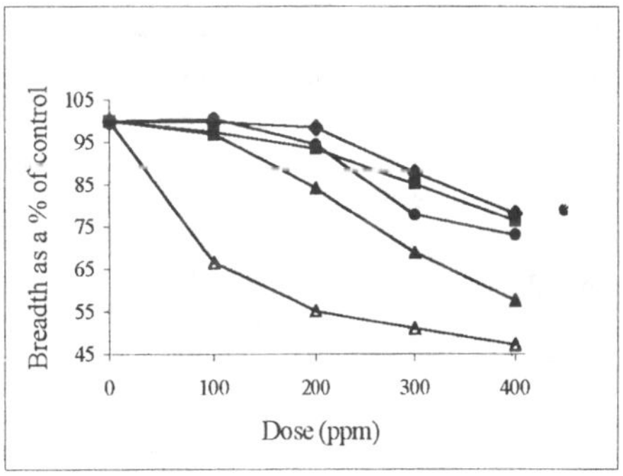

Figs. 1 & 2. Survival and growth of *M. persicae* nymphs after 9 days on artificial diets containing cystatins.

Similar effects on growth and survival were obtained with nymphs of *M. euphorbiae*. After 9 days on artificial diet containing CEWc at 400ppm the surviving nymphs were approximately 35% smaller than those on control diets. Survival was also significantly reduced at concentrations greater than 300ppm.

The survival and fecundity of *M. persicae* feeding on artificial diets containing cystatins

Rates of development, survival time and fecundity of *M. persicae* adults were all reduced on diets containing OcIΔ86 or CEWc. The mean number of days from birth to reproduction was 8.24 days for control aphids, 10.42 days for aphids reared on OcIΔ86 and 11.0 days for aphids reared on CEWc. The survival of aphids on CEWc declined rapidly from day 8 onwards. By day 10, 50% of these aphids were dead. Fifty per cent of the aphids reared on OcIΔ86 had died by day 14, some of the remaining adults survived until day 22 of the study. The total number of nymphs produced on control diet was 215 compared to 91 on OcIΔ86 diet and 17 on CEWc diet. The number of nymphs produced per female per day peaked on day 11 in all the treatments. The values were 1.45 and 1.6 for the control and OcIΔ86 respectively, and 0.63 for CEWc. There were insufficient aphids surviving on CEWc diets to calculate the r_m value but the values for OcIΔ86 and control diets are shown in Table 1.

Table 1. Adult performance of *M. persicae* reared on control diet or diet containing OcIΔ86

Diet	r_m female/female/day	prereproductive time (days)	generation time (T)	doubling time (DT)	Total fecundity (nymphs)
Control	0.21	8.2	11.16	3.30	215
OcIΔ86	0.14	10.4	14.12	5.02	91

The survival and fecundity of *M. persicae* feeding on transgenic plants

No adverse effects were observed on the rate of development, fecundity or doubling time when aphids were caged on plants expressing CEWc (0.4% total soluble protein). The number of nymphs produced per female per day peaked at 3.93 and 4.09 on control and D6/2 plants, respectively. The reproductive peak occurred 7 days earlier on D6/2 than on control plants. The differences in reproductive dynamics are reflected in the calculated r_m values for aphids on the two plant lines (Table 2).

Table 2. Adult performance of *M. persicae* reared on transgenic or control potatoes.

Plant	r_m female/female/day	prereproductive time (days)	generation time (T)	doubling time (DT)
Control	0.299	8.2	11.11	2.32
D6/2	0.330	7.6	10.35	2.10

Aphid populations on transgenic plants in the field

Cool, wet weather during the 1998 growing season resulted in low aphid population densities. Consequently, the counts for *M. persicae* and *M. euphorbiae* were pooled for analysis. Fig. 4

summarises the cumulative number of aphids per 5 plants recorded during the period July-October.

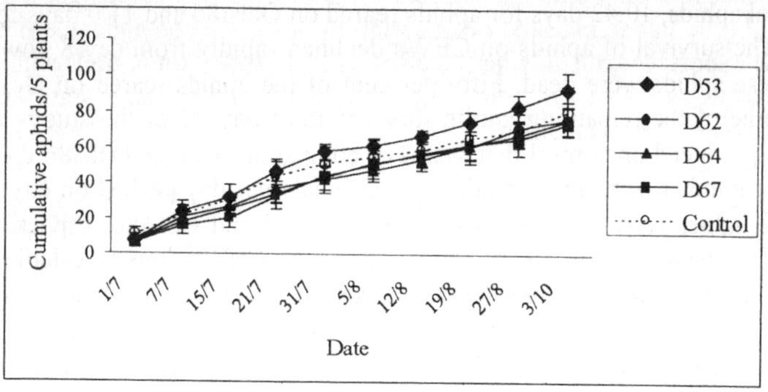

Fig. 4. The cumulative number of aphids on cv Desiree potato plants not expressing (control) or expressing CEWc (D53, D62, D64 & D67). Values were recorded over a 10 week sampling period.

DISCUSSION

Aphids such as *M. persicae* and *M. euphorbiae* represent important non-target insects in the potato-nematode system. Previous studies have shown that although the digestive tract of aphids does not contain detectable amounts of endo-protease activity, the ingestion of some serine PIs can result in reduced growth and survival (Rahbe *et al.*, 1995). Cereal aphids fed PIs from potato also showed reduced survival and fecundity compared to aphids on control diets (Tran *et al.*, 1997). Our studies have shown that cysteine PIs from a range of plant and animal sources also influence aphid growth, survival and fecundity in artificial diet assays. The variation in toxicity between the different cystatins may reflect their different actvity spectra. Maize cystatin has a wider inhibition spectrum than OcI and inhibits cathepsins H and L more efficiently than OcI (Abe *et al.*, 1994).

The observation that the rate of development, fecundity and doubling time of *M. persicae* were not affected when aphids were caged on transgenic plants establishes that tissue specific expression is important in determining effects of PIs on different invertebrates. Levels of expression in the root system of 0.4% total soluble protein were sufficient to provide nematode control (Urwin *et al.*, 1997) but concentrations in the phloem were insufficient to reproduce the adverse effects on aphids shown in the artificial diet assays.

The fact that PI-transformed plants do not stop aphids feeding on the plants immediately has implications for other components of the ecosystem. *M. persicae* and *M. euphorbiae* represent important prey items for several of the predators and parasites that are known to occur in potato crops. Aphid honeydew can also represent an important resource for some predators (Majerus, 1994). Artificial diet studies have shown that PI is present in the honeydew of aphids feeding on diet containing PIs (pers obs.). It is possible that aphid natural enemies may be exposed to cystatins expressed in transgenic plants via either of these two routes. Recent studies have shown that two important predatory insects can be adversely affected by ingestion of cystatins. Walker *et al.* (1998) showed that proteolytic activity of the larval midguts of *Adalia bipunctata* was almost totally inhibited by E64 or CEWc. Ashouri *et al.* (1998) have shown that fecundity and egg eclosion were effected when a pentatomid predator of Colorado potato beetle larvae ingested oryzacystatin (Oc-I), via contaminated prey. These results highlight the importance of studying the effects of PI expression in transgenic plants at all trophic levels and we are currently examining the interaction between cystatin-expressing potatoes, aphids and their natural enemies.

ACKNOWLEDGEMENTS

We would like to thank Dr. P. Urwin and Dr. E. Zubko for scientific support and the Scottish Office Agriculture Environment and Fisheries Department for financial support.

REFERENCES

Abe M; K. Abe K; Iwabuchi K; Domoto C; Arai S (1994). Corn cystatin-I expressed in *Escherichi coli* - investigation of its inhibitory profile and occurrence in corn kernels. *Journal of Biochemistry* **116**(3): 488-492.

Ashouri A S; Overney S; Michaud D; Cloutier C (1998). Fitness and feeding are affected in the two spotted stinkbug, by the cysteine proteinase inhibitor, Oryzacystatin I. *Archives of Insect Biochemsitry & Physiology* **38** 74-83

Dadd R H; Mittler T E (1966). Permanent culture of an aphid on a totally synthetic diet. *Experimentia* **12** 832-833.

Gustafson D I (1993). Pesticides in drinking water. Chapel Hill, N. Carolina, USA, 241pp.

Hilder V A; Gatehouse A M R; Sheerman S E; Barker R F; Boulter D (1987). A novel mechanism of insect resistance engineered into tobacco. *Nature* **330** 160-163.

Leplé J C; Bonadé-Bottino M; Augustin S; Pilate G; (1995). Toxicity to *Chrysomela tremulae* (Coleoptera, Chrysomelidae) of transgenic poplars expressing a cysteine proteinase-inhibitor. *Molecular Breeding* **1**(4): 319-328.

Majerus M E N (1994). *Ladybirds*. Harper Collins, London, pp367.

Rahbe Y N; Sauvion N; Febvay G; Peumans W J (1995). Toxicity of lectins and processing of ingested proteins in the pea aphid *Acyrthosiphon pisum*. *Entomologia Experimentalis Et Applicata* **76**(2): 143-155.

Tran P; Cheesbrough T; Keickheffer R W (1997). Plant proteinase inhibitors are potential anticereal aphid compounds. *Journal of Economic Entomology* **90**(6): 1672-1677.

Urwin P E; Lilley C J; M. J. McPherson; Atkinson H J (1997). Resistance to both cyst and root knot nematodes conferred by transgenic *Arabidopsis* expressing a modified cystatin. *The Plant Journal* 12 455-461

Walker A J; Ford L; Majerus M E N; Geoghegan I E; Birch N; Gatehouse J A; Gatehouse A M R (1998). Characterisation of the mid-gut digestive proteinase activity of the two-spot ladybird, *Adalia bipuncata*, and its sensitivity to proteinase inhibitors. *Insect Biochemistry & Molecular Biology* 28 173-180.

Wyatt I J; White P F (1977). Simple estimation of intrinsic increase rates for aphids and tetranychid mites. *Journal of Applied Ecology* **14** 757-766.

Xu D; Xue Q; McElroy D; MawalY; Hilder V A; Wu R (1996). Constitutive expression of a cowpea trypsin-inhibitor gene, CpTi, in transgenic rice plants confers resistance to 2 major rice insect pests. *Molecular Breeding* **2**(2): 167-173.

Honeybees as vectors of GM oilseed rape pollen

G Ramsay, C E Thompson, S Neilson and G R Mackay

Scottish Crop Research Institute, Invergowrie, Dundee, DD2 5DA, UK

ABSTRACT

One hive of honey bees, placed 800 m from a GM OSR crop, was studied to determine the extent of mixing of pollen taking place. In a one hour period, ten types of pollen were collected with OSR, rosebay willowherb and clover predominating. A combination of cytological analysis and PCR tests demonstrated that OSR pollen was spread through pollen loads of different types. This was likely to be due to worker bees switching crops between foraging trips, or surface contamination within the hive between worker bees returning from different crops. The presence of GM pollen in largely non-GM *Brassica* pollen loads, and the generation of mixed GM and non-GM progeny following the pollination of male sterile plants with pollen from single bees, demonstrates directly that bees can readily spread OSR pollen around their foraging area.

INTRODUCTION

Honeybees (*Apis mellifera*) are an important insect pollinator of oilseed rape (OSR, *Brassica napus* and *B. rapa*) in the UK, numerically more common than *Bombus* species in most field crops. In addition, oilseed rape provides beekeepers with the single most productive honeybee forage crop, and this mutual importance is exploited by oilseed rape growers and beekeepers alike to ensure both good seed set and good honey crops. Some farmers now pay beekeepers for pollination services, which is particularly important for growing non-restored male sterile hybrid OSR. Furthermore, the importance of managed honeybee colonies to OSR production is increasing as feral populations of honeybees are lost due to the spread of the parasitic mite *Varroa*.

Honeybees are known by beekeepers to forage regularly in a radius of about 2 km from the hive, leading to the 'three feet or three miles' rule when moving hives to avoid losing bees returning over familiar territory to the old site of the hive. OSR has a relatively high concentration of sugars in the nectar, and can attract bees from a longer distance. One professional beekeeper, for example, has reported bees in one colony placed in the Aberdeenshire hills flying to a rape field 5 km away (M. McGregor, pers. com.).

Individual honeybees are often regarded as being faithful to one plant type on a particular foraging trip or over days, although switches between forage species have been documented (Proctor, Yeo and Lack 1996). Although studies have been made on bee pollination in regard to quantifying dye analogue (Cresswell *et al.* 1995) or pollen (Ohsawa & Namai 1988) spread from flower to flower during single foraging trips, we are not aware of studies of the mixing and spread of OSR pollen at the level of the colony. In this paper we present some observations on the foraging behaviour of one colony placed 600 m from a commercial spring-sown GM OSR trial, describe the apparent inconstancy of individual bee foraging behaviour and discuss this particularly in relation to the mixing of pollen that a colony may achieve over its foraging area.

MATERIALS AND METHODS

In mid-July 1998, a small colony of honey bees with brood on 6 frames, which was building up from a June swarm, was moved to a site 800 m from a GM OSR field of about 11 ha in Perthshire. The GM OSR was carrying a thioesterase gene from *Umbellularia californica* and an *npt*II marker gene conferring kanamycin resistance. A pollen sampler containing a screen with 4.5 mm diameter perforations was placed in the hive entrance and used to dislodge and collect pollen loads from returning worker honey bees. Pollen loads are masses of pollen grains, collected by grooming and mixing with nectar before transferring onto the hind legs where combs of bristles on a flattened segment (the pollen basket) aid the storage of the pollen for transport back to the colony. Pollen loads were sorted into colour classes, weighed, air dried and stored at room temperature. They were carefully dissected to sample five parts of the load from the position nearest the comb on the leg (determined by the puckering of the load at this point) to the most recent and outermost part. Pollen type was determined where possible by comparison with a colour guide to pollen loads (Kirk 1994) and verified by microscopic observation.

Samples were taken of worker bees emerging from the hive by trapping single bees in a clean plastic centrifuge tube on the hive wall as they emerged. This was performed on the same occasions as pollen sampling, within 30 min of the removal of the pollen sampler from the hive. Bees were stored at -20°C, and used for pollinating male sterile plants or for microscopic examination.

DNA was extracted from pollen loads and from the seedling progeny of pollinations made on male sterile plants and subjected to anchored-SSR PCR fingerprinting or PCR using primers to the *Umbellularia californica* thioesterase and *npt*II transgenes (see Thompson *et al*, these proceedings).

Some additional casual observations were made on a large established hive on the institute farm in an area with spring OSR within 1 km and no GM OSR within flying distance.

RESULTS AND DISCUSSION

Fingerprinting of pollen loads determines their origin

Fingerprints of pollen loads using anchored-SSR PCR (Charters *et al.* 1996) were used to determine the origin of pollen collected at an SCRI hive in an area where no GM OSR was grown. In this case, comparing pollen load patterns with field-collected leaf samples clearly shows the origin of the pollen samples as being cv. Ostara (Fig. 1). The nearest field of Ostara was 900 m from the hive. This method will be used in more extensive studies to examine the number of OSR sources which one colony may be working at any one time.

Pollen load samples

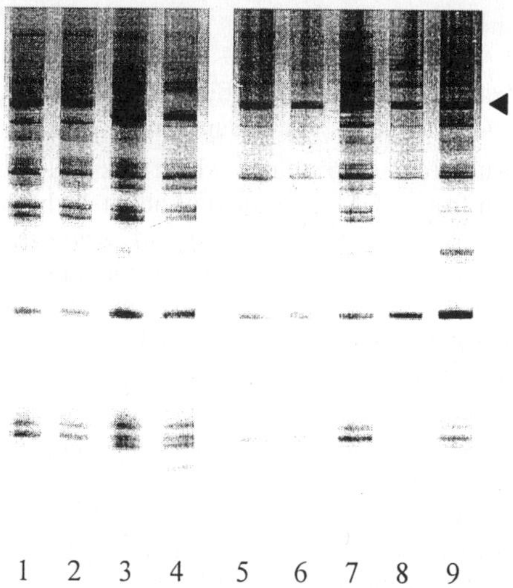

Figure 1. Anchored-SSR PCR fingerprints of pollen loads collected at a hive at SCRI compared to PCR fingerprints of leaf samples from candidate fields. 1-2, leaf samples of cv. Ostara at 900 m from the hive; 3-4, leaf samples of cv. Hyola 401 at 4 km from the hive; 5-9, five separate pollen loads from SCRI hives. An arrowhead marks a diagnostic band.

The data presented in table 1 were collected on 30 July 1998, while the GMO crop was still in full flower, and other OSR crops (the nearest at 500 m distance) were coming out of flower. In one hour, 483 pollen loads were collected from workers returning to the hive placed 800 m from the GMO OSR crop under study. Ten classes were discriminated by the colour of the fresh pollen load (Table 1). *Brassica* pollen accounted for 70% of the pollen loads, with rosebay willowherb (*Chamaenerion angustifolium*) and clover (*Trifolium repens*) making up the majority of the remainder. As OSR was the only *Brassica* type present locally in quantity, we have assumed that all *Brassica* pollen is OSR. Although this represents only one short period of activity of one hive, similar proportions of OSR and non-OSR pollen grains were noted in samples collected earlier in the same month at a hive at SCRI. Close inspection of the pollen loads revealed that although each one is largely composed of one type of pollen grain, many carried a yellow patch where the load had adhered to the comb on the worker's hind leg.

Table 1. Frequency of pollen loads of different types recovered during one hour of sampling.

	Pollen load type									
	1	2	3	4	5	6	7	8	9	10
no.	338	98	37	3	2	2	2	2	2	1
% of total	70	20	8	1	t	t	t	t	t	t
fresh wt (mg)	2308	448	148	19	23	16	7	7	6	8

1 *Brassica* sp., oilseed rape; 2 *Chamaenerion angustifolium*, rosebay willowherb; 3 *Trifolium repens*, clover; 4-10 various *Asteraceae, Rosaceae* and other taxa; t - trace.

Four pollen loads were dissected and examined in more detail to determine the admixture of other pollen (Table 2). Five central portions were taken from each of five sections of the pollen load, numbered from 1 (closest to the original point of attachment to the worker's leg) to 5 (the most recently collected pollen).

Table 2. The percentages of total pollen grains falling into different types in five sections each of 4 pollen loads of different type.

No	1° pollen type	Patch	Sectn.	Percentage of Pollen Grains							
				bra	rbw	clo	urt	cir	com	oth	*n*
1	rbw	N	1	7	93	-	1	-	-	2	150
			2	13	85	-	-	-	-	1	295
			3	9	89	-	1	-	-	1	330
			4	3	95	-	1	-	-	1	466
			5	10	88	-	1	-	-	-	582
2	rbw	Y	1	65	34	-	1	-	-	-	437
			2	36	63	-	1	-	-	-	212
			3	1	99	-	-	-	1	-	140
			4	13	83	-	-	1	1	1	78
			5	16	84	-	-	-	-	-	114
3	clo	N	1	2	-	98	-	-	-	t	273
			2	t	-	99	t	-	-	-	353
			3	2	-	98	-	-	t	-	284
			4	-	-	100	-	-	t	-	241
			5	1	-	99	-	-	1	-	375
4	bra		1	100	-	-	-	-	-	-	407
			2	100	-	-	-	-	-	-	383
			3	100	-	-	-	-	-	-	374
			4	100	-	-	-	-	-	t	366
			5	100	-	-	-	-	-	-	397

Patch – an area of visually different pollen grains at the attachment point, bra – *Brassica* species, rbw – rosebay willowherb *Chamaenerion angustifolius*, clo – clover *Trifolium repens*, urt – nettle *Urtica* sp., cir – thistle *Cirsium* sp., com – other composite *Asteraceae*, oth – other, t - trace.

All four pollen loads carried more than the main pollen type. Trace amounts of pollen of several species were found, including that of *Urtica* (nettle), an anemophilous type. The higher numbers of *Urtica* pollen grains in the rosebay willowherb pollen loads may arise from aerial contamination at the time of collecting due to the frequent proximity of these two plants in fertile waste ground. There was no particular pattern to the distribution of these types of trace admixtures across each pollen load as they were present across all five parts of the loads sampled. *Brassica* pollen was almost ubiqitous. Molecular studies later confirmed that this was OSR pollen. The greatest amount was found in pollen load no. 2 close to the point of attachment to the comb, where 65% of pollen grains scored were OSR. This pollen load was composed predominantly of rosebay willowherb pollen but carried a noticeable yellow patch. The pattern of OSR pollen in this pollen load strongly suggests that the bee previously visited OSR. The OSR pollen in the oldest part of the pollen load could represent either a portion of the previous load adhering to the comb, or further pollen grains groomed and added to the comb after the main load had been placed in the pollen basket for transport. The spread of OSR pollen grains throughout the pollen load suggests that the bee recovered additional OSR grains still contaminating its body surface while grooming to collect rosebay willowherb grains during a foraging trip. The alternative possibility of bees on foraging trips primarily visiting one flower type but regularly moving to OSR during the trip seems unlikely as no banding was detected in any pollen load. The switching from one crop to another between foraging trips was confirmed by microscopic examination of a sample of 30 workers captured emerging from the hive. Most of the workers had signs of OSR pollen adhering to wings and body hairs, or in caked masses on legs, the face and in the pollen baskets. Pollen grains could be seen at different depths on the baskets, indicating the latest and previous foraging activity. Some had last visited OSR after previous visits to rosebay willowherb. In others the reverse was true, and other possible foraging histories were noted.

Mixing of GM pollen with other pollen

The admixture of OSR pollen in non-OSR pollen loads suggested that mixing was taking place. To investigate the mixing of pollen further, DNA was extracted from several OSR, rosebay willowherb and clover pollen loads. Both pollen load 1 and pollen load 2 (Table 2), predominantly rosebay willowherb but with 8.4% and 26.2% *Brassica* content respectively, gave a positive PCR signal (not shown). Among the OSR pollen loads tested (Fig. 2 lanes 14-19), two of the 6 OSR pollen loads gave strong signals for the thioesterase transgene (Fig. 1) suggesting that these were largely composed of GM pollen. Most of the other OSR pollen loads and the rosebay willowherb (lanes 2-7) and clover (lanes 8-13) pollen loads gave a weak signal for the transgene. In a larger sample of OSR pollen loads, 17/44 did not give any detectable signal for GM OSR. This, together with the cytological determination of OSR pollen contamination throughout the 4 pollen loads examined, demonstrates clearly that the GM pollen was widely dispersed on bees in the colony.

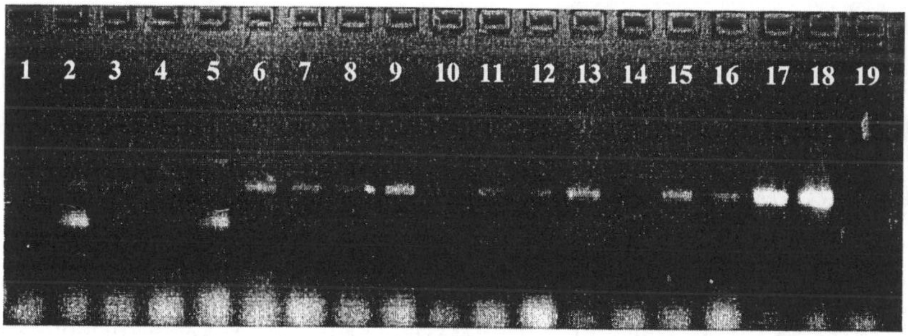

Fig. 2. PCR detection of pollen containing the *Umbellularia californica* thioesterase gene in pollen loads collected from workers returning to the hive. 1, negative control; 2-7, rosebay willowherb pollen loads; 8-13, clover pollen loads; 14-19, oilseed rape pollen loads.

Whether this mixing occurs because of regular switching between foraging on GM rape and other flowers, or by bee to bee spread within the colony was not clear. However one worker returning to an SCRI hive and bearing the typical yellow contamination on the face and elsewhere typical of bees working OSR was found to be carrying 60,000 OSR pollen grains in addition to those in the pollen loads. Workers clean up within the hive and generally emerge free of rape pollen visible to the unaided eye. It seems, therefore, that bee to bee pollen transfer within the hive is inevitable.

Five visually clean bees collected on emergence from the hive near the GM site were used to pollinate flowers on male sterile OSR plants. Two of these failed to induce seed set. The other three all induced fertilisation and the progeny seedlings from each bee contained both transgene carrying and normal individuals. Overall from 62 seedlings screened, 12 carried the transgenes. This demonstrates the mixed nature of the OSR pollen on these bees, and also that this pollen is viable.

The observations presented here illustrate that bees switch from one forage type to another, and that bees carrying many viable OSR pollen grains can be found emerging from a hive. The OSR pollen found on bees leaving a hive can be from mixed sources, indicating either switching between crops and relatively long persistence of the pollen grains on bees, or the picking up of some of the larger numbers of loose pollen grains available on other bees in the close phyical contact in the hive. With most honeybee colonies foraging up to 2 km from the hive, some pollen transfer and fertilisation up to 4 km must be expected. In some circumstances honeybees are capable of much longer foraging trips and these limits will be exceeded.

ACKNOWLEDGEMENTS

Funding of this work by MAFF under the Open Contract CSA4202 is gratefully acknowledged.

REFERENCES

Charters Y M; Robertson A; Wilkinson M J; Ramsay G (1996). PCR analysis of oilseed rape cultivars (*Brassica napus* L. spp. *oleifera*) using 5' anchored simple sequence repeat (SSR) primers. *Theoretical and Applied Genetics* **92**, 442-447.

Cresswell J E; Bassom A P; Bell S A; Collins S J; Kelly T B (1995). Predicted pollen dispersal by honey-bees and three species of bumble bee foraging on oil-seed rape: a comparison of three models. *Functional Ecology* **9**, 829-841.

Kirk W D (1994). A Colour Guide to Pollen Loads of the Honey Bee. International Bee Research Organisation.

Ohsawa R; Namai H (1988). Cross-pollination efficiency of insect pollinators Shimahanaabu *Eristalis cerealis* in rapeseed *Brassica napus* L. *Japanese Journal of Breeding* **38**, 91-102.

Proctor M; Yeo P; Lack A (1996). The Natural History of Pollination. HarperCollins, London.

SESSIONS 7 AND 8

GENE FLOW – THE FUTURE

Chairmen & Session Organisers: Dr P Dale
John Innes Centre, Norwich

and

Dr J Sweet
NIAB, Cambridge

Concerns about gene flow and the implications for the development of monitoring protocols.

J E Hill
Green Alliance, 49 Wellington Street, London WC2E 7BN

ABSTRACT

The possibility that genes will move from crop plants to wild species, as well as to volunteers and to neighbouring non-GM crops, are high on the list of concerns about genetic modification. This paper gives an overview of the way the concerns have been expressed by some influential groups. It then discusses how they might be met by monitoring regimes, and what questions need to be addressed in order to construct such regimes.

INTRODUCTION

Concerns about gene flow are not new. The possibility of gene flow and its possible consequences were amongst the set of concerns about the potential environmental impacts of releases of genetically modified organisms that gave rise to European and UK regulations. The 1989 report from the Royal Commission on Environmental Pollution (RCEP 1989) which was influential in the development of the EU and UK regulatory systems, observed that "The extent to which genes, especially novel genes, may spread is an important uncertainty in assessing the risks in the release of genetically engineered organisms (GEOs)". The report recommended a system of scrutiny that would include examining the likelihood of gene transfer, and put considerable emphasis on the need to develop techniques for monitoring the spread of released GEOs and their introduced genes.

The UK went on to evolve a system of scrutiny, working to the requirements of EU Directive 90/220, that required those proposing to make a release to submit a risk assessment, including answering questions about potential gene flow. For the first 3-4 years of the system, many applications were very cautious in their nature, voluntarily proposing to limit gene flow by means such as preventing flowering of the crop, ensuring large isolation distances, and removing potentially sexually compatible wild species in the vicinity of the release. These measures were not necessarily linked to any identified hazard that could be realised as the result of gene flow - they presumed that gene flow was itself something to be avoided. This trend had three important consequences: it gave the regulatory system the appearance of being highly 'precautionary' in its nature; it simultaneously gave the impression that gene flow was a bad thing *per se*; and it yielded very little useful data about the potential for gene flow in the field.

Thus, when the EU and UK regulatory systems were faced in 1994 with the first application to commercialise a GM crop, the Plant Genetic System's (PGS) hybrid and herbicide-tolerant oilseed rape (PGS 1994) the issue of gene flow came more sharply into focus. The applicants took the view that gene flow to sexually compatible wild relatives of rape could not be ruled out, but that it would be at a low level, and because of the nature of the traits inserted in the rape, would not have deleterious consequences. A majority of the UK Advisory Committee on Releases to the Environment (ACRE) took the same view. A minority of the committee objected to the commercialisation on the grounds that there was still a large measure of uncertainty about the extent of, and consequences of, gene flow (ACRE 1995). The UK Government accepted the majority advice.

To many outside the regulatory system, the UK decision on the PGS application marked a step change in the risk assessment of GMOs - a move from a precautionary approach to one where releases were to be allowed to go ahead with an unacceptable level of uncertainty. Since that point, concerns about gene flow have figured prominently in representations about GMOs. To illustrate this, some of those representations are outlined below.

A 1995 paper from the WorldWide Fund for Nature (WWF 1995) considered primarily what are often referred to as the 'indirect' effects of gene flow: "Transfer of herbicide resistance to weeds, even at low frequency, could clearly create enormous problems for weed control, and reduce the effectiveness of specific herbicides. In such situations, farmers would have fewer options in their weed control practices, and could be forced into using even more environmentally damaging chemicals. Weeds resistant to particular herbicides would have a selective advantage in systems where those herbicides are used".

A 1998 paper issued by Greenpeace International (Fromwald & Strauss 1998) was concerned about uncertainties. It stated: "Current scientific studies and knowledge demonstrate that GE oilseed rape, when commercially released in Europe, may transfer genes to other oilseed rape and wild related species. These hybrids and the GE oilseed rape itself may become a permanent feature of ecosystems and fields. Their overall effects are unpredictable, and once these species are introduced it may take tens or even hundreds of years to recognise their effects".

Like WWF, Greenpeace highlights the possible indirect effects on the use of chemicals: the spread of the tolerance gene to weedy relatives would cause the emergence of glufosinate resistant weeds, therefore creating even further problems of weed control for farmers. The GE oilseed rape and hybrids resulting from cross breeding between the GE oilseed rape and wild related species is also likely to spread onto organic farmers' fields, polluting organically grown products. This could force farmers out of environmentally sound farming methods since organic farming does not allow the use of GE crops, and consumers expect organic products to be free of genetically altered ingredients. Allowing GE oilseed to be grown on a large scale in Europe means undermining environmentally sound farming".

A 1997 briefing document from the Royal Society for the Protection of Birds (RSPB 1997) cites as one of the possible hazards of releasing GMOs into the environment: "The transfer of the introduced genes from the GMO to other species, which subsequently become established as 'pests'". RSPB also states that: "there are many potential effects on bird populations of GMOs which become established 'pests' within the environment including weedy crops or hybrids displacing other plants that are important food sources for birds". RSPB's concerns are more about the identifiable hazards of gene flow than about gene flow *per se*.

The GeneWatch report Genetically Engineered Oilseed Rape: Agricultural Saviour or New Form of Pollution? (GeneWatch 1998a) addresses gene flow as one of the foremost environmental risks associated with genetically engineered organisms (GEOs). The briefing document accompanying the report refers to gene flow as 'genetic pollution' (GeneWatch 1998b). The report notes that it is possible that GEOs will breed with native species, and will transfer introduced genes, and states: "these foreign genes, which could not have entered the native gene pool through natural mechanisms, may become established and alter the genetic diversity of native flora and fauna". The report posits that "in the case of oilseed rape, gene flow to native flora in the UK is inevitable". The report acknowledges that hybridisation events are likely to be low frequency but states that: "the low frequency with which hybridisations take place in experiments cannot be equated with safety. As more varieties of GE oilseed rape are commercialised, the likelihood of harm will be increased"

The report asks the million-dollar question: "if gene introgression occurs - does it matter?". According to GeneWatch there are several causes for concern: "these include matters of immediate practical importance and more intangible concerns about altering genetic diversity so fundamentally and what this means for future generations". The main concerns discussed in the report are that alterations to the genetic makeup of plants could make them 'fitter' and create problem weeds for farmers or alter local ecosystems; and that gene introgression could reduce the fitness of native plants and lead to a reduction in numbers or local extinction of a species. The report points out that "the impacts of gene flow may not be recognised for many decades" and that "once gene introgression has taken place, it will be irreversible".

The GeneWatch report also reviews the advice given in 1994 by the Advisory Committee on Releases to the Environment (ACRE) on the PGS application and challenges two of the key elements of the advice: that gene flow to wild species will be minimal; and that gene flow is unimportant. On the extent of gene flow, the GeneWatch report draws attention to research on hybridisation published since the 1994 evaluation, indicating that hybridisation in the field may be more likely than originally supposed. On the importance of gene flow, the report notes that guidance on 'harm' which ACRE had been involved in developing with the Department of Environment, Transport and the Regions (DETR) "makes the assumption that genetic pollution *per se* is not harmful unless it is connected with a measurable change in species numbers or selected ecosystem parameters". GeneWatch challenges this assumption and states that "because any harm will be irreversible, the decision about acceptable harm should be a matter for wider social debate, particularly given the considerable scientific uncertainty which affects (the) confidence that can be placed in any present day estimate".

GeneWatch called on the Government to withdraw consent to market the PGS oilseed rape and undertake a re-evaluation of the environmental safety aspects, including a systematic examination of the impact of uncertainties on the risk assessment.

Friends of the Earth's 1997 Briefing Sheet on Genetically Engineered Oilseed Rape (FOE 1997) lists ten concerns about Glufosinate Tolerant Oilseed Rape, one of which is that "Glufosinate tolerance will spread to weeds". The briefing states that "Oilseed Rape can cross breed easily with other plants in the Brassica family such as wild turnip and wild radish. Research shows that these too can become herbicide resistant if crossed with the genetically engineered oilseed rape. Therefore the prospect of common weeds becoming agricultural pests (often labelled as 'superweeds') is very real and may lead to more toxic chemicals being used to control them." Friends of the Earth called for a moratorium on the growing of genetically engineered crops until the implications have been fully evaluated.

Friends of the Earth's 1998 court case against DETR attempted to establish that gene flow was possible between a GM maize crop and a nearby organic sweetcorn crop, and would result in unacceptable 'pollution' of the organic crop. The case highlighted the fact that whatever the separation distances, gene flow could never be entirely ruled out or the extent of it accurately predicted, and that the crucial issue was the acceptability of any amount of gene flow.

English Nature's evidence to the House of Lords 1998/1999 enquiry (HOL 1999) had gene introgression at the top of a list of concerns. Uncertainty figures again: "these (introduced) genes have not been exposed to selective pressures in the habitat of the original organism, and may have unpredictable effects if they outcross into wild relatives". The report also states: "Outcrossing to native species is likely to be a rare event, but with increasing acreages of GM crops, will occur more often. We are concerned not only about the possibility of HT (herbicide tolerant) hybrids becoming persistent weeds, and insect resistant hybrids having adverse effects on farmland insect biodiversity, but also that outcrossing risks harm to biodiversity in other semi-natural habitats".

English Nature (EN) notes that: 'There are existing experimental methods of genetically modifying plants which significantly reduce the risk of gene transfer, but these are not being developed as a commercial priority". EN recommends that "if the regulatory process required that the risks of gene transfer were minimised during the developmental state of the product, it would stimulate further research and development of these safeguards in many crops".

Four main categories of concern emerge from these statements:

1. That gene flow constitutes unacceptable 'genetic pollution', whatever the consequences.

2. That gene flow will contaminate organic or other non-GM food crops;

3. That gene flow will occur and will have unpredictable consequences;

4. That gene flow will have identifiable deleterious consequences; such as spread of herbicide tolerance or insect tolerance in a way that has direct or indirect ecological impacts.

RESPONDING TO THE CONCERNS

The concept of 'genetic pollution' as a hazard in itself, independently of its consequences, tends to annoy scientists in the field. They point out that genes already move between species in nature, albeit rarely, and thus even in this GMOs do not present unique considerations. Some have argued that to try to avoid all 'genetic pollution' is to attempt to freeze the environment at a particular point in time and to deny the processes of evolution. Objections to the flow of genes as being 'unnatural' are often labelled as religious or quasi-religious, i.e. more a matter of doctrine than of science.

However, it is important to recognise that genetic manipulation does seem to represent a 'step too far' to some people, even if they can't articulate in strict scientific terms just what that step is (Grove-White et al 1997). It is possible to accept that we routinely intervene in nature, but simultaneously to object to particular interventions, or ones that lead to a change of scale - and the movement of genes between species will certainly increase in scale as a consequence of genetic manipulation and subsequent gene flow. The problem for policy-makers is knowing how many people hold this kind of view, or how many would if they had enough information about the techniques and implications of GM crops. Detailed public consultation is needed to help ascertain this.

The issue of contamination of organic and other non-GM crops has received much more attention. Friends of the Earth and the Soil Association went to court in 1998 on behalf of an organic farmer who feared that the organic status of his crop would be jeopardised by cross-pollination from a GM fodder maize crop. Current organic standards state that "there is no place for GMOs in organic agriculture" which the Soil Association has interpreted as meaning that there should be no cross-pollination and thus contamination of organic crops at all. This would be very difficult to ensure even with large isolation distances, given the possibility that the pollen of some crops can travel for several kilometres, and even the standards for certified seed allow for some contamination.

There is as yet no resolution to this problem, and the case highlighted the fact that there is no mechanism for consultation between the growers of GM and the growers of non-GM crops, in case cross-pollination presents a problem. This may be an issue not just for organic growers, but for those undertaking to provide non-GM sources to retailers, some of whom want to offer own-brand products that are GM-free. It will also be an issue if crops are used as factories for non-food products such as pharmaceuticals and chemical feed-stocks, where it may be important that such substances do not enter the food chain through contamination of neighbouring crops. It would be unwise to put such substances into cross-pollinating crops in the first place, but in case that scenario does arise, there will need to be a regulatory mechanism for ensuring separation.

Unpredictability is also difficult to deal with in the regulatory regime. There are those that argue that avoiding all areas of unpredictability, by constantly invoking the precautionary principle, implies not getting out of bed in the morning. In order to reap benefits we must live with uncertainties, or there will be no progress. The counter-argument is that the regulatory system, and specifically the risk assessments that accompany applications for release consents, fail to acknowledge either the uncertainties or the possible benefits, and certainly do not offer a mechanism for weighing up the two. There is no way of deciding what constitutes an acceptable level of uncertainty in the context of a particular application. Scientific assessments tend to be strong on what is known rather than what is not known, and many critics would be satisfied simply by an honest statement of the gaps in our present knowledge.

DEVELOPING MONITORING PROTOCOLS - QUESTIONS

The fourth category of concerns, that gene flow will lead to direct or indirect damage to the environment of an identified kind, is more amenable to research and monitoring. However, some of the possible impacts are more easily monitored than others. The following are some of the questions that have to be answered before useful monitoring regimes can be constructed.

1. What **kinds** of gene flow do we want to monitor? Flow to wild relatives; to volunteers of the crops; to neighbouring crops? The factors monitored will depend on which of these, or perhaps all of them, are of concern.

2. It is also important to confirm whether the scale, perhaps better expressed as the **frequency** of gene flow is important in itself, or whether the focus should be on the **effects** of gene flow. If the latter, it is debatable how far assumptions about, and confirmation of, frequency is important, since it may be possible to study the effects of, for instance, hybridisation, without knowing how often it happens. On the other hand, there may be cases where the effects are related to the amount of gene flow happening, for instance where the concern is the rapid build-up of volunteer populations carrying introduced genes.

3. If wild relatives are the concern, one of the first considerations will be how much we know about **potential recipients** of introduced genes. The experience with oilseed rape has been that, as research has progressed, it has indicated that more wild relatives are capable of producing hybrids in the field than was thought when the early risk assessments were prepared (Gray & Raybould 1999). This is important because the distribution of those wild species thought to be candidates for hybridisation may determine which areas are targeted for monitoring.

4. What is the **distance** over which we want to monitor? This will be partly determined by what we know about how far pollen can travel, both from and to a crop, as well the issue above about the distribution of related wild species. It will also be strongly driven by the resources available to those monitoring.

5. What should be the size of the **sample** of wild plants to be sure of confirming assumptions about frequency of hybridisation (if frequency is felt to be important). This depends largely on what we think the hybridisation frequency might be, which in turn depends on having made the right assumptions about potential recipients (as in point 2 above). If the hybridisation frequency is assumed to be very low, the sample needs to be very large in order to detect any hybridisation, and confirming the frequency with any degree of accuracy will be difficult. If the hybridisation frequency is assumed to be high, the sample can be smaller. There is also emerging research (Bergelson 1998) that indicates that transgenes may intrinsically be more likely to transfer to near relatives than genes introduced by other means, which complicates assumptions about frequency, but which may also mean that there are more hybrids available for analysis.

6. Having decided the distance and area under scrutiny and the sampling methodology, we have to ask whether or not there are adequate **testing techniques**, and how practical these are to carry out on a large scale. A hybrid of a GM crop and a wild plant may not express the GM trait in a way that can be directly tested - for instance by becoming herbicide tolerant and therefore detectable by applying the herbicide. In such cases molecular analysis will be needed, which is time consuming if a large number of plants are to be tested.

7. In considering the **effects** of gene flow, one of the key concerns is whether introgressed genes will give the plant enhanced **fitness,** making it better able to compete in natural environments, and thus possibly decreasing the populations of other species. Here we need to decide what are the valid measurements of fitness, and over how many generations these need to be monitored to yield a meaningful result. A first generation hybrid may suffer a loss of fitness, which is then enhanced by back-crossing with the original wild-type.

8. In any assessment of effects it will be important to have **base-line data** about species diversity and levels of populations before there is any gene flow. Otherwise it will be impossible to assess how species composition has been altered.

9. Another key concern is whether flow of introduced genes such as those for insect resistance, if they become incorporated into wild species outside the agricultural environment, could affect **insect diversity**, and whether this would be felt higher up the food chain, for instance by affecting food supplies for birds. Here base-line data will be particularly important, not only about populations of plants that could be recipients of the genes and about the insects associated with them, but also about non-target effects, i.e. toxicity to insects that prey on the insects that eat the plants (Hilbeck et al 1998). Information would also be needed about food-webs, i.e. what species are dependent on particular insect species, and how much effect on insect populations would have a knock-on effect on those species.

10. For flow to **volunteers,** it needs to be considered why this is of concern - because volunteer populations will present weed problems in themselves if they take on introduced genes (for instance by becoming herbicide tolerant and therefore not amenable to eradication with that herbicide); that volunteer populations could become a reservoir where a variety of different introduced genes could become 'stacked', producing volunteers with multiple tolerance; or whether the reservoir of introduced genes in volunteer populations could provide a concentrated source of gene flow to wild relatives. The most relevant questions are likely to be those about **distance** to monitor, appropriate **sample size,** and the availability of appropriate **testing techniques.**

11. For flow to **neighbouring crops**, all the questions relevant to volunteers will be important, but an additional consideration will be how much consultation there will need to be with neighbouring farmers to ensure access to the crop to be monitored.

12. For any monitoring regime, a key issue is the **time** over which it is conducted. Effects in the field such as enhanced fitness may take many generations to show up, so it is worth seriously considering whether a three- or even a five-year monitoring programme will yield meaningful results. Monitoring programmes that conclude that nothing is happening will be criticised as lacking credibility and wasting money if they are constructed in timeframes that could not reasonably be expected to detect any effects.

All these questions need to be debated among a wide range of interest groups, not just among the scientifically qualified, before monitoring regimes can be constructed that will satisfy the concerns about gene flow.

REFERENCES

Advisory Committee on Releases to the Environment (ACRE) (1995). *Annual Report No. 2 1994/1995.* Department of the Environment: London.

Bergelson J; Purrington C B; Winchmann G (1998). Promiscuity in transgenic plants. *Nature* 395 p25.

Friends of the Earth (1997). *Briefing Sheet on Genetically Engineered Oilseed Rape*. Friends of the Earth: London.

Fromwald S, Strauss S (May 1998). *Genetically Engineered Oil Seed Rape (Agrevo/PGS): A critical assessment and background information*. Report for Greenpeace International: Amsterdam.

GeneWatch (1998a). *Genetically Engineered Oilseed Rape: Agricultural Saviour or New Form of Pollution? A Report.* GeneWatch: Derbyshire.

GeneWatch (1998b). *Genetically Engineered Oilseed Rape: Agricultural Saviour or New Form of Pollution? Briefing Number 2.* GeneWatch: Derbyshire.

Grove-White R; Macnaghten P; Mayer S; Wynne B (1997). *Uncertain World: Genetically modified organisms, food and public attitudes in Britain.* University of Lancaster.

Hilbeck A; BaumGartner; Fried P M; Bigler F (1998). Effects of transgenic *Bacillus thuringiensis* corn-fed prey on mortality and development time of immature *Chrysoperla carnea* (Neuroptera: Chrysopidae). *Environmental Entomology* 27 (2) 480-487.

House of Lords Select Committee on the European Communities (1999). *EC Regulation of Genetic Modification in Agriculture.* Session 1998-1999 Second Report Minutes of Evidence.

Plant Genetic Systems (PGS) (1994). *A new hybridisation system in oilseed rape (Brassica napus) - application for consent to market genetically modified organisms.* DOE ref: 94/M1/1. Department of the Environment: UK.

Gray A; Raybould A, (1999). *Environmental Risks of Herbicide Tolerant Oilseed Rape - a review of the PGS hybrid oilseed rape.* DETR Research Report Blue Series No.15. (in press).

Royal Commission on Environmental Pollution (RCEP)(1989). *13th Report: The Release of Genetically Engineered Organisms to the Environment*, HMSO, London.

Royal Society for the Protection of Birds (RSPB) (1997). *The Potential Effects of Releasing Genetically Modified Organisms into the Environment.* RSPB: Bedfordshire.

Worldwide Fund for Nature (WWF) (1995). *Genetic Engineering - Examples of Ecological Effects and Inherent Uncertainties.* WWF: Godalming.

Molecular aspects of multiple transgenes and gene flow to crops and wild relatives

I J Senior, P J Dale
John Innes Centre, Colney, Norwich, NR4 7UH, UK

ABSTRACT

The consequences of transgene stacking need to be assessed to provide information to regulatory bodies concerned with genetically modified crops (GMOs). How transgenes will interact depends on the degree of homology between their gene sequences or with resident genes, the copy number present, and their expression. Transgene stacking provides opportunities for plant breeding, by combining traits unobtainable by conventional methods. However, such changes could be transferred to wild populations and impact on the environment. Multiple transgene flow to crop relatives needs careful consideration during biosafety assessment to prevent undesirable combinations becoming established in wild plant species. The design of transgene constructs could play some part in the reduction of environmental impact by targeting gene products more effectively, using inducible promoters and reducing current reliance on constitutive types.

INTRODUCTION

Plant transformation technology has become an established method to introduce new traits into crop plants. This technology is continually being applied to more plant species, with the intention to improve them for human use. The technology is advancing and new ways are being developed to transfer more complex traits and larger DNA fragments to plants, for both research and commercial uses. The first crops developed from transformation technology have already reached the global market place. So far these crops only contain a single construct but, through either deliberate breeding or gene flow, constructs could become stacked together in the same crop plant or in related species. The presence of such multiple transgenes, whether by design or accident, is an important aspect of biosafety assessment. Molecular aspects of gene stacking need to be highlighted to aid evaluations of transgenic crops before they are released into the environment. This paper is intended to address some of the salient molecular issues of multiple gene constructs both in transgenic crops and via gene flow to related crops or wild relatives.

PRESENCE OF MULTIPLE TRANSGENES IN CROP PLANTS.

Plant breeders will want to combine different constructs together to obtain phenotype combinations unavailable by other means. Multiple transgenes will frequently be introduced by sexual hybridisation between single trait elite breeding lines, followed by extensive progeny screening. Table 1 illustrates the possible combinations of traits that may be achieved through breeding. Some of these combinations are more desirable than others, and some have already been produced e.g. herbicide tolerance and male sterility. Combining two or more herbicide tolerance transgenes together is not considered to be an attractive idea, although there are proponents of this approach. Transfer of two herbicide tolerance traits to relatives would make it more difficult to control volunteers or weeds containing these genes. Other combinations are less likely to be disadvantageous if they were to be combined in weeds e.g. male sterility with oil modifications.

Multiple transgenes could also occur through gene flow between sexually compatible transgenic crops planted close together. This is likely to occur at a higher frequency than transfer to wild relatives. Initially at least the probability of the production of multiple transgenes in weedy species is likely to be very low. Availability of related species, along with survival, fitness and fertility of their hybrids will define how quickly transgenes will transfer in the short term, however, in the long term it is likely that multiple transgenes will be found in weedy populations.

STABILITY OF TRANSGENE EXPRESSION

Stable transgene expression is a prerequisite for a commercially viable crop. Usually many independent transformants are assessed in order to find individuals with both appropriate expression levels and transgene stability. Such work has helped to confirm classical genetic evidence of different allelic and non-allelic gene interactions known to affect gene expression (Nap *et al.* 1996). Most of these interactions result in down regulation of gene expression. Gene silencing of transgenes has been a perplexing problem for both researchers and companies alike. Extensive research has shown that transgene silencing is caused by a variety of different mechanisms: position effect, transcriptional and post-transcriptional silencing mechanisms.

Position effect

Analysis of a population of primary transformants reveals a wide range in transgene expression levels, which is often attributed to the 'position effect' - the site of integration of the transgene into the plants genome. These differences occur because of the random way transgenes are integrated into the genome. Transgenes integrating into hypermethylated regions of chromosomes can become methylated and this can inactivate their transcription (Meyer 1995). Introducing matrix attachment region sequences (MARs) along with the desired transgene can moderate expression level variation, due to the site of integration (Mlynarova *et al.* 1994), but have not prevented gene silencing events (Vaucheret *et al.* 1998).

Transcriptional gene silencing

Matzke *et al.* (1989) first observed transcriptional silencing of transgenes, when they sequentially transformed tobacco with two different constructs. These constructs had some sequence homology between each other at the promoter level. Some of the resultant progeny were unexpectedly sensitive to the antibiotic used for selection. Further work showed that this effect could also be observed when two transgene loci were brought together by sexual hybridisation. Analysis of the loci involved in this silencing effect showed multiple copies of one of the transgenes, were present at the locus which could silence any incoming, homologous transgenes (see Matzke *et al.* 1996 for a review).

Post-transcriptional gene silencing

Post-transcriptional gene silencing was first observed with chalcone synthase (*chs*) and dihydroflavonol-4-reductase transgenes when they were introduced into petunia. Instead of over-production of the gene product, suppression of both transgene and endogenous genes occurred. Previously it was known that reversing the coding sequence (antisense) could silence genes, whereas genes introduced in a sense orientation would be over-expressed. It is now established that genes in the sense orientation can also silence expression. This type of gene silencing has been seen in many species and has now become an alternative method to silence plant genes (see Depicker & van Montagu for a review).

Over the last decade considerable research efforts have probed the mechanisms of plant gene silencing (reviewed in Kumpatla *et al.* 1998; van den Boogaart *et al.* 1998). Similar silencing systems are also known in fungi and it is postulated that all genomes have evolved similar defence mechanisms (Kumpatla *et al.* 1998). Gene silencing mechanisms are not fully understood, but it is thought that many different levels of interaction are involved in the recognition of invasive sequences. Recent evidence has suggested that a signalling system is operational, which spreads the silencing command throughout the plant (Voinnet & Baulcombe, 1997). Further evidence to support this signalling theory came from grafting experiments by Palauqui *et al.* (1997), who showed that the silencing phenotype was graft transmissible.

INTERACTIONS OF TRANSGENES WITH THE ENVIRONMENT

Pathogen interaction

Evidence for viral interaction with transgene constructs is well known (Baulcombe & English 1996) and recent data has shown gene silencing can also be initiated by viral infection of transgenic *Brassica napus*. When infected with cauliflower mosaic virus (CaMV), a gradual decline in transgene expression was observed in plants containing transgenes driven by the viral CaMV 35S promoter (Al Kaff *et al.* 1998). Transgenes driven by other non-homologous promoters remained active in infected plants. This indicated the silencing mechanism relied on sequence homology between the virus and transgene to silence gene expression of both.

Heat stress

Transgene silencing can also be induced by changes in temperature. Exposure to both heat and bright light caused *chs* expression to become gradually weaker when petunia plants were grown under field conditions (Meyer *et al.* 1992). Walter *et al.* (1992) found 95% of cells were silenced after suspension cultures of glufosinate ammonium tolerant *Medicago sativa* were given heat shock treatment (37°C for 10 days). Heat induced transgene inactivation in whole plants was investigated by Broer (1996) using *Nicotiana tabacum*. A reduction of expression was detectable at 35°C, while all tested plants showed damage if exposed to 40°C for 10 days, decreasing the temperature back to 25°C restored the transgene phenotype. Further work by Neumann *et al.* (1997) showed temperature sensitivity or tolerance of transgenes was heritable. The phenomena did not correlate with copy number or to the allelic state.

Transfer and cultivation

Brandle *et al.* (1995) found that seeds sown directly into the field retained their transgene expression, but those transplanted from the glasshouse into the field suffered silencing. Other silencing events have been found to depend on environmental conditions (van der Krol *et al.* 1990; Hart *et al.* 1992; de Carvalho Niebel *et al.* 1995) while others are developmentally controlled (de Carvalho *et al.* 1992; Dorlhac de Borne *et al.* 1994).

COMBINING TRANSGENES

Plant breeders will want to combine different transgenic elite breeding lines (e.g. herbicide tolerance with quality traits) to produce new transgenic cultivars. This process is similar to conventional plant breeding where ranks of characters are combined by hybridisation, and selection sorts out the superior combinations. Stacking transgenes could potentially result in transgene silencing if the constructs have high levels of sequence homology either between themselves or with endogenous genes. Silencing can also occur where single copies of highly expressed transgenes are present (Elmayan & Vaucheret 1996). However, silencing is more likely where multiple constructs are located at a single locus. Strong silencing loci have been found which completely silence any incoming construct carrying homology to the first. Separation of the two in the next generation can still result in lower expression than control plants as the imprinting of the loci is retained for several generations (see Matzke *et al.* 1996) Such effects in breeding lines would result in the combination being discarded. High performing lines that survive the rigorous testing and selection procedure are likely to become commercial cultivars.

Accidental stacking

Little data is available on the effects of single transgenes in wild populations and even less for multiple transfers. Examining the literature for single combinations shows that transgenes will remain active in new hybrids and that they can be passed on through the generations (Chevre *et al.* 1998). This should also be the case for multiple transgenes, with the expectation that different constructs will segregate in a Mendelian manner in most instances. Only transgenes that are closely linked will remain together over the following generations. It is expected that progeny from crop x wild species hybrids may have a range of phenotypes for the introduced genes. Some gene silencing could occur, but in most instances we would expect expression. After the transfer of transgenes to wild relatives, they will be under the control of a new genetic background. This could influence transgene expression and the environmental impact it has. Dormancy characteristics of the crop x weed hybrid will also influence how long seeds containing transgenes can survive in the seed bank.

CONTROLLING TRANSGENE ESCAPE

In many instances the risk assessment will conclude that gene movement from transgenic varieties is no more hazardous than from conventionally bred varieties. In other cases, such as the production of specialist products (pharmaceutical substances and special oils) it may be necessary to control transgene movement. Various methods are emerging to control transgene escape from the crop to related plants. Some of these methods are more appealing than others, but all could influence transgene spread.

Transformation of plastids

This is a developing field where plastids are targeted for transformation instead of the nuclear genomic DNA (Dix & Kavanagh 1995). Transformation of plastids has been achieved for several model species, notably tobacco (Daniell *et al.* 1998) and *Arabidopsis thaliana* (Sikdar *et al.* 1998). As plastids have their own DNA which is mostly maternally inherited, any integrated transgenes are very unlikely to be transferred into relatives via pollen, although this does not preclude the reverse hybridisation from occurring (e.g. the crop pollinated by the weed species).

F1 hybrids, male sterility

Many workers have been engaged in obtaining male sterility in crops. This has been achieved by both conventional and transgenic approaches. The advantages of such systems lie in the ability to produce F1 hybrids, which can be more productive than normal cultivars. Hybrids offer productivity advantages to farms but prevent farmers from saving their own seeds. The rigors of hybrid seed production will make it less likely that seeds for sowing will become genetically contaminated. Male sterility has also been considered as an isolating mechanism to reduce gene flow e.g. forest trees.

'Killer' constructs

Mechanisms by which fertile seed is prevented from being produced are being evaluated. These could potentially halt the spread of transgenic feral plants. Introgression of such traits into wild relatives of crops could influence ability to reproduce, but there is likely to be intense selection against individual hybrids expressing this transgene.

Construct design

The way a construct is designed could influence the spread or activity of transgenes. Consideration of the type of promoter to be used, and in what context, should become an integral part of construct design. It is clear from the work of Al Kaff *et al.* (1998), that the use of pathogen derived sequences in their natural plant hosts can result in transgene silencing. The use of tissue specific promoters rather than constitutive promoters may reduce some of the potential impact of gene transfer. For example, modified oil composition can be precisely targeted to seeds or particular parts of the seed, with all other plant parts retaining a 'wild type' phenotype.

Inducible promoters have now been isolated (Gatz & Lenk 1998), and could provide an efficient way to control transgene expression in field crops. For example, crops containing herbicide tolerance transgene driven by such promoters could be sprayed with the inducer, followed by the herbicide. This would control both unrelated weeds and non-transgenic weeds. Following harvest, a second round of spraying could occur with the herbicide only. Any crop volunteers or transformed weeds will be destroyed along with their non-transgenic counterparts as the transgene would be silent.

DISCUSSION

The introgression of multiple transgenes into related wild populations will occur for some of our transgenic crop plants. The genes that are transferred will determine the significance of such events. It should be remembered that plant breeding has already produced many novel combinations of genes for e.g. disease resistance, insect resistance and tolerance to environmental stresses, all of which have had the opportunity to be transferred to wild relatives. Effects of transgenes in wild populations are likely to be the same or similar to those seen for conventional plant breeding traits. If they convey selective advantage or are neutral in their effect, it is possible that they will be assimilated into the wild species genotype. Observations on the effects of some transgenes have been shown to result in poor survival (Sweet *et al.* 1997), while other traits (Bt gene) have given greater advantage in an ecological niche and therefore could become established in the wild population (Stewart *et al.* 1997). Factors which could reduce the impact of gene stacking include; construct design, use of targeted constructs, less reliance on constitutive expression of traits and greater use of inducible promoters. These changes should help to reduce the rate of introgression resulting from selection and subsequent expression of multiple transgenes in the environment.

ACKNOWLEDGEMENTS

We would like to thank Department of Environment, Transport and the Regions and the BBSRC for their financial support.

REFERENCES

Al Kaff, N S; Covey S N; Kreike M M; Page A M; Pinder R; Dale P J (1998). Transcriptional and posttranscriptional plant gene silencing in response to a pathogen. *Science* **279**, 2113-2115.

Baulcombe D C; English, J J (1996). Ectopic pairing of homologous DNA and post transcriptional gene silencing in transgenic plants. *Current Opinion in Biotechnology* **7**, 173-180.

Brandle J E; McHugh; S G; James L; Labbe H; Miki B L (1995). Instability of transgene expression in field grown tobacco carrying the *crs1-1* gene for sulfonylurea herbicide resistance. *Bio/ Technology* **13**, 994-998.

Broer, I (1996).Stress inactivation of foreign genes in transgenic plants. *Field CropsResearch* **45**, 19-25.

Chevre A M; Eber F; Baranger A; Hureau G; Barret P; Picault H; & Renard M (1998). Characterisation of backcross generations obtained under field conditions from oilseed rape wild radish F1 interspecific hybrids: an assessment of transgene dispersal. *Theoretical & Applied Genetics* **97**, 90-98.

Daniell H; Datta R; Varma S; Gray S; Lee S B (1998). Containment of herbicide resistance through the genetic engineering of the chloroplast genome. *Nature Biotechnology* **16**, 345-348.

De Carvalho F; Gheysen G; Kushnir S; Van Montagu M; Inze D; Castresana C (1992). Suppression of β-1,3-glucanase transgene expression in homozygous plants. *EMBO Journal* **11**, 2595-2602.

De Carvalho Niebel F; Frendo P; Van Montagu; Cornelissen M (1995). Post-transcriptional cosuppression of β-1,3-glucanase genes does not affect accumulation of transgene nuclear mRNA. *The Plant Cell* 7, 347-358.

Depicker A; Van Montagu M (1997). Post-transcriptional gene silencing in plants. *Current Opinion in Cell Biology* **9**, 373-382

Dix, P J; Kavanagh T A (1995). Transforming the plastome: genetic markers and DNA delivery systems. *Euphytica* **85**, 29-34.

Dorlhac De Borne F; Vincentz M; Chupeau Y; Vaucheret, H (1994).Cosuppression of nitrate reductase host gene & transgenes in transgenic tobacco plants. *Molecular & General Genetics* **243**,613-21

Elmayan, T; Vaucheret H (1996). Expression of single copies of a strongly expressed 35S transgene can be silenced post-transcriptionally. *The Plant Journal* **9**, 787-797.

Gatz C; Lenk I (1998). Promoters that respond to chemical inducers. *Trends in Plant Science* **3**, 352-358.

Hart C M; Fischer B; Neuhaus J M; Meins F (1992). Regulated inactivation of homologous gene expression in transgenic *Nicotiana sylvestris* plants containing a defence related tobacco chitinase gene. *Molecular & General Genetics* **235**, 179-188.

Kumpatla S P; Chandrasekharan M B; Iyer L M; Li G; Hall T C (1998). Genome intruder scanning modulation systems & transgene silencing. *Trends Plant Science* **3**, 97-104.

Matzke M A; Park Y D; Papp I; Vaucheret H; Matzke A J M (1996). Features of promoter homology-dependent gene silencing in transgenic plants. In *Mechanisms and applications of gene silencing*, D Grierson; G W Lycett; G A Tucker, eds (Nottingham: Nottingham University Press), pp. 33-42.

Matzke M A; Primig M; Trnovsky J; Matzke A J M (1989). Reversible methylation and inactivation of marker genes in sequentially transformed tobacco plants. *EMBO Journal* **8**, 643-649.

Meyer P (1995). Understanding and controlling transgene expression. *Tibtech* **13**, 332-337.

Meyer P; Linn F; Heidmann I; Meyer H; Niedenhof I; Saedler H (1992). Endogenous and environmental factors influence 35S promoter methylation of a maize A1 gene construct in transgenic *Petunia* and its colour phenotype. *Molecular & General Genetics* **231**, 345-352.

Mlynarova L; Loonen A; Heldens J; Jansen R C; Keizer P; Stiekema W J; Nap, J P (1994). Reduced position effect in mature transgenic plants conferred by the chicken lysozyme matrix-associated region. *The Plant Cell* **6**, 417-426.

Nap J P; Mlynarova L; Steikema W J (1996). From transgene expression to public acceptance of transgenic plants: a matter of predictability. *Field Crops Research* **45**, 5-10.

Neumann K; Droge-Laser W; Kohne S; Broer I (1997). Heat treatment results in loss of transgene encoded activities in several tobacco lines. *Plant Physiology* **115**, 939-947.

Palauqui J P; Elmayan T; Pollien J M; Vaucheret H (1997). Systemic acquired silencing: transgene specific post-transcriptional silencing is transmitted by grafting from silenced stocks to non silenced scions. *EMBO Journal* **16**, 4738-4745.

Sikdar S R; Serino G; Chaudhuri S; Maliga P (1998). Plastid transformation in *Arabidopis thaliana*. *Plant Cell Reports* **18**, 20-24.

Stewart C N; All J N; Raymer P L; Ramachandran S (1997). Increased fitness of transgenic insecticidal rapeseed under insect selection pressure. *Molecular Ecology* **6**, 773-779.

Sweet J B; Shepperson R; Thomas J E; Simpson E (1997). The impact of releases of genetically modified herbicide tolerant oilseed rape in the UK. *Brighton Crop Protection Conference* **1**, 291-302.

Van Der Krol A R; Mur L A; Beld M; Mol J N M; Stuitje A R (1990). Flavonoid genes in petunia: addition of a limited number of gene copies may lead to a suppression of gene expression. *The Plant Cell* **2**, 291-299.

Van Den Boogaart T; Lomonossoff G P; Davies J W (1998). Can we explain RNA-mediated virus resistance by homology-dependent gene silencing? **11**, 717-723.

Vaucheret H; Elmayan T; Thierry D; Van Der Geest A; Hall T; Conner A J; Mlynarova L; Nap J P (1998). Flank matrix attachment regions (MARs) from chicken, bean, yeast or tobacco do not prevent homology-dependent *trans*-silencing in transgenic tobacco plants. *Molecular & General Genetics* **259**, 388-392.

Voinnet O; Baulcombe D C (1997). Systemic signalling in gene silencing. *Nature* **389**, 553.

Walter C; Broer I; Hillemann D; Puhler A (1992). High frequency, heat treatment induced inactivation of the phosphinothricin resistance gene in transgenic single cell suspension cultures of *Medicago sativa*. *Molecular & General Genetics* **235**, 189-196.

Table 1. Deliberate stacking of transgenes in plant varieties

Characters*	Viral	Fungal	Bacterial	Insect	Breeding System	Stress	Morphology	Propagation	Herbicide	Pharmaceuticals	Protein	Starch	Oil
Viral resistance	✓✓	-	-	-	-	-	-	-	-	-	-	-	-
Fungal resistance	✓✓	✓	-	-	-	-	-	-	-	-	-	-	-
Bacterial resistance	✓	✓	✓	-	-	-	-	-	-	-	-	-	-
Insect resistance	✓✓	✓	✓	✓	-	-	-	-	-	-	-	-	-
Breeding system	✓	✓	✓	✓	✓	-	-	-	-	-	-	-	-
Stress tolerance	✓	✓	✓	✓	✓	✓	-	-	-	-	-	-	-
Morphology	✓	✓	✓	✓	✓	✓	✓	-	-	-	-	-	-
Propagation	✓	✓	✓	✓	✓	✓	✓	✓	-	-	-	-	-
Herbicide tolerance	✓	✓	✓	✓	✓✓	✓	✓	✓	✓**	-	-	-	-
Pharmaceuticals	✓	✓	✓	✓	✓	✓	✓	✓	✓	✓	-	-	-
Protein	✓	✓	✓	✓	✓	✓	✓	✓	✓	X	✓	-	-
Starch	✓	✓	✓	✓	✓	✓	✓	✓	✓	X	X	✓	-
Oil	✓	✓	✓	✓	✓	✓	✓	✓	✓	X	X	X	✓

✓ = Stacking probable; ✓✓ = Stacking already practised; ✓** = Stacking not recommended; X = Stacking unlikely

* Examples: Viral resistance - coat protein, replicase; Fungal resistance - chitinase, glucanase, lysozyme; Bacterial resistance - cecropin; Insects resistance - Bt, trypsin inhibitor, lectin; Breeding System - male sterility, self compatibility, self incompatibility; Stress tolerance - drought, cold, heat, salt; Morphology - flowering, dwarf, root form, pod shattering, canopy structure; Propagation - seed number, seed dormancy, vegetative propagation; Herbicide tolerance - glufosinate, glyphosate, bromoxynil; Pharmaceuticals - hirudin, anticancer, antibodies, vaccines; Protein - storage proteins; Starch - quantity and structure; Oil - quantity and type.

Management of transgenic crops within the cropping system

J Champolivier (1), J Gasquez (2), A Messéan (1) & M Richard-Molard (3)
(1) *CETIOM, 174 Ave Victor Hugo, Paris, France*
(2) *INRA, Laboratoire de malherbologie, rue Sully, Dijon, France*
(3) *ITB, 45 Rue de Naples, Paris, France*

ABSTRACT

With the development of Genetically Modified Organisms (GMO) in agriculture, new concerns about crop management have been addressed. Apart from the evaluation carried out with the regulation process before marketing, observations under current agricultural practices are required in order to build suitable agronomic management and design a monitoring system. A multi-year and multi-crop monitoring study has been carrying out in France since 1995 and preliminary results suggest that a more integrated crop management should be required.

INTRODUCTION

After about 15 years of transgenic research carried out by public research teams as well as private companies, the first marketing releases occurred in North America in 1995, while in Europe the first applications are still under discussion. Tobacco resistant to bromoxynil, imports of glyphosate resistant soybean, insect resistant corn through the Bt strategy and some herbicide tolerant rapeseed sum up the current European status for GMO marketing. Several other applications for marketing clearance have been submitted.

Corn, sugar beet and rapeseed are the main crops for which genetic modification has been applied. While several traits have been introduced (oil quality, disease and insect resistance) and are under development, herbicide resistance has been developed extensively and three systems are near marketing : glyphosate and glufosinate resistance for the three crops and bromoxynil resistance for rapeseed.

Development of transgenic plants raises several questions, most of them are not specific to recombinant DNA techniques : ethical concerns, relationship between science, society and organization of collective expertise, marketing of transgenic plants with new rules, protection of biotechnology and patent policy, food and feed safety of these novel plants, environmental and agronomic concerns. With respect to these last concerns, the evaluation has to be performed on a case-by-case basis. The risk assessment of gene flow must take into account the specific trait introduced (e.g. herbicide resistance vs oil quality), the biology of the plant (open vs self pollination, seed dormancy) and the agricultural context (cropping systems, spatial organization of the crops, agricultural practices, ...).

Herbicide resistance is not only one of the first traits for which marketing clearance has been applied but it is also an adequate model to carry out the risk assessment of crop management of transgenic plants. In this paper the main criteria for consideration in herbicide tolerant crops and the effect of their use in cropping systems will be reviewed. Rapeseed and sugar beet provide a good example of the principles involved (Richard-Molard & Gestat de Garambe, 1998).

RISK ASSESSMENT

For several years, the main question with respect to modified rapeseed and sugar beet was : will the transgene be disseminated outside the field and be transferred to other plants and, especially, to weeds ? From many studies carried out by different scientific teams, it can be concluded that transgenes will disseminate and can lead to outcrossing with weeds. Although interspecific crosses between rapeseed and related wild species lead to less fertile plants, they can produce a small quantity of seeds (Kerlan *et al.*, 1992).

As we know that transgenes will disseminate, the question is now : So what ? Could the consequences of such a dissemination be managed ? With respect to long-term effects, no experiments are available for assessing the transgene behaviour. In order to estimate gene flow, simulations using genetic models are performed. These models generally represent the gene transfer from a field towards the wild species located at field edges and take into account various parameters such as the gene migration rate, its dominance level or the competitivity of the hybrid. Long-term behaviour appears to be difficult to predict as the model is highly dependent on specific events. It is thus necessary to take into account the spatial and temporal variability.

On the other hand, we can look for markers already introduced into rapeseed in the past and to survey their behaviour in the non-cultivated areas. Such a survey is being performed in various regions of France, where we are intending to detect the introgression of traits like "low-erucic" in wild species.

Gene flow

In the case of rapeseed, gene flow can occur through two different ways :

* the pollen, either towards rapeseed plants (intraspecific crosses) or towards wild relatives which are quite numerous (interspecific crosses) ;
* the seeds, through volunteers in subsequent crops or seed dissemination during transportation.

The long-term effect of such phenomena on farmers' crop management of transgenic plants and the design of adequate agricultural practices are assessed by carrying out several types of studies :

* Modeling the gene flow. Models of gene flow between two adjacent fields have been designed (Reboud, 1992 ; Lavigne *et al.*, 1994) and are being improved by taking into account crop rotations, spatial patterns of crops and agricultural practices.
* Specific studies about outcrossing have been performed in order to estimate pollination distances and interspecific crosses (Jorgensen and Andersen, 1996 ; Kerlan *et al.*, 1992 ; Eber *et al.*, 1994 ; Baranger *et al.*, 1995). Spontaneous hybridization of rape with wild mustard, wild hoary mustard and wild radish has been demonstrated to occur when using a male-sterile oilseed rape cultivar as the pollen recipient, i.e. without pollen competition (Chèvre *et al.*, 1996). Lefol *et al.* (1996b) have showed that the reciprocal cross can occur in the field with hoary mustard, but it was not been observed with wild mustard to date (Lefol *et al.*, 1996a).

Recently, and for the first time, the possibility of producing interspecific hybrids in the field between oilseed rape and wild radish as the seed parent have been reported (Darmency *et al.*, 1998). Two hybrids were obtained from 59 wild radish plants grown at low density in the field, but none when wild radish was grown at high density. The germination rate of these hybrids is low and their fitness is reduced.

Other studies have been performed in North America and first large scale releases already took place there. However, climatic and agricultural conditions are quite different in Europe : shorter rotations (every two years in some European regions), winter sown type rather than spring sown types, different kinds of wild relatives. Thus, it appears to be rather difficult to extrapolate data from North America for assessing the gene flow and agronomic impact of herbicide tolerant crops.

Sugar beet is a biannual crop, but some bolter beets can produce seeds from the first year. They are hybrids between crop and wild forms, which have been observing in the seed production area for 25 years (Desplanque *et al.*, 1996). These weed beets are controlled in the current cropping systems by respecting recommendation about lifting bolters. The development of non-selective herbicide tolerant varieties may induce the appearance of a tolerant weed flora, which can develop from bolter beet and volunteer beet.

For the maize, there is no possibility of outcrossing with wild relatives in Europe. But the transgene escape remains possible by pollination between varieties.

A multi-crop and multi-year monitoring study

In order to assess the effect of such outcrossing under agricultural conditions, in 1995 the French technical institutes CETIOM, AGPM, ITB and ITCF, designed and implemented a monitoring study for various transgenic crops on three platforms located in different regions of France : Champagne, Burgundy and Midi-Pyrénées (South-West). Each platform consists of a 6 ha field where transgenic corn, rapeseed and sugar beet are cropped with the usual local cropping system (see Figure 1). The transgenic traits are as follows :

* glufosinate and glyphosate resistance in corn, rapeseed and sugar beet ;

* bromoxynil resistance in rapeseed and corn borer tolerance (using the Bt system) in corn.

A 500 meter area around the field was defined and monitored in order to assess the spatial impact of transgenic crops.

This multi-year experiment aimed mainly at :

* assessing the impact of transgenic crops when cultivated together in the same field area ;

* designing the weed control of volunteers in subsequent crops which are resistant to the same herbicide (e.g. glyphosate-resistant rapeseed volunteers in the subsequent sugar beet resistant to glyphosate) ;

* evaluating the multiple resistance rate when cropping two adjacent rapeseed fields with two different herbicide resistances ;

* estimating the interspecific outcrossing towards the wild relatives under real and local conditions and

* estimating the cost-benefit of herbicide resistance technology with respect to conventional techniques.

Figure 1. Example of a cropping system in Burgundy (1996-97) using transgenic traits.

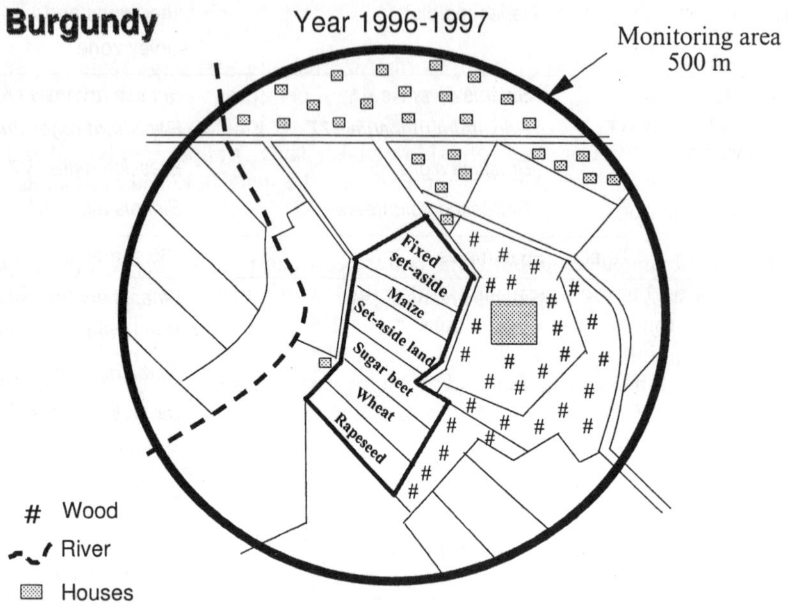

Example of cropping system, Burgundy

Year 1995-1996	Set-aside land	Rapeseed	Wheat	Sugar-beet	Maize	Fixed set-aside land (trefoil)
Year 1996-1997	Rapeseed	Wheat	Sugar-beet	Wheat	Maize	Fixed set-aside land (trefoil)
Year 1997-1998	Set-aside land	Sugar-beet	Set-aside land	Rapeseed	Maize	Fixed set-aside land (trefoil)

Table 1. Identification of wild relatives within the monitoring area from 1996 to 1998.

Location		Weed species and number of samples	
		Rapeseed plot	Other crops and survey zone
Midi-Pyrénées		*Sinapis arvensis - 1*	*Sinapis arvensis - 6*
		Rapistrum rugosum - 77	*Rapistrum rugosum - 1*
		Brassica nigra - 5	*Brassica nigra - 77*
		Rapessed volunteers - 13	*Sinapis alba - 51*
	Total	96 samples	135 samples
Burgundy		*Sinapis arvensis - 54*	*Sinapis arvensis - 58*
			Rapeseed volunteers - 14
			Arabidopsis thaliana - 1
			Capsella bursa pastoris - 13
			Sisymbrium officinale - 1
			Thlaspi arvense - 1
			Barbarea intermedia - 1
			Alliara petiolata - 18
	Total	54 samples	107 samples
Champagne-Ardennes		*Sisymbrium officinale - 20*	*Sinapis arvensis - 28*
		Capsella bursa pastoris - 15	*Sinapis alba - 4*
		Calepina irregularis - 9	*Raphanus raphanistrum - 1*
		Rapeseed voulunteers - 38	*Capsella bursa pastoris - 13*
		Sinapis arvensis - 6	*Rapeseed volunteers - 22*
			Sisymbrium officinale - 2
			Calepina irregularis - 7
			Thlaspi arvense - 4
			Cardamine hirsuta - 4
			Alliara petiolata - 20
	Total	88 samples	105 samples
Total		238 samples	347 samples

Outcrossing with wild relatives

Within the monitoring area, each wild relative plant of rapeseed was located and surveyed until seed maturity. The flowering period was observed and compared with the flowering periods of the transgenic rapeseed crops. Seeds were sampled for assessing the herbicide resistance which was checked by spraying herbicides after re-sowing. Table 1 gives the occurrence of wild relatives observed during three years of the study (1996-1998) : a plot represents one or several plant(-s) located at the same place.

RESULTS

Preliminary results indicated that no herbicide resistance with wild mustard and other mustard species occurred during the first two years. Unfortunately, wild radish was not present in our situations and a specific location site has been implemented in 1998 in the South-West region. frequency of outcrossing will continue to be studied in the subsequent years and will allow us to increase the precision of this frequency.

There was some weed beets in our situations. The frequency of outcrossing with the sugar beet varieties can be estimated from 0.07 to 0.2%. The distribution of resistant seeds let believe that insects are involved in the pollination. This hypothesis will be tested in the further years.

Figure 2. Average dispersal curves for the three location sites

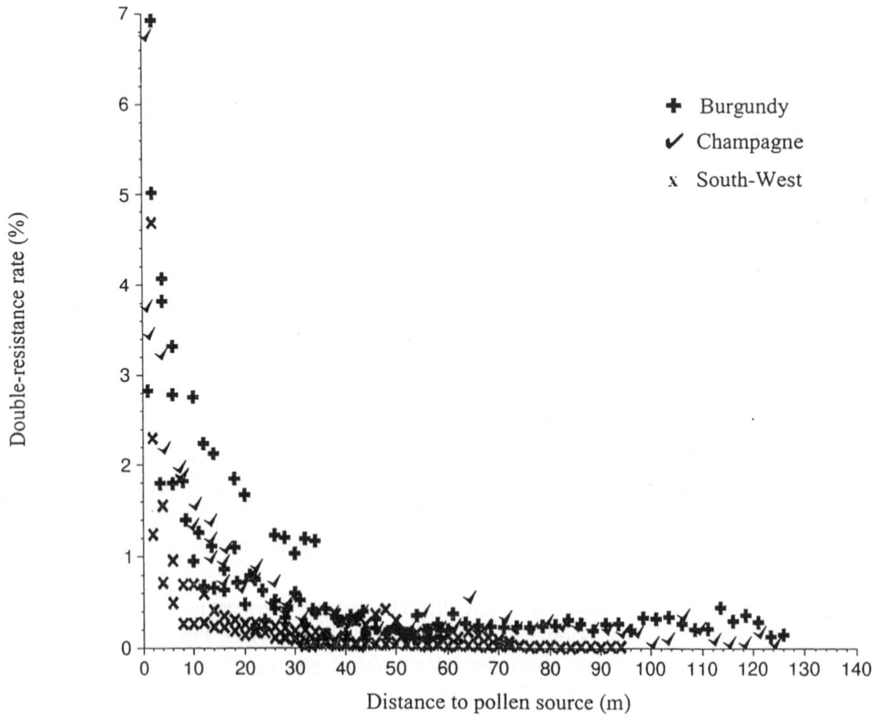

Multiple resistance and dispersal of pollen

The three herbicide resistant rapeseed varieties were cropped in adjacent fields and double resistant plants were detected in two different ways :

* by applying the herbicides on volunteers whose emergence occurred after harvesting ;
* by sampling seeds and re-sowing using a specific design of experiments and direct application.

Both methods gave similar results with respect to the rate of double resistance. Although the results were dependent upon the variety, the average rate of double resistance can be estimated under our specific conditions : about 2 % at a one meter distance, 0.2 % at 20 meters and less than 0.01 % at 65 meters.* (Figure 2)

The pollen dispersal of sugar beet outside the field was assayed by introducing male-sterile plants at various distances from the field, to biologically trap pollen. At maturity, seed production was recorded, seeds collected and tested for resistance by screening a portion of the sampled seed with herbicide. The analysis of seed production of the male-sterile plants showed a rapid decrease of the pollen flow with the distance (- 85% for 30 meters).

Asynchrony of flowering between the two varieties of maize limited the study of gene flow. The first results showed that the seed production on emasculated plants at 150 m from the field dropped to about 10%.

CONCLUSIONS

Although further data are still required, these results seem to indicate that multiple resistance should probably be the major concern for farmers rather than interspecific crosses. Suitable agronomic practices should be proposed in order to provide a sustainable use for transgenic crops such as herbicide tolerant oilseed rape.

Preliminary results obtained during the first years of our project confirmed what was expected from previous studies. Thus results have been obtained under current farmer practices and provided data which will be used to fit simulation models for gene flow. While further data will be obtained for three more years, adequate crop management practices can already be discussed and recommendations provided for public decision-making.

REFERENCES

Baranger A ; Chèvre A M ; Eber F ; Renard M. (1995). Effect of Oilseed rape genotype on the spontaneous hybridization rate with a weedy species : an assessment of transgene dispersal. *Theoretical and Applied Genetics.*, **91**, 956-963.

Darmency H ; Lefol E ; Fleury A (1998). Spontaneous Hybridizations Between oilseed rape and wild radish. *Molecular Ecology*, **7**, 1467-1473.

Desplanque B ; Boudry P ; Broomberg K ; Van Dijk H (1996). Flux géniques entre betteraves sauvages et cultivées : des populations de betteraves mauvaises herbes peuvent-elles s'installer rapidement ? *Xème colloque international sur la biologie des mauvaises herbes, 11-13 septembre, Dijon, ANPP,* 239-245.

Eber F ; Chèvre A M ; Baranger A ; Vallée P ; Tanguy X ; Renard M (1994). Spontaneous Hybridization between a male sterile oilseed rape and two weeds. in : *Theoretical and Applied Genetics*, **88**, 362-368.

Jorgensen R B ; Andersen B (1996). Spontaneous Hybridization between oilseed rape (*Brassica napus*) and weedy relatives. *Acta Horticulturae,* **407**, 193-200.

Kerlan MC ; Chèvre A M ; Eber F ; Baranger A ; Renard M. (1992). Risk Assessment of outcrossing transgenic rapeseed to related species : I. Interspecific hybrid production under optimal conditions with emphasis on pollination and fertilization. *Euphytica,* **62**, 145-153.

Lavigne C ; Godelle B ; Reboud X ; Gouyon P H (1996). A Method to determine the mean pollen dispersal of individual plants growing within a large pollen source. *Theoretical and Applied Genetics,* **93**, 1319-1326.

Lefol E ; Danielou V ; Darmency H (1996a). Predicting Hybridization Between Transgenic oilseed rape and wild mustard. *Field Crops Research,* **45**, 153-161.

Lefol E ; Fleury A ; Darmency H (1996b). Gene Dispersal From Transgenic Crops. Ii. Hybridization between oilseed rape and the hoary mustard. *Sexual Plant Reproduction,* **9**, 189-196.

Messéan A (1995). Release of genetically modified rapeseed : so what ? In : *Rapeseed : today and tomorrow*, Proceedings of the 9th International Rapeseed Congress, Cambridge, 4-7 July 1995, **Vol.4**, **L-3**, 1335-1337.

Messéan A ; Reau R ; Wagner D (1995). La charte environnement du colza énergétique en France. In *Rapeseed : today and tomorrow*, Proceedings of the 9th International Rapeseed Congress, Cambridge, 4-7 July 1995. Vol 1, C-7, 235-237.

Mikkelsen T R ; Andersen B ; Jorgensen R B (1996). The Risk Of Crop Transgene spread. Nature, 380, 31.

Reboud X (1992). Les Risques associés aux manipulations génétiques : le cas de la résistance aux herbicides. *Thèse INAPG,* 62 p.

Richard-Molard M ; Gestat de Garambe Th (1998). Utilisation de variétés tolérantes à un herbicide non sélectif. Conséquences sur le système de culture. *61ème Congrès de l'IIRB, février 1998*, pp. 269-288.

Vigouroux Y ; Darmency H ; Gestat de Garambe Th (under press). Gene flow between sugar beet and weed beet. *British Crop Protection Council (Gene Flow and Agriculture) 12-14 april 1999 – Staffordshire (UK)*

Assessing the impact and consequences of the release and commercialisation of genetically modified crops

J B Sweet, C E Norris, E Simpson, J E Thomas
National Institute of Agricultural Botany, Cambridge, UK

ABSTRACT

Genetically modified (GM) crops will have impacts on agriculture and both the agricultural and natural environment. Analysing the consequences for the agricultural environment requires study of the characteristics of the GM crop and its hybridising relatives, and study of the management systems involved in growing the GM crop and other crops grown in rotation with it. GM crops may also have impacts on uncultivated land and natural environments. Thus risk assessments are concentrating on whether the genetically modified characteristics of a GM crop are likely to change the behaviour of the plants in their environments to the extent that ecological balances are altered. This paper discusses approaches to risk assesments and reviews results of risk assessments of GM Brassica crops.

INTRODUCION

Genetically modified (GM) crops have potential benefits for growers, processors and consumers. However many will have impacts on agriculture and both the agricultural and natural environment while others will have food quality and safety implications.

Pest, disease and herbicide resistant crops will require different (often reduced) pesticide and herbicide inputs in order to exploit these novel traits. These modified management systems will themselves have an impact on agricultural environments. Thus analysing the consequences for the agricultural environment requires study of the characteristics of the GM crop and its hybridising relatives, and study of the management systems involved in growing the GM crop and other crops grown in rotation with it. For example GM herbicide tolerant (HT) crops will be treated with different herbicides, with different activity spectra, at different crop development growth stages. Thus the effects on botanical diversity in the GMHT crop and in subsequent crops will be a product of the GM crop and the herbicide treatment.

GM crops may also have impacts on uncultivated land and natural environments. These environments may be affected by characteristics of crop and wild species induced by novel genetic constructs and their products. Thus risk assessments concentrate on whether the genetically modified characteristics of a GM crop, and of similarly modified hybridising relatives, are likely to change the behaviour of the plants in their environments to the extent that ecological balances are altered. Risk assessments study both the severity and extent of the hazard or damage as well as the likelihood and frequency at which the damage will occur. Thus the definition:

Risk (Impact) = Frequency x Hazard

GENE FLOW MEASUREMENTS

The impact of transgenes on particular wild species will depend on whether genes will introgress into wild populations and the rate of that introgression. Introgression of a transgene is a product of cross pollination with transgenic relatives, survival of the hybrid and its hybridisation with the wild population to the extent that the gene becomes established in a proportion of the population. If there is sexual incompatability between species the assumption is that no gene transfer will take place and thus the risk is zero. However, in experiments, certain wild Brassica and other crucifer species previously considered incompatible with *B. napus* have shown some ability to hybridise (Scheffler and Dale, 1994). Thus there is the possibility that hybridisation can occur, albeit at very low frequencies, and that transgenes have a route to introgression. Studies to determine the frequency of these interspecific hybrisations are described in other papers in this conference report. However information is also required on whether the hybrids can survive and back cross with their wild parental species so that the gene can become established in the wild population. Studies by Scott and Wilkinson (1998) of *B. rapa* (*B. campestris*) populations growing outside fields of *B. napus* in England indicated that cross pollination frequencies were low (0.4-1.5%) in 7% of *B.rapa* populations surveyed. The remaining 93% of populations contained no hybrid seed. In addition they found that on average less than 2% of all seedlings survive, so that unless the trangene conferred higher survival characteristics, establishment of GM *rapa* x *napus* hybrids would be very poor and introgression of the gene into rapa populations would be very slow. Hybridisation frequencies appear to be much higher where *B. rapa* occurs as a weed in *B. napus* crops (Jorgensen *et al* 1998 and Sweet and Norris, unpublished data). However in the UK there is no indication that *B. rapa* has modified its behaviour or is spreading as a weed in *B. napus* crops as a consequence of hybridisation. The relationship between weedy and wild *B. rapa* is yet to be determined but may be a significant route for introgression of a transgene from oilseed rape.

Chevre *et al* (1998) have shown that hybridisation and backcrossing with *Raphanus* and *Sinapis* can occur at low frequencies under particular conditions. Studies are underway in the UK to determine whether oilseed rape genes can be detected in *Sinapis* spp. (Dale and Moyes, pers comm.) However little is known of the ability and frequency of hybridisation of weedy crucifers eg. *Sinapis* and *Raphanus* with *B. napus* under natural conditions and the survival and reproductive characteristics of the interspecific hybrids. Thus we appear to be far from determining whether natural hybridisation frequencies are zero or close to zero for many related cruciferous spp.

Hybridisation between weedy crucifers and oilseed rape would tend to lead to introgression of advantageous adaptive characters into the weedy populations such as increased waxyness and reduced leaf hairs associated with tolerance to the herbicides currently used on oilseed rape. There are indications that this may have occured in weedy *B. rapa* populations occurring in oilseed rape crops in Denmark (Jorgensen *et al* 1998) and in England (Sweet and Norris, unpublished data). However no morphological modifications associated with enhanced herbicide tolerance has been reported in populations of other weedy crucifers. Some *Sinapis* and *Raphanus* populations appear to have partial tolerance to some oilseed rape herbicides, though whether this is due to gene exhange, evolution or selection from an inherent natural variability is unknown.

There is also a need to study indirect gene flow through intermediary species which are compatible with both oilseed rape and with other wild crucifers. Thus we need to determine whether transgenes are able to introgress from *B. napus* into *B. rapa* and in hence to *Sinapis* or *Raphanus* spp.

There have been and continue to be numerous studies of transgene flow through seed dispersal, pollination and hybridisation within the same species. These are gradually being scaled up to levels that reflect field scale releases and the results confirm those found with non-GM pollen. There are less reports of studies of the persistance and survival of introgressed genes in volunteers. Sweet *et al* (1998), and Simpson *et al,* in this report, studied volunteer populations of winter and spring oilseed rape occuring at release sites in subsequent crops. There are also reports on volunteer GM rape from France (Messean, 1997) but none of the studies have reported increased numbers or fitness of herbicide tolerant volunteers.

Impact of plant species

Many studies have concentrated on measuring frequency phenomena such as gene flow and inter-specific hybridisation without necessarily considering the impact of the transgene when it has dispersed or introgressed into other populations or species. Frequency and hazard are dependant on the characteristics of both the crop that is modified and of the GM trait and so risk assessments require measurement and study of the hazard or impact of both the crop/plant and the trait. In addition the impact of the release of the GM plant will depend on the type and location of the environment into which it is being released. This means that risk assessments are not necessarily transferable from one site, area, region or country to another.

Plants vary in the degree to which they dominate or are invasive of certain environments and in their ability to disperse genes to different populations and species. Thus they will have different environmental impacts when genetically modified and, for any particular country or region, plants can be classified as being high, medium and low impact.

High Impact plants

Plants in this group are hardy, perennial, competitive, open pollinating and prolific, have a wide range of relatives with which they hybridise and an ability to colonise a range of natural and semi natural habitats. Examples of such plants are perennial grasses (eg *Lolium perenne* - perennial ryegrass) and certain indigenous and introduced trees and shrubs which form a significant proportion of forests and woodlands (eg poplar -*Populus* spp.) Modifications of these plants which affect their competitiveness and behaviour could have significant impacts on the ecology of a range of environments.

Medium Impact plants

Plants in this group are open pollinating, hybridise with some wild relatives, prolific and colonise a limited range of habitats. Examples of such plants are oilseed rape, oats, sugar beet and rice all of which have closely related wild relatives with which they hybridise and an ability to colonise disturbed ground. These plants and their close relatives rarely form climax populations except in particular environments such as coastal areas or in disturbed ground.

Low Impact plants

These are usually annual or biennial species, largely self pollinating with few hybridising relatives that are poorly adapted to the area in which they are cultivated. In the UK examples are maize and sunflower.

It is important to appreciate that the impact of plant species will depend upon the environment into which they are being released. Maize and potato are considered low impact plants in England. However in countries of Central and South America where their centres of genetic diversity occur, along with many wild relatives, their impact would be considered very high.

Impact of Transgenes

Transgenes, operating through their expression in plants, will have different impacts on environments. Since genes often operate uniquely it is not easy to classify transgenes as having high or low impact. In addition their impact is also dependant upon the nature of the receiving environment.

High Impact Transgenes

Generally genetic modifications which improve the fitness of GM plants by increasing their reproduction, competitiveness, invasiveness and/or persistance will have the greatest environmental impact. Thus transformations which significantly increase plant productivity and overcome constraints and stresses such as pests, diseases, drought etc will have the highest impact. Thus very high yielding and vigorous GM plants with enhanced and broad spectrum pest, disease and stress tolerance will have the greatest impact.

Many pest and disease resistance genes will have effects on non-target species either directly through gene products which destroy or debilitate non-targets or indirectly by altering relationships between pests and beneficials. It is important that these non-target effects are thoroughly understood before commercialisation progresses.

Low Impact Transgenes

These are genes that do not noticeably enhance the fitness of the modified plant or of other organisms so that they have minimum ecological impact. Examples would be herbicide tolerance and genes that modify seed composition eg high lauric acid genes in oilseed rape, high starch genes in potato. However it is important to verify that these genes do not significantly increase seed tuber overwinter survival through enhanced frost resistance, or dormancy characteristics of oilseed rape seed so that it has enhanced soil survival characteristics.

AGRICULTURAL IMPACT

Genetic modification can have a range of impacts on agricultural systems and hence will require specific management. Genetic modification can alter the nature of crop volunteers in subsequent crops and the GM trait can disperse to other crops and weeds through cross pollination and seed dispersal. Low impact genes such as herbicide tolerance, which have little impact on natural environments, become highly significant in the agricultural environment because of the changes in herbicides required for their management. These herbicides will differ in the effect they have on plant and other species diversity in cropped fields. These aspects are now the subject of several research projects at NIAB and other European institutes.

Deployment of high impact genes such as pest and disease resistance will result in reductions and changes in pesticide usage and thus offer opportunities to enhance diversity in cropped fields, especially if the trangene products are very specific to selected pest species. However it is important that the selection pressures they impose on pests and diseases do not encourage the development of virulent races of pests and pathogens which are resistant to the genes and require additional pesticide treatments..

Impacts, Consequences and Monitoring

There are considerable concerns in the UK and Europe about the environmental effects of releases and the indirect effects through agriculture of the commercialisation of GM crops. Risk assessments conducted for regulatory purposes tend to concentrate on the direct effects of the GM crop and its relatives on the natural environment. However it is now becoming apparent that the agricultural consequences of the deployment and management of GM crops could also have significant impacts on the environment in regions where a very high proportion of the total land area is managed by man. Plans are being developed for monitoring the early years of the commercialisation of each GM crop so that its impact on both agriculture and the environment can be evaluated in farm scale releases.

ACKNOWLEDGEMENTS

NIAB acknowledges MAFF, DETR, AgrEvo, PGS, Monsanto and other biotechnology companies for their support of studies of the environmental impacts and consequences for agriculture of releases of GM crops.

REFERENCES

Chevre A-M, Eber F, Baranger A, Renard M, Boucherie R, Broucqsault L M, & Bouchet Y. (1998) Risk assessment on crucifer species. *Acta Horticulturae* **459**, 219 - 224.

Dale P J & Senior I J (1997) The possible impacts of multiple transgenes. *Proceedings of the 1997 Brighton Conference* 265-272.

Harding K; Harris P S (1994). Risk assessments of the release of genetically modified plants: a review. MAFF 54pp.

Jorgensen R B, Andersen B & Thure P. (1998) Introgression of crop genes from oilseed rape (*Brassica napus*) to related wild species - an avenue for the escape of engineered genes. *Acta Horticulturae* **459**, 211- 217.

Messean A (1997) Management of herbicide tolerant crops in Europe. *Proceedings of the 1997 Brighton Conference* 947-954..

Rogers H J; Matharu B; Parkes H C (1996) Monitoring releases of transgenic plants: theoretical and practical considerations. *BCPC Symposium Proceedings* **65**, 39-45.

Rudelsheim P; Huybrechts I (1995). A case study of hybrid oilseed rape: from conception to marketing. *Report of the 4th NIAB International Forum: Genetically Modified Crop Cultivars,* 10pp.

Scheffler J A & Dale P J. (1994) Opportunities for gene transfer from transgenic oilseed rape to related species. *Transgenic Research,* **3**, 263 - 278.

Sweet J B; Shepperson R (1996). Monitoring commercial releases of genetically modified oilseed rape. *Proceedings of 10th International Weed Biology Conference*, Dijon, 1996, 217 - 222.

Sweet J B & Shepperson R.(1998). The impact of releases of genetically modified herbicide tolerant oilseed rape in the UK. *Acta Horticulturae* **459**, 225-234

Gene flow and the practical management of genetically modified crops in the UK

J H Orson
Morley Research Centre, Morley St. Botolph, Wymondham, Norfolk, NR18 9DB, UK

J F Oldfield
Raynham Farms Co. Ltd., East Raynham, Fakenham, Norwich, Norfolk, NR21 7EH, UK

ABSTRACT

Crops that are tolerant to the non-selective herbicides glyphosate or glufosinate have been used as examples to describe the possible impact of gene flow on the on-farm management of genetically modified crops in the UK. There are many practical, economic and environmental advantages that could accrue from their adoption. This is provided that there is consumer and regulatory acceptance of these crops and that any practical disadvantages of gene flow are limited. Their management to minimise the impact of gene flow will require a disciplined approach to record keeping, seed handling, crop management and crop handling and storage. It is vital that all those on the farm, including office and support staff, should be fully informed of the procedures required to handle the seed, grow the crop and handle and store the produce. It is concluded that the ease with which gene flow is managed in order to meet grower, consumer and regulatory requirements will help to determine the extent of the adoption of herbicide tolerant crops and any consequent economic and environmental benefits.

BACKGROUND

The implications of the introduction of genetically modified crops are likely to be very profound. The first introductions into the UK will probably be oilseed rape, sugar beet, fodder beet and maize that have been modified to be tolerant to the non-selective herbicides glyphosate or glufosinate. These are used as examples in this paper on the on-farm implications of gene flow, particularly as only the guidelines for the management of herbicide tolerant crops are available at the time of writing.

UK arable production is becoming more exposed to world markets. This trend will continue with the support measures for combinable crops that are proposed under Agenda 2000. UK wheat and oilseed rape growers are becoming internationally competitive because they have invested wisely in technologies such as mechanisation, plant breeding, pesticides and chemical fertilisers. These have resulted in a fuller exploitation of a climate and soils that can sustain high yields (Orson, 1997). It is because of this experience of technology determining international competitiveness that some concern is being expressed that Europe is dragging its feet over the introduction of genetically modified crops.

Whilst UK sugar production from beet is competitive within Europe, the production from cane is significantly more competitive. Therefore, there will have to be a dramatic increase in the efficiency of European sugar beet production in the event of less protected markets.

Forage maize production in the UK has increased rapidly in recent years because of its value in rations for ruminants. Fodder beet production has been limited by the increased separation of livestock and arable production. Most livestock farms do not have the expertise to manage the complexity of weed control using conventional herbicides.

The lower product prices over the last two seasons are resulting in radical changes on arable farms. Variable costs have been minimised and fixed costs reduced by thoroughly reviewing every cost, increasing the scale of production, and machinery and labour rationalisation and sharing. There is now little scope for further reductions in the variable costs of producing conventionally bred cultivars of arable and forage crops.

THE ON-FARM MANAGEMENT OF HERBICIDE TOLERANT CROPS

The management of herbicide tolerant crops will depend on several factors:

- The conditions of registration and management guidelines from the provider of the seed and/or the buyer of the produce.
- The reasons for adopting herbicide tolerant crops.
- The consequences of gene flow.

Gene flow, in the context of this paper, is defined as the transfer of herbicide tolerance to other crops of the same or related species and/or to weeds in pollen and also the movement of herbicide tolerance in seed and produce. It is acknowledged that the conditions for registration and guidelines for crop management are designed to minimise gene flow and that gene flow could also have undesirable environmental consequences.

Conditions of registration and management guidelines

The management guidelines for herbicide tolerant crops (British Society of Plant Breeders, personal communication) will result in great care having to be taken on the farm. The most demanding situation is likely to be where a small-seeded herbicide tolerant crop, such as oilseed rape, is being grown on the same farm as 'conventionally bred' cultivars. This will be particularly so where there is a requirement to segregate the produce of modified crops and to minimise the spread of herbicide tolerance to other fields. In such a situation, at least initially, the extent of the management required to ensure segregation might be determined by the standards imposed for 'GM-free' produce by buyers rather than those standards imposed by the regulatory process.

Management guidelines may involve some or all of the following, depending on regulatory requirements, crop species, objectives of the grower, the standard of segregation required and end-user requirements:

- Recognition of isolation distances.
- Recording of receipt and safe and separate storage of seed and its protection from vermin.
- Preventing seed spills during transport about the farm, re-sealing of bags containing unused seed, cleaning out seed drills and the cleaning up, recording and monitoring of seed spills.
- Eliminating seed production by hand roguing.
- Avoiding leakage of the produce of genetically modified crops during transport. Trailers will have to be sealed and probably covered. Handling equipment, such as grain elevators,

will have to be cleaned prior to handling 'conventional' produce and in-field spills and spills in the vicinity of the stores will have to be cleaned up and recorded.

- Matching the area of production to storage facilities to ensure separate storage from the produce of conventional varieties. This may cause difficulty in some cases when there is a significant isolation distance required between commercial herbicide tolerant crops and commercial conventionally bred crops.
- Thoroughly cleaning stores and handling equipment after use.
- Restricting movement of herbicide tolerant seed in harvesting and cultivation equipment. The problems associated with the prevention of the movement of the seed of small-seeded crops, such as oilseed rape, around the farm are of great practical significance. Cleaning out combines and cleaning tyres and cultivation equipment is very time consuming and conflicts with the commercial necessity of minimising the costs of labour and machinery.
- Adopting practices to minimise the number of volunteers and their control in subsequent crops.
- Keeping meticulous and accurate field records and providing a robust and effective audit trail of records to prove that the above has been carried out. This task should not be underestimated and sound preparation will be necessary to ensure that all aspects of the management of these crops are carried out and recorded to the required standards at a time when labour may be under significant time pressure.
- Providing information on all aspects of the management of genetically modified crops and the possible consequences of failing to meet management guidelines to every person who works on the farm, including office and other support staff. This is absolutely essential and, in the view of the authors of this paper, the responsibility of those selling the seed.

Isolation distances from neighbouring crops on adjoining farms may be a cause of difficulty and discussions between farmers should start at the earliest possible stage. There could be friction between neighbouring farmers where an end-user specifies greater isolation distances than the regulatory requirements in order to claim its products are 'GM-free'. It has been suggested that where isolation requirements are not very demanding, an isolation strip could be sown with a conventional cultivar, but the separate requirement for weed control is not compatible with the farmers' requirement to reduce labour and machinery on farms. However, such a strip would also serve as an area to help purge combines of herbicide tolerant seeds; the success of this will depend on the level of gene flow into this area.

The reasons for adopting herbicide tolerant crops

Meeting specific agronomic requirements

Initially, it is envisaged that the crops will be introduced to meet specific agronomic requirements that could include:

- Control of perennial weeds.
- Control of troublesome 'annual' weeds such as volunteer potatoes and weed beet that are expensive and difficult to control in sugar beet and complexes of broad-leaved weeds which can be difficult and/or expensive to control in oilseed rape.
- To assist in preventing or managing herbicide resistant weeds.

These objectives may also influence the time when glyphosate and glufosinate is applied to the herbicide tolerant crop. For instance, it may be preferable to use selective herbicides in glyphosate tolerant crops until some specific target weeds, notably perennial broad-leaved weeds such as creeping thistle, have reached a development stage when they will transport the glyphosate to the perenniating organs.

Increasing profits

Farmers will do their own cost/benefit analysis when deciding whether to adopt herbicide tolerant crops. Herbicide tolerant crops will have to produce, at a minimum, similar levels of profit to conventionally bred crops. The indirect benefits to take into account include reduced weed control costs through the rotation as well as reductions of fixed costs due to increased flexibility in the time of application of herbicides when compared to conventionally bred cultivars. The additional costs will include higher seed costs and/or technology fees as well as those involved in complying with management guidelines.

Herbicides are currently a very significant variable cost in sugar and fodder beet and herbicide tolerance offers the prospect of significant reductions in the cost, ease and flexibility of weed control. Also, independent research has suggested that herbicide tolerant crops of beet treated with glyphosate are higher yielding than when treated with selective herbicides (Wevers, 1998). Herbicide tolerance may also result in an increase in the area of fodder beet grown in livestock areas because less expertise is required for weed control. All this, plus the fact that pollen production should be eliminated by removing flowering plants i.e. bolters during the season, suggests that there could be rapid adoption of herbicide tolerant beet crops. This is based on the assumption that there will be consumer acceptance of the technology and that herbicide tolerant cultivars will have a similar yield potential to conventionally bred cultivars.

Improving the environmental value of arable land

One of the major objections to the adoption of herbicide tolerant crops is that there will be fewer weeds in the crop resulting in less biodiversity. On the other hand, the ability to manage, both easily and cheaply, weeds in every crop within the rotation will provide more opportunity to leave uncompetitive populations to grow, flower and set seed thus benefiting in-crop biodiversity. Farmers could also adopt strategies in herbicide tolerant beet to encourage biodiversity such as delaying weed control or leaving narrow unsprayed strips in the field to meet prescribed targets every year. With conventional herbicides, there is a variation from year to year in the number of weeds at harvest. However, weed numbers are usually very low because farmers have to adopt an insurance approach with conventional herbicides because the levels of control are less predictable than with glyphosate and glufosinate.

There are other environmental advantages that can be envisaged:

- Preventing resistance to selective herbicides and controlling 'difficult weeds' will reduce herbicide usage through the rotation.
- Reducing the threat of resistance to selective herbicides and increasing the options for perennial weed control may result in the re-introduction of non-plough tillage with its attendant benefits of reduced costs, reduced erosion and an increase in over-wintered stubble. The latter will also improve the food chain for farmland birds.

- Controlling perennial weeds that have spread from habitats on headlands that have been introduced and managed in order to encourage biodiversity. This would remove one of the major concerns about allowing perennial weeds to thrive on the edge of fields.
- Glyphosate and glufosinate are more environmentally benign than some herbicides. Their use in maize will reduce the use of atrazine. Atrazine still occurs regularly in water.

The consequences of gene flow

Volunteers

The implications of herbicide resistant volunteers will have to be judged on an individual herbicide and crop species basis. For instance, glyphosate tolerant volunteer potatoes will be very difficult to control through the rotation because one of the most effective control measures is pre-harvest applications of this herbicide in a range of crops, notably cereals.

The volunteers of the crops in question should not cause a significant control problem later in the rotation. They are equally susceptible to conventional herbicides used in other crops and should not cause significant problems in between-crop management. However, the increased priority for their control may influence herbicide choice and timing in the other crops in the rotation.

It is essential that the number of volunteers are minimised through careful harvesting, control of pollen production from herbicide tolerant sugar or fodder beet and care with post-harvest cultivations.

Sugar beet is a biennial and if drilled too early will vernalise and bolt i.e. set seed. Hence, herbicide tolerant crops should not be sown too early in the spring. It should be noted that the current method for controlling high numbers of bolters is to use a rope wick to apply glyphosate. This option will not be available in a glyphosate-tolerant crop. Another form of herbicide tolerance could be used in the subsequent sugar beet crop to help ensure that volunteers tolerant to the herbicide used in the previous tolerant crop are controlled

The seed shed from oilseed rape will be reduced where stem canker has been controlled and the crop matures evenly, favouring the production of herbicide tolerant crops in fields with little variation in soil types. It should be noted that glyphosate or glufosinate would not be available to desiccate the crop before harvest where the crop is tolerant to one of these herbicides, although diquat may be used on either modification. Research has shown that leaving shed seed of oilseed rape on the soil surface for three to four weeks after harvest will dramatically reduce the number of volunteers in following crops (Pekrun *et al.*, 1998).

The consequences of gene flow to other species

Extensive gene flow to botanically similar weeds will reduce the efficiency of weed control of the non-selective herbicides but it must be pointed out that farmers will not be worse off than they were before the introduction of herbicide tolerant crops. This is because most weeds, including some 'problem' weeds, will remain susceptible to the non-selective herbicides and farmers can always revert back to selective herbicides. However, those species which are botanically similar to the crop are the ones which are often the most difficult to control with selective herbicides. There is also the possibility that selective herbicides may be withdrawn from the market because of herbicide tolerance restricting their sales. The inability to control the weeds that have become resistant to the non-selective herbicides may also cause limited problems in between-crop management.

It is most unlikely that weeds will develop resistance to glyphosate or glufosinate. There are instances of glyphosate resistance in *Lolium rigiditum* in Australia (Powles *et al.*, 1998). However, it should be pointed out that this species has a propensity to develop herbicide resistance and the usage of glyphosate involved in these instances will not be matched in this country, unless every crop is treated with the herbicide.

CONCLUSIONS

There will be advantages from the limited adoption of herbicide tolerant crops where there is the requirement to control perennial weeds or difficult weeds in specific fields. However, to exploit fully any economic and environmental advantages will require a very significant adoption of these crops. Assuming there is end-user and regulatory acceptance, the ease with which gene flow is managed will not only help to determine the extent of their adoption and any consequent economic benefits but also the extent of any environmental benefits. Therefore, it is likely that the environmental advantages from the adoption of herbicide tolerant sugar and fodder beet and herbicide tolerant maize, where gene flow should be more easily controlled, will be more rapidly achieved.

This paper has discussed the implications of gene flow on the crop management of herbicide tolerant crops. It is clear that these have to be considered on a case by case basis. Similarly other transgenes in individual crop species and their potential for and impact on gene flow would also have to be considered on a case by case basis.

REFERENCES

Orson J H (1997). Pesticide management in UK wheat production - optimisation and pragmatism. *Brighton Weed Control Conference,* pp. 445-452

Pekrun C; Hewitt J D J; Lutman P J W (1998). Cultural control of volunteer oilseed rape (*Brassica napus*). *Journal of Agricultural Science (Cambridge)* **130**, 155-163.

Powles S B; Lorraine-Colwill D F; Dellow J J; Preston C (1998). Evolved resistance to glyphosate in rigid ryegrass (*Lolium rigidum*) in Australia. *Weed Science* **46**, 604-607.

Wevers J D A (1998). Agronomic and environmental aspects of herbicide-resistant sugar beet in the Netherlands. *Aspects of Applied Biology 52, Protection and production of sugar beet and potatoes*, pp. 393-399.

POSTER PAPERS

Monitoring weediness and persistence of genetically modified oilseed rape (*Brassica napus*) in the UK

C E Norris, E C Simpson, J B Sweet, J E Thomas
The National Institute of Agricultural Botany, Huntingdon Road, Cambridge, CB3 OLE, UK

ABSTRACT

Results from monitoring the incidence and persistence of genetically modified (GM) oilseed rape volunteers at National List sites and at eight large scale release sites throughout the UK show that volunteer numbers in subsequent crops can vary considerably from site to site. At some sites volunteers were fully controlled in subsequent crops. At other sites viable GM seeds persisted in the soil, producing volunteer populations for up to three years after the release. Where GM and non GM varieties were grown together, the proportion of transgenic volunteers in subsequent crops appears to be lower than the original percentage grown. Initial results from this study suggest that weediness and invasiveness of oilseed rape volunteers is not enhanced by the specific genetic modifications studied.

INTRODUCTION

Some of the first large scale releases of genetically modified (GM) oilseed rape in the UK have been monitored by NIAB since 1995. The work has been commissioned by the Department of the Environment, Transport and the Regions (DETR) and by the Ministry of Agriculture, Fisheries and Food (MAFF) to provide more data on the agricultural and environmental consequences of growing GM oilseed rape.

This study investigates whether genetically modified (GM) oilseed rape is likely to become more persistent in agricultural environments and more invasive of natural habitats, by studying the frequencies and consequences of intra-specific gene flow and seed dispersal.

Oilseed rape seeds from conventionally bred varieties have been shown to persist in the soil for several years (Lutman & Lopez-Granados, 1998) and can cause significant volunteer problems in broadleaved crops (Knott, 1995). Cross-pollination between oilseed rape varieties occurs readily at short distances and infrequently at long range (Simpson *et al.*, in press). The persistence and 'escape' of transgenes could arise by pollen flow and seed dispersal, however several comparative experiments have provided no conclusive evidence to suggest that transgenic rape is any more persistent or invasive than conventional types (Sweet *et al.*, 1997, Booth *et al.*, 1996, Crawley *et al.*, 1993).

The incidence and persistence of volunteers and feral populations following GM releases has been examined at a range of sites across the UK. These sites have grown either herbicide tolerant oilseed rape (glufosinate tolerant or glyphosate tolerant), or oilseed rape with a seed oil modification for high lauric acid.

MATERIALS AND METHODS

A total of eight sites with areas of between 1ha and 38ha were monitored for this study, three of which had their first GM releases in 1995. Monitoring of high lauric acid oilseed rape release sites commenced in 1998. All sites will continue to be studied for a minimum of two years.

Each site was visited monthly from April to November. The incidence of volunteers and feral rape plants in these fields and in the surrounding field margins and roadsides was recorded. Volunteers were counted in the cereal crops following the various releases and where possible after the harvest of following cereal crops. Volunteer and feral plants were tested for herbicide tolerance using a non-destructive assay disk test (Sweet *et al.,* 1997). Leaf tissue samples were also taken from plants to confirm presence or absence of the transgenes by polymerase chain reaction (PCR) using transgene specific primers. DNA was isolated from plant tissue using a DNeasy extraction kit obtained from Qiagen Ltd. UK. PCR reaction conditions were supplied by the biotechnology companies responsible for the production of each transgenic variety grown at the sites. Samples collected in 1998 are currently being tested by PCR.

Seeds were taken from volunteers or feral plants flowering concurrently with, and within 400m of the herbicide tolerant GM crop. Seedlings were screened for tolerance to either glyphosate or glufosinate (Simpson *et al.*, in press).

Post release monitoring of National List (NL) GM oilseed rape trials was conducted on two occasions, the first in November and a second in April to determine volunteer survival.

RESULTS

Numbers of volunteers recorded at National List sites were low, although only one year of post release monitoring has been conducted. Most sites recorded no volunteers and only one site recorded more than one plant per square metre. Numbers of volunteers generally decreased between autumn and spring monitoring visits.

No volunteers were seen throughout the 1998 growing season at the three sites where high laurate GM oilseed rape was grown in 1997.

Numbers of volunteers found at sites where herbicide tolerant GM oilseed rape was grown (Tables 1-4) varied from zero to several thousand. Numbers tended to be low in the crop following the GM rape, and to be more prevalent in the second crop post GM release. At the Cambridgeshire site 77 volunteers were recorded in 1997, in the winter wheat crop following the release. In 1998 over 2400 volunteers were counted in the same field, two years after the GM trial had been harvested. In contrast, in the cereal crops following the GM release no volunteers were recorded in two other fields where GM trials had been grown. Between harvest of the cereal crop and the drilling of the following crop of peas (approximately six months), an estimate of over 15000 volunteers was made in one of these fields. At the Devon site over 600 volunteers were counted in the release field three years post release.

At the Cambridgeshire site herbicide tolerant plants were found both in following crops and in semi-natural locations. (Tables 1-4). Phenotypic testing with glufosinate consistently detected lower numbers of herbicide tolerant plants compared to non-tolerant plants at this site, although the proportions at each release field had originally all grown between 45% and 100% GM plants. Testing of seeds from a roadside population of feral rape at this site at 150m distance from the flowering GM trial (Table 4) showed no tolerant plants, although an experiment using male sterile bait plants at 100m distance in the same location gave 8% tolerant seedlings (Simpson *et al.*, 1999).

Table 1. Testing of volunteers within GM crop field.

Site	Area of rape (ha)	Approx % GM	Year tested	Crop	Number tested	%tolerant
Cambs 1995/96	5.2	48%	1996/97	w.wheat	77	6.5 #
			1997/98	w.barley	90	23.3
			1998	post harvest	142	16.9
Cambs 1996	1.0	100%	1997/98	w.oats	0	0.0
			1998	post harvest	87	42.5
Cambs 1996/97	7.2	69%	1997/98	w.wheat	0	0.0
			1998	post harvest	19	10.5
N.I.A.B. 1996/97	2.8	49%	1997/98	w.wheat	16	0.0
			1998	post harvest	0	0.0

Results confirmed by PCR

Table 2. Herbicide screening of seeds taken from volunteers within GM crop area.

Site	Area of rape (ha)	Approx % GM	Year Tested	Crop	Number tested	%tolerant
Cambs 1995/96	5.2	48%	1998	w.wheat	3650	6.0
Cambs 1996/97	7.2	69%	1997	w.wheat	3696	7.6
Devon 1995	1.0	100%	1997	forage rape	615	100.0

Table 3. Testing of volunteers outside GM crop fields.

Site	Location	Year Tested	Number tested	% tolerant
Cambs	adjacent field	1997	541	0.0
	roadside	1998	92	16.3
	adjacent field	1998	135	40.0
Norfolk	adjacent field	1998	160	0.0

Table 4. Herbicide screening of rape seeds taken from outside GM crop area.

Site	**Location**	**Distance from GM rape**	**Year tested**	**Number tested**	**% tolerant**
Cambs	adjacent field	20m	1997	7998	0.0*
	roadside	20m	1997	6758	0.1
	roadside	150m	1998	3947	0.0
Norfolk	adjacent field	50m	1998	2004	0.0
Berkshire	adjacent crop	10m	1998	14421	0.4

DISCUSSION

The very low incidence of volunteers found at National List sites shows that GM herbicide tolerant oilseed rape does not appear to increase problems of volunteer management in subsequent crops. Furthermore, the gene flow data from the NL GM winter rape trials (Simpson *et al.*, 1998) showed that a proportion of plants grown from seed sampled from GM plots are hybrids expressing tolerance to both glufosinate and glyphosate. There is no indication at present that these multiple tolerant hybrids are more difficult to control in following crops than conventional or single tolerant rape varieties.

The absence of oilseed rape volunteers at the high laurate sites shows that at these sites weed management was extremely good in the following wheat crop, and the GM oilseed rape presented no more of a problem than conventional volunteers. Post harvest conditions in 1997 were ideal for fast germination of oilseed rape seeds prior to cultivation which reduced numbers of seeds returning to the soil seed bank. Post harvest cultivation and weather conditions affect seed return to the soil and may influence the numbers of volunteers seen one year after harvest of the rape crop (Pekrun & Lutman, 1998; Bowerman,1993). These sites will continue to be monitored as different post rape harvest conditions at different sites may produce more volunteers in subsequent crops. However these results indicate that the high laurate trait does not appear to cause problems in control.

Monitoring volunteer incidence indicates that large numbers of oilseed rape seeds are persisting in the soil for up to three years post GM release at some sites (Cambs and Devon). Previous studies show that rape seeds can survive in soil for several years (Lutman & Pekrun 1998, Lutman 1993) due to environmentally induced secondary dormancy.

Plant Genetic Systems releases of glufosinate tolerant winter and spring rape at the Cambridgeshire site show that the numbers of GM compared to non-GM volunteer plants found both in following crops and in semi-natural situations were low. Previous work looking at the survival and persistence of GM rape lines reflects the situation reported here (Sweet *et al.*, 1997, Booth *et al.*, 1996, *Crawley et al.*, 1993). The differences in proportions of GM to non-GM plants detected could however be due to a number of factors. Loss of expression of the transgene in subsequent volunteer populations may occur (Metz *et al.*, 1997). Results of

PCR tests on samples from susceptible volunteers will determine whether the transgene is present. There is also the possibility that some GM seed was shed at a different time to non GM seed prior to harvest, affecting the ratio of transgenic seed returning to the soil seed bank. Different proportions of GM to non GM seed may have germinated immediately after harvest, or the GM volunteers may be less fit or viable than their non transgenic counterparts. Incidence of GM herbicide tolerant rape plants in these volunteer populations suggest that weediness and invasiveness is not enhanced by this specific genetic modification. Although some transgenic plants were found outside the initial GM area at one site, these were all eliminated by normal cultivation methods. Spread of the transgene outside the crop area was probably due to seed spillage in this case. Pollen flow can occur as indicated by the bait plant experiment (Simpson *et al*, in press), but monitoring detected no cross-pollination of feral plants. This may have been due to pollen competition from neighbouring plants, and from self-fertilization.

In conclusion, results of monitoring to date do not indicate that herbicide tolerant and high laurate GM oilseed rape varieties are likely to be any more weedy in an agricultural environment than conventionally bred oilseed rape. Transgenic volunteers with these traits do not appear to be more invasive of habitats outside the crop than their non transgenic counterparts. Future monitoring of these sites will give an indication of the longevity and viability of transgenic seeds in the soil several years after the initial release. This information will be useful for predicting effects of future GM traits on persistence of oilseed rape volunteers.

ACKNOWLEDGEMENTS

The authors wish to thank the Department of the Environment, Transport and the Regions and the Ministry of Agriculture, Fisheries and Food for funding this research, Plant Genetic Systems, AgrEvo, Monsanto, Calgene and John K King & Sons for permission to monitor their sites, and for providing diagnostic primer sequences. Monitoring and harvesting of National List trials was carried out by trials officers at the National Institute of Agricultural Botany.

REFERENCES

Booth E J; Walker K C; Whytock G P; Sovero M (1996) Assessment of the ecological consequences of introducing transgenic rapeseed. *4th ESA Congress, Book of Abstracts (Persistence of oil- modified oilseed rape, Sinapis arvensis and Brassica nigra),* pp 144-145.

Crawley M J; Hails R S; Rees M; Kohn D; Buxton J (1993) Ecology of transgenic oilseed rape in natural habitats. *Nature,* **363**, 620-623.

Knott C M (1995) Control of volunteer rape in peas. *Proceedings Brighton Crop Protection Conference - Weeds*, **2**, 889-894.

Lutman P J W & Lopez-Granados F (1998) The persistence of seeds of oilseed rape (*Brassica napus*). *Aspects of Applied Biology,* **51**,147-152.

Metz P L J ; Jacobsen E; Stiekema W J (1997) Occasional loss of expression of phosphinothricin tolerance in sexual offspring of transgenic oilseed rape (*Brassica napus* L.). *Euphytica,* **98**, 189-196.

Simpson E C; Norris C E; Law J R; Thomas J E; Sweet J B (in press) Gene flow in genetically modified herbicide tolerant oilseed rape (*Brassica napus*) in the UK. *BCPC Proceedings. In press*

Sweet J B; Shepperson R; Thomas J E; Simpson E C (1997) The impact of releases of genetically modified herbicide tolerant oilseed rape in the UK. *Proceedings Brighton Crop Protection Conference - Weeds*, **4**, 291-302

Millet as a model-crop to assess the impact of gene flow toward weed populations

H Darmency, L Assémat
Laboratoire de Malherbologie, INRA, BV 1540, 21034, Dijon, France

T Wang
Institute of Crop Germplasm Ressources, CAAS, 30 Baishiqiao road, 100081, Beijing, China

ABSTRACT

Data on pollen dispersal and interspecific crosses between foxtail millet, an autogamous summer cereal, and the green foxtail, are reported. Progeny of hybrids are identified as the giant green foxtail, a botanical variety of green foxtail called "major". Growth and reproduction of the two wild types are compared as well as their competition effect against maize. The fitness associated with a herbicide resistance gene transferred to millet is studied. The results are used to make inferences about the impact of gene flow from the crop to its wild relative.

INTRODUCTION

Biotechnologically derived herbicide-resistant crops are purported to be major steps in weed control. However, one of the main scientific and public concerns is that recombinant genes may escape from agronomic species and enter wild plants, via spontaneous hybridization between crops and wild plants, and create "super weeds". The literature provides little information on the probability of such hybridizations, and even less on the survival and behaviour of hybrids and descendants. Indeed, the basic question is not only the possible occurrence of hybrids, but the conditions that could lead to the spread of their descendants in arable fields and wild habitats, and the final impact of these plants for farmers and the environment.

A well-documented case is that of foxtail millet (*Setaria italica*), an autogamous summer cereal, which can intercross spontaneously with the green foxtail (*S. viridis*). Resulting hybrids and descendants are identified as the already known giant green foxtail, a botanical variety of the wild species (*S. viridis* var. *major*). These plants have a bigger vegetative development than green foxtail and reproduce abundantly. They can be found as weeds in cultivated fields as well as in waste places. Surveys show that they are currently released in fields of foxtail millet, for instance in the area of Angers (France), or in the northern half of China where more than three million hectares are still grown. However, in spite of the recurrent gene flow working in these areas from antiquity times, the frequency of the "*major*" variety is not so high as it would be expected on the basis of their apparent developmental advantage. This suggests that "*major*" plants could be unadapted. Since herbicide resistant varieties of millet are highly desired and available now, this must be quantified in order to predict the spread of herbicide resistance genes in the wild species.

The paper summarizes the main biological features regulating the appearance of the hybrid "*major*" plants. Then growth, reproductive and competitive ability of typical green foxtail and giant green foxtail are compared to estimate their relative weediness. The effects on growth and reproduction of a herbicide resistance gene transferred to the crop is also investigated on nearly isogenic lines in order to check if it is a neutral change in the absence of herbicide. The results are used to indicate the likely impact of gene flow from the crop to its wild relative.

POLLEN FLOW AND HYBRIDIZATION

Foxtail millet and green foxtail are two autogamous species. Outcrossing rate between plants growing 20-40 cm apart ranges from 0 to 2.2 % in both species (Till-Bottraud *et al.*, 1992). We carried out an experiment to study the effect of the distance between a pollen donor and a target plant. One m^2 plots of a green-pigmented variety were sown at different distances along eight lines radiating from a 20 m^2 plot of a red-pigmented variety. The total area of the green-pigmented plants was 120 m^2 spaced over a 1.1 ha circular field. Red pigmentation of seedlings was used as the dominant marker to identify hybrids among the progeny of green-pigmented target plants.

The farthest hybrids were detected at 24 m (Table 1). Of course, a larger size pollen donor plot should improve the chances of observing pollen migrating farther, and also higher hybrid frequency at each distance. We conclude that, even with an autogamous crop, risks of long distance pollination cannot be excluded. More data on pollen flow, as obtained with male sterile plants as target, are reported in Wang *et al.* (1997).

Table 1. Percentage of hybrids between two varieties of foxtail millet according to distance (mean, and maximum value at the risk P=0.05).

Hybrid	Distance (m)								
(%)	0.5	1	2	4	8	12	18	24	>30
Mean	0.71	0.57	0.32	0.10	0.033	0.009	0.005	0.001	0
Max	0.74	0.60	0.34	0.12	0.039	0.012	0.009	0.004	0.001

The above results can apply to interspecific crosses as no peculiar barrier to interspecific hybridization seems to occur. Plants of green foxtail within a millet field (inter-rows=0.75 m; 10 plants per linear m) produce on average 0.2% of hybrids (Darmency *et al.*, 1987; Till-Bottraud *et al.*, 1992). Reciprocal crosses occur at a lower rate when both species are grown at the same density, 0.002%, and in addition most of the seeds are harvested and exported out of the field, which finally reduces again the actual hybrid output on the field.

Interspecific hybrids are 76% sterile, but further selfed generations behave normally (Darmency & Pernès, 1985). Multivariate analysis of morphological and reproductive traits of the F2 generation of hybrids and various "*major*" plants collected in the wild showed overlapping of the two groups and clear-cut difference from green foxtail and millet. The similarity of hybrid descendants with the "*major*" variety was confirmed by the finding, in "*major*" plants collected in France, of some segregation for characters that make the difference between green foxtail and millet (Darmency & Pernès, 1987, Till-Bottraud *et al.*, 1992).

COMPARISON OF WILD TYPE TO "*MAJOR*"

Seedlings development in growth cabinet

Emergence and growth of young seedlings is a key period that often determines the intensity of weediness of a species against crops. Two populations of green foxtail and two of giant green foxtail, in both cases one population belonging to China and the other from France, were germinated and grown in a growth cabinet at 25°C during 16 h day (230 μmol.photon.m^{-2}.s^{-2}) and 20°C during 8 h in the dark (30 plants each population). A cultivar of foxtail millet was also used. The development (height and weight) of the "*major*" was intermediate between that of the green foxtail and that of the millet (Table 2). This suggests higher potential for competition of the "*major*" type.

Table 2. Characteristics of seedlings of a cultivar of millet, two samples of green foxtail (minor) and two of giant green foxtail (*major*).

Type	No. of days to 3rd leaf	Height at 19 days (cm)	Height at 31 days (cm)	Weight at 31 days (g)
Millet	15.7 d	4.4 a	9.0 b	14.0 a
Major 1	16.7 c	3.2 b	8.1 b	8.1 b
Major 2	16.7 c	3.5 b	10.2 a	7.8 b
Minor 1	17.2 bc	2.0 c	5.9 c	4.5 c
Minor 2	18.4 a	1.6 c	5.4 c	4.4 c

a-d values different at P = 0.05

Experiment in a maize field

A field competition experiment was conducted at the INRA experimental farm to compare the growth of green foxtail and giant green foxtail in a maize field. As the soil must be kept weed free to protect the experiment from interferences with other weeds, and additional weed control techniques such as hoeing disturb the soil too much, atrazine-resistant assessions of the weeds were used, and the maize sprayed with atrazine. The two weeds (origin France) and maize (cv. Dea) were grown pure or with a second species. Maize was sown at 7.4 plants/m^2 with an inter-row of 0.75 m. The same day, plots 3.75 m x 0.75 m were sown with *Setaria* seeds and the mean emergence density was 47 plants/plot for the "minor" green foxtail and 184 for the giant, "*major*" type. The time of emergence was recorded for each plant and all plants were located on a map. The size of 33 days old seedlings of maize, 40 days for *Setaria*, were measured. Each plant was harvested separately, observed for morphological traits, then dried and weighed, four months after sowing for *Setaria*, five for maize. Both design and result interpretation were performed as in Assémat & Allirand (1995) and Assémat *et al.* (1995).

Table 3. Early size (L, cm) and reproductive dry weight (DW, g/plant) for small, medium and big plants in pure and mixture plots (see text).

Species		Small plants		Medium plants		Big plants	
		L	DW	L	DW	L	DW
Maize	Pure	35.6	130.3	38.4	160.8	40.0	176.8
	+ *Major*	34.8	122.0	38.8	163.2	40.8	181.5
	+ Minor	33.1	120.7	39.6	163.4	44.7	186.8
Major	Pure	5.3	0.15	6.9	1.66	7.9	5.95
	+ Maize	3.3	0.08	5.7	0.38	7.0	0.88
Minor	Pure	3.8	0.20	4.1	1.47	5.3	5.96
	+ Maize	2.6	0.04	4.2	0.22	5.7	0.83

Main results concerning plant reproductive output are listed in Table 3. Within each plot, plants were split in 3 groups according to total dry weight (small plants less the 1st quartile, then medium plants, and big plants higher than 3rd quartile). Both size at 33 days and reproductive dry weight increased with total biomass. A large variability was found in both *Setaria* for the reproductive output, up to a ratio of 1 for small plants to 20 for big plants, much less for the size at 33 days. In this experiment, the competitive effects of "*major*" and "minor" types on crop yield was low and significant against small plants only. The competitive effect of maize on both *Setaria* was very important for all types of plants. A higher reproductive investment for "minor" was suspected for most plants, especially because vegetative biomass was much lower than for "*major*". The strong link between results of competition and early size (which, in fact, is measuring emergence earliness within the population) has to be inferred in order to make clear ecological differences between "*major*" and "minor" types.

EFFECTS OF A RESISTANCE GENE

Three different herbicide resistances have been recently obtained in foxtail millet through classical breeding. Atrazine- (Darmency & Pernès, 1985), trifluralin- (Wang *et al.*, 1996) and sethoxydim- (Wang & Darmency, 1997) resistant lines will solve the weed control problem in millet fields. However, secondary effects of the resistance genes are possible. For instance, atrazine resistance resulted in millet varieties having lower potential yield, but yield penalty was compensated by better weed control (Darmency & Pernès, 1989). In contrast, results on BC2 descendants of sethoxydim resistant and susceptible lines showed higher grain yield with the resistant type in the absence of herbicide (Wang & Darmency, 1996).

We show here data on the growth and yield of more isogenic materials, BC7, resistant and susceptible to sethoxydim. Millet was sown as pure lines in a randomized block design (5 replicates) in the field. No significant differences were recorded for vegetative characters and most reproductive ones (Table 4). The weight of 1000 grains was lower for the resistant plants. Since grain weight per plant seemed to be higher, this made higher number of grains produced by the resistant plants. These results confirm previous records on BC2. Apart from a possible pleiotropic effect of the resistance gene on seed production, this would suggest a close linkage with a gene controlling seed output in the original resistant parent. Therefore, transfer of this group of linked genes to wild plants, even in the absence of herbicide use, could be advantageous and enhance fitness and weediness of green foxtail populations.

Table 4. Characteristics of sethoxydim resistant and susceptible BC7 lines of millet (Per plant values, Test of Tuckey at P=0.05).

Trait	Resistant	Susceptible	Difference
Flowering time (day)	83.1	83.4	NS
Height (cm)	156.0	167.0	NS
Flag length (cm)	28.8	31.8	NS
Tiller number	3.6	3.4	NS
Spike length (cm)	22.6	21.3	NS
Vegetative weight (g)	34.6	34.3	NS
Grain weight (g)	37.9	32.2	NS
Weight of 1000 grains (g)	2.9	3.2	0.03

CONCLUSIONS AND PERSPECTIVES

The case of millet illustrates the ancient and repeated occurrence of hybrids between a crop and a wild relative. Gene flow is continuous and leads to hybrid descendants that could establish in the field and waste habitats long before the use of transgenes. However, their apparent more vigorous morphology does not make them more competitive weeds against maize. The reproductive to vegetative dry weight ratio is higher for the common green foxtail, which could explain the low frequency of "*major*" in field populations. In contrast, when "*major*" plants will originate in crosses with a herbicide resistant millet, they could pick up the herbicide resistance that provides a new adaptive value when sprayed with the herbicide. Therefore, it is likely that the dispersal of herbicide resistance genes would change the present situation. In addition, the resistance genes transferred to millet can also have side effects, such as that observed above on the seed output, which can enhance fitness of the weeds.

All these features show that accurate investigations on the biology of the target weeds and wild plants are needed in order to predict the impact of gene flow between transgenic crops and their wild relatives. The case of herbicide resistance is easy to study because the selection pressure, the herbicide is under the control of the researcher or the farmer. More complex are the cases of stress and pest resistances. As well as knowledge on weed biology and actual selection pressures in the fields, there are working plans to reduce gene flow. Transformation and breeding strategies of the crop could be improved. For instance, reciprocal crosses in millet have not the same chance, so that a resistance gene located on the chloroplast DNA of the crop would be safer. Linkage to genes unadaptive or detrimental to wild plants, such as those encoding for less seed produced and no seed shedding, or multigenic and recessive traits that would not be expressed in hybrids, could delay the spread of transgenes.

Further studies are carried out in the framework of a EC-China project involving both millet and wheat. These aim to provide guidelines for both the breeders and the farmers in order to use safely and durably herbicide resistant crops. As for the work on millet, they focus on the presence of the "*major*" type (i.e. past introgression) versus the epidemiology of the spread of resistance genes in the vicinity of fields where resistant millet is grown now in China (i.e. current gene flow). In addition, molecular markers are being developed to study the genetic structure of wild populations as well as cytological investigation using *in situ* chromosome painting to check the presence in wild plants of chromosomes of the crop.

Foxtail millet is also a good model to study the role of the mode of inheritance of the resistance gene, and the effect of the location of the gene on chromosomes to prevent or delay such a gene escape. For instance, the three resistances obtained in millet are differently inherited:- cytoplasmic (atrazine), nuclear recessive (trifluralin) and nuclear dominant (sethoxydim), which provides the opportunity to validate computer models with data from artificial populations involving green foxtail and hybrids submitted to various treatments. This approach is expected to show which kind of gene or population management is best suited to delay the establishment of the resistance genes within a population.

REFERENCES

Assémat L; Allirand J M (1995). Une méthode d'analyse de la compétition pour la lumière dans un maïs enherbé. *16° Conférence du Columa*, Reims, ANPP, 306-314.

Assémat L; Champroux G; Ney B (1995). Measuring light interception by weeds in maize crop. In: *Ecophysiology of tropical intercropping*, eds H Sinoquet & Cruz, pp. 109-114. INRA, Paris.

Darmency H; Pernès J (1985). Use of wild *Setaria viridis* (L.) Beauv. to improve triazine resistance in cultivated S. *italica* (L.) by hybridization. *Weed Research,* **25**, 175-179.

Darmency H; Pernès J (1987). An inheritance study of domestication in foxtail millet using interspecific crosses. *Plant Breeding,* **99**, 30-33.

Darmency H; Zangre G R; Pernès J (1987). The wild-weed-crop complex in *Setaria*: a hybridization study. *Genetica,* **75**, 103-107.

Darmency H; Pernès J (1989). Agronomic performance of a triazine resistant foxtail millet (*Setaria italica* (L.) Beauv.). *Weed Research,* **29**, 147-150

Till-Bottraud I; Reboud X; Brabant P; Lefranc M; Rherissi B; Vedel F; Darmency H (1992). Outcrossing and hybridization in wild and cultivated foxtail millets: consequences for the release of trangenic crops. *Theoretical and Applied Genetics,* **83**, 940-946.

Wang T; Darmency H (1996). Comparison of growth and yield of foxtail millet (*Setaria italica*) resistant and susceptible to acetyl-coenzyme A carboxylase inhibiting herbicides. *X° Colloque International sur la Biologie des Mauvaises Herbes, Dijon,* ANPP, 203-210.

Wang T; Darmency H (1997). Inheritance of sethoxydim resistance in foxtail millet, Setaria italica (L.) Beauv. *Euphytica,* **94**, 69-73.

Wang T; Chen H B; Reboud X; Darmency H (1997). Pollen-mediated gene flow in an autogamous crop: foxtail millet (*Setaria italica*). *Plant Breeding,* **116**, 579-583.

Wang T; Fleury A; Ma J; Darmency H (1996). Genetic control of dinitroaniline resistance in foxtail millet (*Setaria italica*). *Journal of Heredity,* **87**, 423-426.

Origin of wild beet and gene flow between *Beta vulgaris* and *B. macrocarpa* in California

D Bartsch
Biology V (Ecology), Aachen University of Technology RWTH, 52056 Aachen, Germany

J Clegg, N C Ellstrand
Botany & Plant Sciences, University of California, Riverside, CA 92521-0124, USA

ABSTRACT

We evaluated genetic variation in 52 germplasm accessions of wild and domesticated beets (genus *Beta* L) and examined the origin of wild beet accessions in California using allozyme analysis. UPGMA relationship analysis showed overall that domesticated and wild beets form genetically coherent groups. Wild beets in California have two different taxonomic origins from European *Beta vulgaris* or from *Beta macrocarpa*. Population level pattern of allozyme variation for wild California beets related to *B. vulgaris* suggest those populations evolved from naturalized populations of the cultivated *B. vulgaris* ssp. *vulgaris* which had hybridized to varying degrees with the sea beets *B. vulgaris* ssp. *maritima*. Wild California beets related to *B macrocarpa* are essentially genetically identical to European accessions. In addition, we found substantial evidence for hybridization and introgression of *B. vulgaris* alleles in Californian *B. macrocarpa*.

INTRODUCTION

The main target of genetic engineering in the genus *Beta* is sugar beet, *Beta vulgaris* ssp. *vulgaris* var. *altissima* DÖLL, which is ecologically limited due to its low competitiveness and susceptibility to several plant pathogens and phytophagous animals. Nearly all beets are wind-pollinated. Thus, gene flow between sympatric cultivated and compatible wild forms is unavoidable (Boudry et al. 1993, Raybould & Gray 1993, Bartsch & Schmidt 1997). To address biosafety concerns for beet in California three questions should be answered: 1. Where are potential areas of sympatry of wild and cultivated beet in California? 2. Can we use isozyme marker to assign wild beets of California species and subspecies? 3. Is there genetic evidence of gene flow between wild and cultivated beets in California?

MATERIAL AND METHODS

For this study, allozyme diversity was assayed from 52 wild and cultivated accessions of the genus *Beta*. Samples were obtained from seed companies, from international plant genetic resource collections or from collecting directly from wild and cultivated California populations. Accessions were selected so that most of the geographical range of wild and cultivated beets in Europe and California was represented (Table 1). Starch gel electrophoresis was performed on crude protein extracts of young leaf tissue. Approximately 100 mg of tissue from each individual was ground in 0.5 ml extraction buffer [0.1 M Tris-HCl pH 7, 4% polyvinylpyrrolidone (PVP), 0.1% dithiothreitol (DTT), and 0.1% ascorbic acid]. Gel buffers are cited below, and stain techniques are as described by Devlin & Ellstrand (1989) and Wendel & Weeden (1989).

Tab. 1: Species and accessions of beet surveyed in this study: N_i = number of individuals examined (Standard accessions recommended by WBN, World Beta Network)

No.	Species	Accession	Origin	Location	N_i
1	*B. vulgaris*	FC172	USA	(WBN standard)	15
2	- Sugar beet	KWS-2N1009	Germany		30
3		KWS-Kavetina	Germany		49
4		KWS-Rizor	Germany		48
5		KWS-246	Italy		39
6		KWS-247	Italy		41
7		Betaseed-4035	California		25
8		Betaseed-4581	California		24
9		Betaseed-4776	California		31
10		Spreckels-HH103	California		29
11		Spreckels-IV2R	California		30
12		Spreckels-NB2	California		34
13		Spreckels-SS781	California		85
14		Spreckels-VB7R	California		26
15	- Swiss chard	Dark Green	USA		70
16		Chard Fordhook	USA		21
17		Chard Lucullus	USA		21
18		Chard Rhubarb	USA		39
19	- Red beet	W300C	USA	(WBN standard)	32
20		Burpee	USA		78
21		Detroit Dark Red	USA		16
22		Red Ball	USA		25
23		Tall Top	USA		29
24	- Sea beet	PI 504266	France	Corsica, Ajaccio	26
25		PI 540575	France	Gironde County	21
26		PI 540588	France	Charante Marit. County	30
27		HHU	Germany	Oldenburg, Bot. Garden	85
28		RWTH – 1	Greece	Chalkidiki	7
29		RWTH – 2	Greece	Crete	21
30		RWTH – 3	Greece	Peleponnes	20
31		PI518398	Ireland	Kerry County, Dingle	37
32		BGRC 54228	Ireland	(WBN standard)	26
33		RWTH collection	Italy	Ravenna County, Cervia	57
34		RWTH collection	Italy	Goriza County, Grado	10
35		RWTH collection	Italy	Venice County, Venice	60
36		PI518310	UK	East Sussex County	48
37		UCR – 01	California	Contra Costa County	15
38		UCR – 02	California	Alameda County	20
39		UCR – 03	California	Santa Clara County	37
41		UCR – 05	California	San Benito County	17
42		UCR – 06	California	Los Angeles County	25
43		UCR – 07	California	Riverside County	40
44		UCR – 08	California	San Diego County	25
45		UCR – 09	California	Santa Barbara County	40
46	*B. macrocarpa*	UCR 10; UCR 13	California	Imperial County	506
47		UCR – 11	California	Ventura County	30
48		UCR – 12	California	Los Angeles County	40
49		BGRC 53034	Israel	Athlistean Plain	26
50		BGRC 57644	Cyprus	Larnaca	13
51		BGRC 57664	Spain	Cartagena	70
52		BGRC 57676	Spain	Granada	15

Our nine enzyme systems revealed a minimum of 13: aspartate amino transferase (*Aat1*, *Aat2*; E.C. 2.6.1.1), aconitase (*Aco*; E.C. 4.2.1.3), glutamate dehydrogenase (*Gdh*; E.C. 1.4.1.2), leucine aminopeptidase (*Lap*; E.C. 3.4.11.1), NAD^+ malate dehydrogenase (*Mdh1*, *Mdh2*;

E.C. 1.1.1.37), phosphoglucomutase (*Pgm1*, *Pgm2*; E.C. 5.4.2.2), shikimate dehydrogenase (*Skd*; E.C. 1.1.1.25), triose phosphate isomerase (*Tpi1*, *Tpi2*; E.C. 5.3.1.1), and uridine diphosphoglucose pyrophosphorylase (*Udp*; E.C. 2.4.1.1). To resolve these isozymes, we used three different electrophoretic buffer systems: tris-EDTA-borate pH 8.8 (Heywood, 1980) for *Gdh, Lap,* and *Udp*, lithium-borate pH 8.0 (Rieseberg & Soltis 1989) for *Aat, Pgm,* and *Tpi*, and morpholine-Citrate pH 7.0 (O'Malley et al., 1980) for *Aco, Mdh,* and *Skd.* Genetic interpretations of allozyme variation patterns were based on previously published reports for *Beta* (Abe & Tsuda, 1987, Nagamine et al., 1989, Letschert 1993, Raybould et al. 1996).

Banding patterns of polymorphic loci were congruent with typical angiosperm zymograms, and were interpreted accordingly to Weeden & Wendel (1989). Loci encoding the less anodally migrating allozyme for each enzyme system were designated "1", with additional loci numbered sequentially in order of increasing mobility. This nomenclature is inverted relative to locus/allele designations for allozymes of Letschert (1993). Data analysis was carried out with POPGENE (http://www.ualberta.ca/~fyeh/index.html)

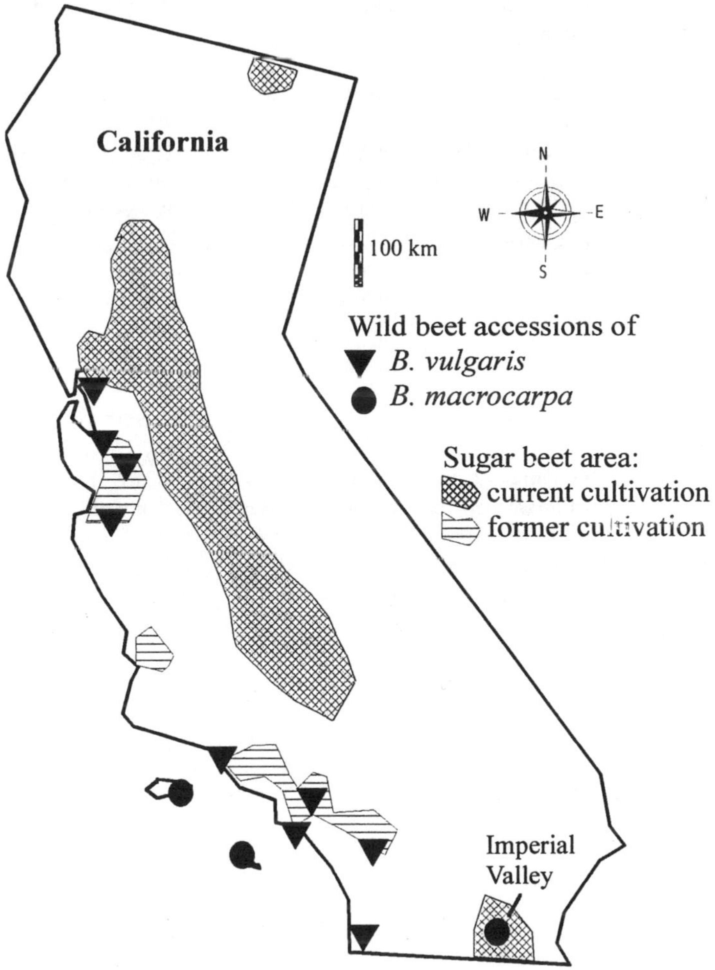

Figure 1. Sugar beet cultivation areas and wild beet accessions in California

RESULTS AND DISCUSSION

Wild beet populations in California were found in several locations (Figure 1). We compiled the different Californian wild beet accessions into groups of *B. macrocarpa* and *B. vulgaris*. The latter group clusters closely with the sea beet group (Figure 2), strongly supporting the hypothesis that at least some of the wild beets originated directly from sea beet. However, the morphologies of the collections demonstrated both sea beet and Swiss chard ancestors. Our isozyme study suggests that also some small populations of vegetable *B. vulgaris* have escaped cultivation in inland areas. The weed beet *B. macrocarpa* was found in sugar beet fields in the Imperial Valley and at coastal sites of Southwest-Californian islands.

In the Imperial Valley, McFarlane (1975) reported phenotypic observations of hybridization between wild and cultivated beets. In this area sugar beet is grown in winter culture, and vernalization of the biennial plants is a common phenomenon due to moderate cold winter temperature. An examination based on 15 sugar beet fields (representing an area of approximately 2 million m^2) showed a sugar beet bolting rate of 0.6 plants/m^2. This rate seems to be higher than typical in this area, probably due to an extraordinary cool winter 1997/98 with periods of low freeze in some parts of the area.

The density of the annual weed *B. macrocarpa* is in the range of 2.7 plants/m^2 (representing an area of approximately 1 million m^2 sugar beet plantation examined). Although the annual *B. macrocarpa* usually flowers earlier than sugar beet bolters, a flowering time overlap could be detected in May 1998. Based on 9 specific isozymes (Table 2), introgression in this area could be detected at a rate of 2 % wild beet individuals (13 of 594 examined Californian plants), which were morphologically similar to *B. macrocarpa*, but had isozyme alleles specific to *B. vulgaris*.

CONCLUSION

There is genetic evidence that Californian wild beets belong to two different taxa, and have at least three different origins. We found wild beet evolved from (1) escaped Swiss chard or Red beet, (2) *B. macrocarpa*, presumably introduced from Spain, and (3) hybridization of *B. vulgaris* with introduced *B. macrocarpa*. Although wild sea beet probably played some role in the origin of California wild beets, our genetic information is insufficient to determine that extent to which hybridization of cultivated beet with sea beet and/or direct introduction of sea beet from Europe contributed to contemporary *B. vulgaris*-type wild beets in California.

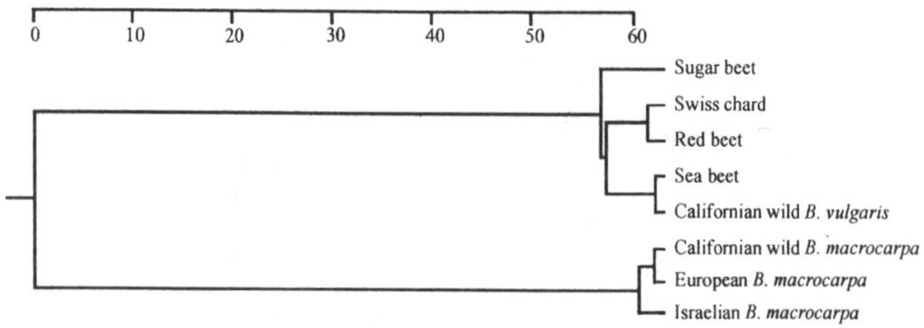

Figure 2. UPGMA dendrogram based Nei's (1978) genetic distance method of allele frequencies for 13 isozymes of *B. vulgaris* (see Table 1).

Engineered cultivars grown in California, especially the Imperial Valley, would have an increased probability of gene escape to wild relatives if their bolting properties were similar or higher than tendencies of the current cultivars planted in this area. Because of this increased tendency for beets to bolt in this area, the impact of an engineered trait escaping into wild populations would have to be assessed.

Table 2. Mean allele frequencies for 13 isozymes of genus *Beta*. Isozyme alleles specific to *B. vulgaris* which were found in 2% of Californian *B. macrocarpa* individuals, are printed bolt. (Origin of *B. macrocarpa*: I = California, II = Europe, III = Israel)

		Sugar beet	Swiss chard	Red beet	Sea beet	Wild beet California	*B.macro-carpa*-I	*B.macro-carpa*-II	*B.macro-carpa*-III
Mdh1	-1	0.765	0.525	0.827	0.885	0.760	0.692	0.726	0.961
	-2	0.157	0.443	0.005	0.023	0.152	**0.001**	0.000	0.000
	-3	0.078	0.032	0.168	0.092	0.088	0.307	0.274	0.039
Mdh2	-1	0.536	0.000	0.004	0.019	0.080	**0.006**	0.000	0.000
	-2	0.464	0.993	0.965	0.924	0.912	**0.002**	0.000	0.000
	-3	0.000	0.000	0.000	0.000	0.000	0.992	1.000	1.000
	-4	0.000	0.007	0.031	0.057	0.000	0.000	0.000	0.000
Aco	-1	0.914	0.176	0.254	0.728	0.612	1.000	1.000	1.000
	-2	0.086	0.824	0.746	0.272	0.388	0.000	0.000	0.000
Skd	-1	0.036	0.000	0.061	0.055	0.024	**0.045**	0.000	0.000
	-2	0.005	0.000	0.010	0.082	0.024	0.946	0.988	1.000
	-3	0.826	0.973	0.822	0.781	0.841	0.009	0.012	0.000
	-4	0.133	0.027	0.107	0.082	0.111	0.000	0.000	0.000
Lap	-1	0.390	0.725	0.551	0.821	0.791	**0.001**	0.000	0.000
	-2	0.610	0.275	0.449	0.179	0.210	**0.008**	0.000	0.000
	-3	0.000	0.000	0.000	0.000	0.000	0.991	1.000	1.000
Tpi1	-1	1.000	1.000	1.000	1.000	1.000	1.000	1.000	1.000
Tpi2	-1	0.000	0.000	0.000	0.064	0.000	0.000	0.000	0.000
	-2	0.971	1.000	0.911	0.878	1.000	0.000	0.000	0.000
	-3	0.029	0.000	0.089	0.057	0.000	0.000	0.000	0.000
Aat1	-1	0.001	0.000	0.000	0.000	0.000	0.000	0.000	0.000
	-2	0.533	0.500	0.496	0.570	0.515	0.005	0.000	0.096
	-3	0.000	0.000	0.004	0.210	0.067	0.995	1.000	0.904
	-4	0.466	0.500	0.500	0.220	0.415	0.000	0.000	0.000
Aat2	-1	0.000	0.000	0.000	0.000	1.000	1.000	0.985	1.000
	-2	0.000	0.000	0.000	0.000	0.000	0.000	0.015	0.000
Udp	-1	0.556	0.482	0.831	0.180	0.520	0.000	0.000	0.000
	-2	0.444	0.518	0.169	0.820	0.480	1.000	1.000	1.000
Pgm1.	-1	0.000	0.000	0.000	0.052	0.000	0.984	0.988	1.000
	-2	0.998	1.000	1.000	0.930	0.995	0.016	0.012	0.000
	-3	0.002	0.000	0.000	0.019	0.005	0.000	0.000	0.000
Pgm2	-1	0.000	0.000	0.000	0.013	0.000	0.992	0.933	0.519
	-2	0.266	0.028	0.009	0.109	0.069	0.003	0.067	0.000
	-3	0.686	0.964	0.991	0.828	0.902	0.005	0.000	0.481
	-4	0.048	0.008	0.000	0.050	0.029	**0.001**	0.000	0.000
Gdh	-1	0.824	0.933	1.000	0.742	0.198	**0.002**	0.000	0.000
	-2	0.176	0.067	0.000	0.215	0.000	**0.008**	0.000	0.000
	-3	0.000	0.000	0.000	0.042	0.000	0.989	1.000	1.000

ACKNOWLEDGEMENTS

We thank the following for assistance during early phases of this project: K Meyerholz (KWS Einbeck), J H Griffin, D C Melin (Spreckels-Holly Sugar Corporation), J R Stander (Betaseed), L Panella (Fort Collins), R Levellen and G Wisler (USDA-ARS Salinas), B Ertter (UC Berkeley), S. Chaney (Channel Islands National Park), D. Cudney, C Koehler, A Sanders, R Whitkus (UC Riverside), S Boyd (Rancho Santa Ana Botanic Garden), M Hoefs (Wrigley Botanic Garden Avalon), P Schuyler (Catalina Island Conservancy), D Wilken, S Junak (Santa Barbara Botanic Garden), M Beauchamp (Pacific Southwest Biological Services Inc.), J Gibson (San Diego Natural History Museum), W L Bretz (UC Irvine Natural Reserve System), C Morak, M Pohl-Orf, I Schuphan (RWTH Aachen), L Frese (FAL Braunschweig), S Kaffka and R Norris (UC Davis). This study was supported by University of California Competitive Grant (#1997-980069).

REFERENCES

Abe J; Tsuda CH (1987). Genetic analysis for allozyme variation in the section *Vulgares*, genus *Beta*. Japanese Journal of Breeding **37**, 253-261.

Bartsch D; Schmidt M (1997). Influence of sugar beet breeding on populations of *Beta vulgaris* ssp. *maritima* in Italy. *Journal of Vegetation Science* **8**, 81-84.

Boudry P; Mörchen M; Saumitou-Laprade P; Vernet P; Van Dijk H (1993). The origin and evolution of weed beets: consequences for the breeding and release of herbicide-resistant transgenic sugar-beets. *Theoretical and Applied Genetics* **87**, 471-478.

Devlin B; Ellstrand NC (1989). Transmission genetics of isozyme loci in *Raphanus sativus* (Brassicaceae): stress depended non-Mendelian segregation. *American Journal of Botany* **76**, 40-46.

Heywood JS (1980). *Genetic correlates of edaphic differentiation and endemism in Gaillardia*. PhD thesis. University of Texas, Austin.

Letschert JPW (1993) *Beta* section *Beta*: biogeographical patterns of variation and taxonomy. *Wageningen Agricultural University Papers* **93**,: 153pp.

McFarlane JS (1975). Naturally occuring hybrids between sugarbeet and *Beta macrocarpa* in the Imperial Valley of California. *Journal of the American Society of Sugar Beet Technology* (Fort Collins) **18**, 245-251.

Nagamine T; Catty JP; Ford-Lloyd BV (1989). Phenotypic polymorphism and allele differentiation of allozymes in fodder beet, multigerm sugar beet and monogerm sugar beet. *Theoretical and Applied Genetics* **77**, 711-720.

O'Malley DM; Wheeler NE; Guries RP (1980). A manual for starch gel electrophoresis. *University of Wisconsin Madison Department of Forestry Staff Paper Series* **11**.

Raybould, AF; Gray AJ (1993). Genetically modified crops and hybridization with wild elatives: a UK perspective. *Journal of Applied Ecology* **30**, 199-219.

Rieseberg LH; Soltis DE (1989). Assessing the utility of allozyme number for determining ploidal level: evidence from *Helianthus* and *Heliomeris* (Asteraceae). *Aliso* **12**, 277-286.

Weeden NF; Wendel JF (1989). Genetics of plant allozymes. In: *Allozymes in plant biology*, eds DE Soltis & PS Soltis PS. pp 46-72. Dioscorides Press.

Wendel JF, Weeden NF (1989) Visualization and interpretation of plant allozymes. In: *Allozymes in plant biology*, eds DE Soltis & PS Soltis PS. pp 5-45. Dioscorides Press

Evaluating the risk of transgene spread from *Brassica napus* to related species

R Pinder, N Al-Kaff, M Kreike, P Dale
John Innes Centre, Norwich, NR4 7UH UK.

ABSTRACT

There is a need to evaluate the likelihood of the sexual spread of transgenes from genetically modified crops to wild species. We attempt to estimate this by crossing a transgenic (hemizygous) line of oilseed rape to weed species with which it has varying degrees of sexual compatibility. The offspring were selfed and backcrossed in order to examine the inheritance of the transgene in subsequent generations. We found that while the transgene was inherited in a normal Mendelian fashion for the first generation, most of the offspring were infertile. However, analysis of those that are fertile could be very informative in estimating the risk of transgene spread in different genetic backgrounds.

INTRODUCTION

Oilseed rape (*Brassica napus oleifera* AACC 2n=38) is an important oil and fodder crop in the UK. There is increasing commercial pressure to develop genetically modified lines, with for example, herbicide tolerance, disease resistance, and improved oil quality. *Brassica napus* is classed as self-fertile, but shows a degree of out-breeding. It is both insect and wind pollinated. *Sinapis arvensis* (SarSar 2n=18) is classed as an out-breeder. Chromosome pairing studies show it has some chromosomal homoeology to *B. napus*, as does *Raphanus raphanistrum* (RrRr 2n=18) (Mizushima 1980). *Sinapis arvensis* and *Raphanus raphanistrum* have been shown to be weeds in areas of *B. napus* cultivation. *B. rapa* (AA 2n=20) and *B. juncea* (AABB 2n =36) can also be found as weeds (Chevre *et al.* 1996). Thus these plants are potential hybridization partners with *B. napus* . The possibility of hybridization between *B. napus* and some of its more or less compatible relatives has been reviewed (Scheffler and Dale, 1994). There is a possibility that the 'escape' of such transgenes into the wider environment could lead to the development of, for example, herbicide-tolerant weeds that are able to survive treatment by particular herbicides. There may also be other effects from the presence of such transgenes in the gene pool of weed species. These four annual species were all used with *B. napus* (also an annual) in a crossing programme designed to generate both backcrossed and selfed hybrid lines. An embryo rescue procedure was used to obtain hybrids between *B.napus* and the poorly compatible species (*S. arvensis* and *R. raphanistrum*); and chromosome number was determined for these hybrids.

MATERIALS AND METHODS

All plants were grown in a containment glasshouse. All hybridizations were made by hand-pollination of emasculated stigmas. Self-seeding was performed by bagging the entire plant.
Plants used in the crossing program: At least four ecotypes of each of the above species were collected from different sources. (Table 1). The *B. napus* (dihaploid line W10) contained a single, stable copy of the *bar* and *npt II* genes, both were driven by the cauliflower mosaic virus 35S

promoter, and terminated by the octopine synthase terminator. The number of seed obtained per cross, as well as the proportion of viable offspring containing the transgene, was measured.

Table 1. Code number and source of the wild species used in the crossing program with transgenic *B. napus*.

Plant species	Chromosome number	Ecotype code number	Sources
Brassica juncea	(2n=36)	NAK1	Dr. R. Mithen, JIC, UK.
		NAK2	Dr. R. Jorgensen, NL., Denmark
107*		NAK3	C. Gomez-Campo, Spain
108*		NAK4	C. Gomez-Campo, Spain.
Brassica rapa	(2n=20)	NAK5	Dr. R. Jorgensen, NL., Denmark
133*		NAK6	C. Gomez-Campo, Spain
		NAK7	Dr. R. Mithen, JIC, UK.
		NAK8	Dr. R. Mithen, JIC, UK.
		NAK25	Dr. A. Meikle, SAC, UK.
Brassica napus	(2n=38)	NAK9	C. Gomez-Campo, Spain
		NAK11	Dr. Philip Dale , JIC, UK.
		NAK12	Dr. Philip Dale , JIC, UK.
		NAK13	Dr. Philip Dale , JIC, UK.
Raphanus raphanistrum	(2n=18)	NAK14	Herbiseed Ltd., UK
511*		NAK15	C. Gomez-Campo, Spain
512*		NAK16	C. Gomez-Campo, Spain
513*		NAK17	C. Gomez-Campo, Spain
Sinapis arvensis	(2n=18)	NAK18	Herbiseed Ltd., UK
		NAK19	Dr. R. Jorgensen, NL., Denmark
		NAK20	Dr. R. Jorgensen, NL., Denmark
537*		NAK21	C. Gomez-Campo, Spain
538*		NAK22	C. Gomez-Campo, Spain
		NAK23	Dr. A. Meikle, SAC, UK
		NAK24	Dr. A. Meikle, SAC, UK

* Germplasm collection catalogue number, Madrid, 1995

Embryo rescue: Hybrids generated by crossing with ecotype numbers 14 to 25 were subjected to embryo-rescue. Whole pods were removed from plants 5-10 days after pollination, then sterilised in 20% sodium hypochlorite solution (BDH) containing 5 drops of Tween 20 (Sigma) per 20 ml for 5 minutes. Pods were then washed three times with single-distilled water for 5 minutes. The pod pedicle was embedded in Nitsch medium containing 2.14 g Nitsch medium concentrate (Imperial Laboratories), 0.3 g casein hydrolysate (Oxoid), 15 g sucrose (BDH), 15 g glucose (BDH) pH 5.8 with KOH, made up to 1 litre, then 2.5 g Gelrite (Imperial Laboratories). Developed ovules were removed to MS medium containing 4.41 g MS medium (Imperial Laboratories), 10 g sucrose (BDH), pH 5.8 with KOH, made up to 1 litre, then 2.5 g Gelrite. Pods and ovules were grown under artificial light, with a 16 h photoperiod at temperature of 25 °C during light and 20 °C during dark. Developed plants were moved to Jiffy pots and transferred to the glasshouse.

Transgene detection: Initially, presence of the transgene was assayed by polymerase chain reaction for both the *npt II* and *bar* genes, but recently this has been superseded by the much faster *bar* colour assay, which relies on the expression of the *bar* gene to allow ammonia metabolism in a pH sensitive indicating medium. (D'Halluin *et al.* 1992).

Bar colour assay: 5mm leaf discs were placed in microtitre plates, on liquid MS medium containing 3% sucrose (BDH), 20 mgL^{-1} phosphinothricin (Duchefa), 50 mgL^{-1} chlorophenol red (Sigma), 25 mgL^{-1} nystatin (Duchefa), 500 mgL^{-1} carbenicillin (Duchefa). The plates were sealed with transparent film, and placed at 30 °C for 16 hours.

Chromosome number: Root tips, excised in the morning, were incubated at 18 °C in 0.05 % colchicine (Sigma), then fixed in 3:1 ethanol: acetic acid for 3-7 days at –20 °C. Fixed root tips were digested in 2% cellulase (Onozuka RS), 1.5% macerozyme R200, 0.3% pectolyase Y-23 (Genetic Research Instruments) in 1mM EDTA (BDH) for 45 minutes at 37 °C, then spread on slides in a drop of fixative (Fukui *et al.* 1991). After staining with Giemsa (BDH) for 2 hours, the slides were examined at X 400 magnification. At least 10 metaphase nuclei per slide were counted in order to estimate chromosome number.

RESULTS

The results of the initial pollinations are presented in Table 2.

Chromosome numbers were determined for a small number of hybrids between *B. napus* x *Raphanus raphanistrum*, *B. napus* x *Sinapis arvensis* and *B. napus* x *B. rapa*. Each of the hybrids tested had an estimated chromosome number of 28, apart from one from *B. napus* x *Raphanus raphanistrum* (NAK14), which had 56 chromosomes.

Percentage of F1 hybrids giving rise to viable F2 seeds described in Table 3.

No crosses between *B. napus and R. raphanistrum* or between *B. napus* and *S. arvensis* gave fertile hybrids. The hybrids have been maintained vegetatively by taking cuttings for over 2 years. Pollination was allowed in a contained glasshouse, but no seeds were set (Table 3).

Table 2. Results of initial pollinations.

Species	Ecotype crossed with *B. napus*	Number of buds pollinated	Number of hybrids	Efficiency	No. Offspring tested *bar* positive.	No. offspring tested *bar* negative
B. juncea	NAK 1	24	67*	31	11	13
	NAK 2	27	122	46	75	59
	NAK 3	37	187	46	22	27
	NAK 4	16	24*	31	22	5
B. rapa	NAK 5	19	104	60	30	36
	NAK 6	17	64	41	34	20
	NAK 7	37	156	40	34	17
	NAK 8	8	23*	12	1	5
	NAK 25	65	13*	98	10	3
B. napus	NAK 11	22	166	80		
	NAK 12	43	450*	87	4	3
	NAK 13	14	64	80	9	9
R. raphanistrum	NAK 14	81	1*	40	1	0
	NAK 15	68	4*	21	4	0
S. arvensis	NAK 18	38	0	79		
	NAK 19	71	0	42		
	NAK 20	80	1*	40		
	NAK 22	30	0	40		
	NAK 23	81	1*	21	0	1
	NAK 24	101	0	34		

-* Sample too small for chi-squared analysis.

- Number of putative hybrid seeds generated is given for samples 1-13; number of plants generated by embryo rescue is given for plants 14-25. Efficiency is defined as percentage of pollinated buds producing pods containing either seeds (1-13) or embryos suitable for embryo rescue (14-25).

Table 3. Percentage of F1 hybrids giving rise to viable F2 seeds.

B. napus x weed hybrid	% giving F2	*B. napus* x weed hybrid	% giving F2
NAK 1 (*B. juncea*)	4	NAK 11(*napus*)	nt
NAK 2 (*B. juncea*)	2	NAK 12(*napus*)	4
NAK 3 (*B. juncea*)	10	NAK 13(*napus*)	34
NAK 4 (*B. juncea*)	25	NAK 14 (*R. raphanistrum*)	0
NAK 5 (*B. rapa*)	30*	NAK 15 (*R. raphanistrum*)	0
NAK 6 (*B. rapa*)	nt	NAK 23 (*S. arvensis*)	0
NAK 7 (*B. rapa*)	33*	NAK 25 (*B. rapa*)	32
NAK 8 (*B. rapa*)	nt		

- *Samples gave at least 3 plants which produced F3 generation.
- nt: not tested.
-
-

DISCUSSION

Hybridizations between distantly related species:

Our poor success in generating *B.napus x S.arvensis* hybrids was not unexpected. Kerlan *et al.* (1992) had some success in generating such hybrids but only when using *B.napus* as the female parent. Lefol *et al.* (1996) estimated the probability of generating a natural hybrid between wild *S. arvensis* and *B. napus,* based on flowering period overlap, hybridization rate and hybrid reproduction. They concluded that, for the Burgundy region of France this was about 1 in 10^{10} pollinations and, extrapolating for the whole of Western Europe, that this would result in about 1000 such hybrids per year. The success of those hybrids will depend on their sexual fertility which, from the evidence available, is likely to be very low.

Hybridization with more closely related species:

Mikkelsen *et al.* (1996a) report that crossing transgenic *B. napus* with *B. rapa* can create fertile, transgenic offspring after 2 generations of hybridization and backcrossing. This is confirmed by our results with the *B. rapa* and *B. juncea* hybrids. The selfing data show that for 3 ecotypes (one B. *juncea* and two *B. rapa*) there is evidence that such hybrids might persist. Backcross data are still being generated and at present there is no indication that the transgene is inherited at a higher rate than expected. Indeed, in one *B. napus* x *B. juncea* hybrid the transgene was entirely lost in the backcross population (data not shown). This might imply that the transgene is located on the C-genome, and has been lost in meiotic pairing. This result is in contrast to a comparable study by Frello *et al.* (1995) who found the transgene in approximately half of such progeny. They concluded that gene transfer from *B. napus* to *B. juncea* is possible. Mikkelsen *et al.* (1996b) hypothesized that 'safe' areas of the chromosome might be identified for transgene insertion. This could reduce the opportunity for transgene introgression to from *B. napus* to *B. rapa* and *B. juncea*, but they concluded from the analysis of transmission of 33 predominantly C-genome-linked *B. napus* RAPD markers from *B. napus* x *B. juncea* hybrids back to *B. juncea,* that this was unlikely. However, Metz *et al.* (1997) provide evidence to support this hypothesis. On backcrossing a transgenic *B. napus* x *B. rapa* hybrid to *B. rapa,* there was a low frequency of transmission of the transgene. They concluded that the transgene was located on the C-genome, and that location on either the A or the C genome would influence the probability of transfer to related species.

CONCLUSION

Further work needs to be done to evaluate the risk of transgene spread from *B napus* to related species. It is likely that if transgene flow occurs, it will be to the more closely related species, *B. rapa* and *B. juncea.* The persistence of a herbicide tolerance gene in the absence of herbicide selection pressure is undetermined. There is evidence that the specific chromosomal location of the transgene will affect the persistence of the transgene in a population. Long term backcrossing data would be useful to estimate the persistence of a transgene in a wild population.

ACKNOWLEDGMENTS

We are grateful to Mr Mike Hill and Mr Damien Alger for assistance in the glasshouse. The project was funded by MAFF, project number CSA 2915.

REFERENCES

Chevre A M; Eber F; Kerlan M C; Barret P; Festoc G; Vallee P; Renard M (1996). Interspecific gene flow as a component of risk assessment for transgenic brassicas. *Acta Horticulturae* **407,** 169-179.

D'Halluin K D; Block M; Denecke J; Janssens J; Leemans J; Reynaerts A; Botterman J (1992). The bar gene as a selectable and screenable marker in plant engineering. *Methods in Enzymology* **216**, 415-426.

Frello S; Hansen K R; Jensen J; Jorgensen R B (1995). Inheritance of rapeseed (*Brassica napus*) specific RAPD markers and a transgene in the cross *B. juncea* x *(B. juncea x B. napus)*. *Theoretical and Applied Genetics* **91**, 236-241.

Fukui K; Iijima K (1991). Somatic chromosome map of rice by imaging methods. *Theoretical and Applied Genetics* **81**, 589-596.

Kerlan M C; Chevre A M; Eber F; Baranger A, Renard M (1992). Risk assessment of outcrossing of transgenic rapeseed to related species. *Euphytica* **62**, 145-153.

Lefol E; Danielou V; Darmency H (1996). Predicting hybridization between transgenic oilseed rape and wild mustard. *Field Crops Research* **45**, 153-161

Mikkelsen T R; Andersen B; Jorgensen R B (1996a). The risk of crop transgene spread. *Nature* **380**, 3.

Mikkelsen T R; Jensen J; Jorgensen R K (1996b). Inheritance of oilseed rape *(Brassica napus)* RAPD markers in a backcross progeny with *B. campestris*. *Theoretical and Applied Genetics* **95,** 442-450.

Mizushima U (1980). Genome analysis in *Brassica* and allied genera. In: *Brassica crops and wild allies*, eds S Tsunoda; K Hinata; C Gomez-Campo, pp 89-108. Japan Scientific Societies Press, Tokyo.

Scheffler J; Dale P J (1994). Opportunities for gene transfer from transgenic oilseed rape (*Brassica napus*) to related species. *Transgenic Research* **3**, 263-278.

Estimation of allele frequencies for *Bacillus thuringiensis* resistance in diamondback moth, *Plutella xylostella* and cotton bollworm, *Helicoverpa armigera* : An isofemale line (F_2) approach

M Ahmad, R Roush
Department of Applied & Molecular Ecology, The University of Adelaide, Waite Campus, Glen Osmond, SA 5064, Australia

ABSTRACT

The detection of resistance while still at low frequencies has been identified as a regulatory issue for transgenic crops. Recently, it has been assumed that the frequency of resistance in unselected populations is about 10^{-2}-10^{-4}. Using a leaf-dip bioassay method, 4 of 943 (=3772 alleles) isofemale lines (IFLs) were detected carriers of low level resistance to *Bacillus thuringiensis* (Bt) toxins in diamondback moth, *Plutella xylostella*. The frequency of alleles for Bt resistance found in this study is much lower than estimated elsewhere. Currently, transgenic cotton plants expressing Cry1Ac protein are being used to screen for Bt resistance alleles in field populations of cotton bollworm, *Helicoverpa armigera.* We are still looking to detect a resistant IFL after screening 347 lines. This also supports our hypothesis that the frequency of Bt resistance alleles in natural populations is lower than reported recently for other bollworms.

INTRODUCTION

The current decade has been earmarked as an era of transgenic crops. Transgenic plants offer a novel approach to insect pest control especially against those pests which normally escape sprays. It is widely believed that transgenic crops have the potential to replace chemicals to a large extent, eg., transgenic cotton can reduce overall insecticide use world-wide by at least 10-15% (Fitt *et al.*, 1994; Roush, 1994).

The most critical concern for these transgenic crops and *Bacillus thuringiensis* (Bt) sprays is the evolution of resistance in the targeted insect pests. Diamondback moth (DBM), *Plutella xylostella*, a major pest of cruciferous vegetables, and cotton bollworm (CBW), *Helicoverpa armigera*, a major pest of cotton, are major targets of Bt sprays and transgenic crops, respectively. Resistance to Bt toxins in field populations of DBM has already been documented from many parts of the world (Tang *et al.*, 1997; Tabashnik *et al.*, 1992; Hama *et al.*, 1992). Transgenic cotton may exert higher selection pressure in target insect pests due to continuous expression of insecticidal proteins. Therefore, planting transgenic cotton on a large scale has the potential of selecting resistance to Bt toxins in insect pests (Kaiser, 1996). Wide-spread evolution of resistance to Bt toxins would eliminate them as a valuable pest control tool (Roush & Shelton, 1997).

Before selection, the resistance alleles are carried mainly in heterozygous (SR) individuals in accordance with Hardy-Weinberg law (p^2: 2pq: q^2) (Dobzhansky, 1955). Because routine bioassays are inefficient to detect recessive alleles in the heterozygous form, eg., many cases of Bt resistance (Tabashnik *et al.*, 1997), any technique that has the potential to improve the precision of detecting resistance would be very attractive.

The frequency of resistance alleles is one of the key factors in the effectiveness of the more promising resistance management tactics for Bt crops (Roush, 1997). Change in allele frequencies is also the key indicator of the effectiveness of a resistance management program. Most resistance management models have been based on theoretical estimates of initial allele frequencies because of scanty empirical data (Gould, 1998). The initial allele frequency for Bt resistance has become highly controversial with recently published estimates as high as 10^{-3} (Gould *et al.*, 1997; Tabashnik *et al.*, 1997). Since the alleles conferring resistance to Bt are predominantly recessive, we used an IFL technique in which insects are inbred to an F_2 to estimate the frequencies of alleles for Bt resistance in the field populations. This technique increases homozygosity and allows ready detection of recessive alleles. (Andow & Alstad, 1998).

The sample sizes needed for detection of resistance and its frequency are in inverse proportion. For example, at a frequency of 10% or more, a sample size of 100 individuals can detect resistance easily (at 95% probability level). A sample size of approximately 300 would be required for detecting a resistant individual at a frequency of 1%. When resistance frequencies are 0.1% or less, it would require sample size in thousands (Roush & Miller, 1986).

The main objective of this work was to estimate the initial frequency of alleles conferring resistance to Bt endotoxins in field populations of DBM and CBW. The implications of these results will be highly relevant for devising future resistance management strategies for biopesticides based on Bt endotoxins in general, and transgenic crops in particular.

METHODS AND MATERIALS

Isofemale lines

IFLs (single pair mating) of DBM and CBW were established by adults reared in the lab from field-collected eggs, larvae and pupae. The setting up of IFLs can be summarized as below:

Generation I	SS	X	SR	(field population, in single pairs)
		↓		
Generation II (F_1)	SR	X	SR	(mating of siblings in mass crosses)
		↓		
Generation III (F_2)	RR	SS	SR	(in lines carrying resistance, 1/16 (6%) offspring are RR)

The "SS" refers to homozygous susceptible, "SR" to heterozygotes, and "RR" to homozygous resistant. The single pair crosses of DBM were set up in clear plastic bag cages especially designed for rearing IFLs. The plastic bags were shifted when F_1 moths emerged and wrapped around another fresh rape-seedling pot for oviposition. The larvae (F_2) were reared until 2nd - 3rd instar (~5 day old) and used for the bioassay studies.

In order to establish IFLs of CBW, one day-old male moths were paired with 2-3 day-old females in a rearing cage measuring 50x60x75 cm^3. Moths while mating were transferred gently into one litre plastic containers. Eggs were collected on kitchen paper towel and the F_1 larvae were reared on artificial diet. The moths were enclosed in cages made of clear plastic bags with tiny holes for aeration.

Plants

Canola (oilseed rape) and Chinese cabbage plants were grown in plastic pots (17.5 cm diameter) in the glass house. Usually, 7-10 day-old canola seedlings and four to six week-old Chinese cabbage plants were used for rearing and bioassays, respectively. Broccoli, *Brassica oleracea* subsp. *italica* (Green Comet-non-Bt and transgenic broccoli, expressing the Cry1Ac protein of Bt), plants were used in bioassays at Cornell University (New York). A lab resistant (Res) and a lab susceptible (G88) strain were also bioassayed for confirmation of resistance in IFLs detected from South Australia (SA).

Bt-cotton and non-Bt cotton were grown in a growth room for bioassaying CBW neonates (F_2). We used 4-6 week old plants which normally have an expression level of Cry proteins high enough to kill susceptible homozygotes (SS) and heterozygotes (SR).

Screening of IFL (F_2)

A leaf-dip bioassay method was adopted for screening Bt resistance alleles in DBM. Delfin WG (Sandoz®) containing 53,000 *Spodoptera* units per mg Bt Berliner subsp. *kurstaki* was used for screening IFL's of DBM (with a single diagnostic dose as well as multidose bioassays). On an average, 50 larvae (5 day old) from each IFL (F_2) were tested for Bt resistance. The surviving 4 of 943 IFLs from SA were tested again by using leaf-dip bioassays with a range of doses. The dose mortality data were analysed using POLO-PC (Russel *et al.,* 1977). These 4 IFLs were further tested on broccoli (expressing Cry1Ac) at Cornell University, New York.

CBW eggs from the caged F_1 adults were laid directly onto the transgenic cotton plants. When the neonates hatched, they fed directly on the plant and took a dose of Bt toxin. The survivors (if any) were saved for further confirmation of resistance.

RESULTS & DISCUSSION

Screening of diamondback moth

Each IFL allowed us to test four alleles since DBM is diploid. We tested 943 IFLs (=3772 alleles) using a diagnostic dose (@1ppm Bt *kurstaki*). The survivors were reared further and tested again to confirm the presence of resistance alleles. Twelve IFLs survived the diagnostic dose initially. In order to eliminate false positives, we tested them again using the same diagnostic dose and consequently eight families were discarded.

Further testing of surviving IFL's of DBM

A range of doses were used to confirm the presence or absence of any major resistance genes among the surviving four families (IFL-448, IFL-563, IFL-574, IFL-712). The probit regression analysis of mortality data indicate a low level of resistance among the four IFLs detected from SA (Table 1; Figure 1). The resistance ratios calculated on the basis of LC_{50}s also indicate a low level of resistance. This was further confirmed when these IFLs were bioassayed with transgenic broccoli plants along with a highly resistant population (Res) (Tang *et al.*, 1997) and a susceptible strain (G88) at Cornell University. None of the IFLs from SA, even though they showed a low level of resistance in leaf-dip bioassays, could survive on the transgenic broccoli. The resistant population (Res) survived successfully (av. 87% and 93% for neonate and third instar larvae) on the transgenic broccoli (Figure 2). This has further strengthened our hypothesis that the frequency of alleles conferring high levels of Bt resistance is much lower than 10^{3} in natural populations.

Table 1. Probit regression analysis of IFLs of DBM detected as carriers of low level of resistance from SA by using leaf dip bioassays.

Population	n[a]	Slope ± SE	LC_{50} mg a.i./l	LC_{90} mg a.i./l	RR[b]
IFL-448[c]	480	0.674 ± 0.07	0.12 (0.066-.205)	9.6 (4-30)	24
IFL-563	480	0.437 ± 0.06	0.082 (0.011-0.286)	71 (9-1314)	16.4
IFL-574	480	0.419 ± 0.06	0.07 (0.005-0.298)	80 (7-189429)	14
IFL-712	480	0.641 ± 0.07	0.11 (0.035-0.267)	11 (3-111)	22
GF[d]	480	0.568 ± 0.7	0.417 (0.17-0.964)	75 (19-929)	83.4
TW[e]	240	0.804 ± 0.21	0.013 (0-0.045)	0.5 (0.2-1.2)	2.6
LP[f]	240	0.547 ± 0.7	0.005 (0.002-0.010	1.1 (0.4-5.4)	1

[a] Number of larvae tested; [b] Ratio is LC_{50} divided by the lab susceptible population; [c] Isofemale line number; [d] Lab resistant population (originally from Gatton field areas, Queensland); [e] Field population (from Two Wells, SA); [f] SA lab susceptible population.

The evolution of resistance can be delayed significantly by adopting resistance management strategies. The low frequencies of Bt resistance alleles detected in DBM are quite promising for the future use of Bt sprays and for transgenic crops in SA.

Screening of cotton bollworm

We are still looking to detect a resistant family after testing 347 single families on transgenic cotton. The results obtained so far suggest that the frequency of alleles conferring Bt resistance seems to be much lower than has been reported recently in other bollworms (eg., *Heliothis virescens*). This experiment is still under way.

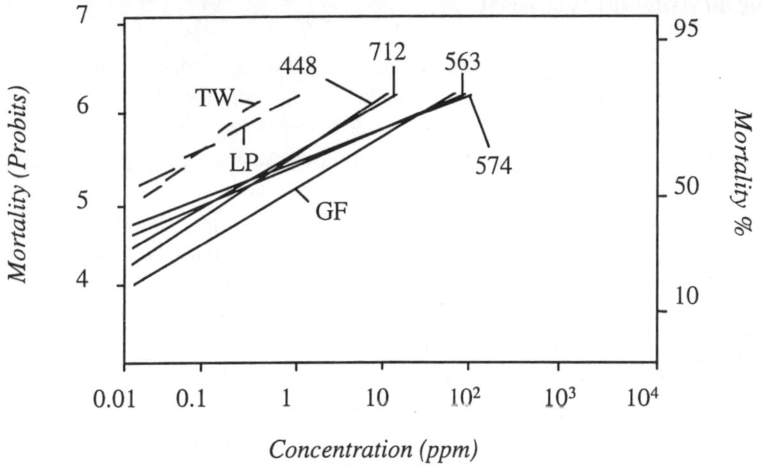

Figure 1. Genotypic responses of IFLs of DBM (448, 563, 574 and 712); lab susceptible (LP), resistant (GF) and a field population (TW) from SA.

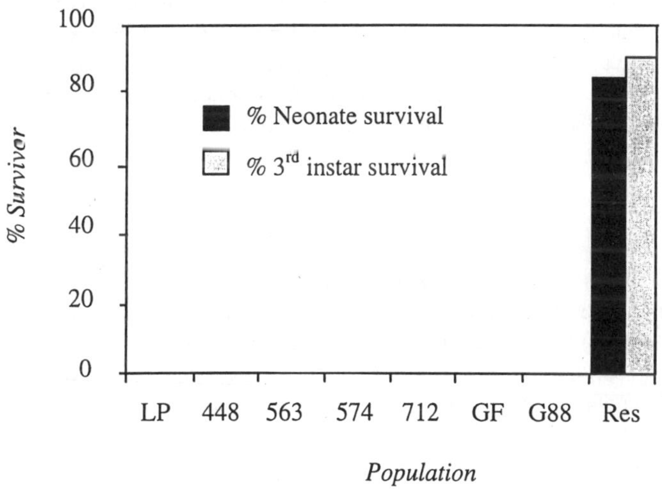

Figure 2. Response of neonate and third instar larvae of DBM to transgenic broccoli (Cry1Ac). LP refers to SA lab susceptible population; 448, 563, 574, and 712 are IFLs; GF, SA resistant lab population; G88, New York lab susceptible strain; Res, New York lab resistant strain (originally from Florida).

ACKNOWLEDGEMENTS

We thank G. Fitt, C. Tann, N. Forrester (New South Wales) and M. Cahill (Queensland) for supplying *H. armigera*.

REFERENCES

Andow D A; Alstad D N (1998). F_2 screen for rare resistance alleles. *Journal of Economic Entomology,* **91**, 572-578.

Dobzhansky, T (1955). *Evolution, genetics, and man*. John Wiley & Sons: New York.

Fitt G P; Mares C L; Llewellyn D J (1994). Field evaluation and potential ecological impact of transgenic cottons (*Gossypium hirsutum*) in Australia. *Biocontrol Science and Technology,* **4**, 535-548.

Gould, F (1998). Sustainability of transgenic insecticidal cultivars: Integrating pest genetics and ecology. *Annual Review of Entomology,* **43**, 701-26.

Gould F; Anderson A; Jones A; Sumerford D; Heckel D G; Lopez J; Micinski S; Leonard R; Laster M (1997). Initial frequency of alleles for resistance to *Bacillus thuringiensis* toxins in field populations of *Heliothis virescens. Proceedings of the National Academy of Sciences of the United States of America,* **94**, 3519-3523.

Hama H; Suzuki K; Tanaka H (1992). Inheritance and stability of resistance to *Bacillus thuringiensis* formulations of the diamondback moth, *Plutella xylostella* (Linnaeus) (Lepidoptera: Yponomeutidae). *Applied Entomology and Zoology,* **27**, 355-362.

Kaiser J (1996). Pests overwhelm Bt cotton crop. *Science,* **273**, 423.

Roush R (1997). Managing resistance to transgenic crops. In: *Advances in insect control: The role of transgenic plants,* eds. N. Carozzi and M. Koziel, pp. 271-295. Taylor & Francis Ltd: London.

Roush R T (1994). Managing pests and their resistance to *Bacillus thuringiensis*: can transgenic crops be better than sprays? *Biocontrol Science and Technology,* **4**, 501-516.

Roush R T; Miller G L (1986). Considerations for design of insecticide resistance monitoring programs. *Journal of Economic Entomology,* **79**, 293-298.

Roush R T; Shelton A M (1997). Assessing the odds - the emergence of resistance to Bt transgenic plants. *Nature Biotechnology,* **15**, 816-817.

Russel R. M; Robertson J L; Savin N E (1977). POLO: a new computer program for probit analysis. *Bulletin of Entomological Society of America,* **23**, 209-213.

Tabashnik B E; Liu Y B; Finson N; Masson L; Heckel D G (1997). One gene in diamondback moth confers resistance to four *Bacillus thuringiensis* toxins. *Proceedings of the National Academy of Sciences of the United States of America,* **94**, 1640-1644.

Tabashnik B E; Schwartz J M; Finson N; Johnson M W (1992). Inheritance of resistance to *Bacillus thuringiensis* in diamondback moth (Lepidoptera: Plutellidae). *Journal of Economic Entomology*, **85**, 1046-1055.

Tang J D; Gilboa S; Roush R T; Shelton A M (1997). Inheritance, stability, and lack-of-fitness costs of field-selected resistance to *Bacillus thuringiensis* in diamondback moth (Lepidoptera: Plutellidae) from Florida. *Journal of Economic Entomology,* **90**, 732-741.